Graduate Texts in Physics

Graduate Texts in Physics

Graduate Texts in Physics publishes core learning/teaching material for graduate- and advanced-level undergraduate courses on topics of current and emerging fields within physics, both pure and applied. These textbooks serve students at the MS- or PhD-level and their instructors as comprehensive sources of principles, definitions, derivations, experiments and applications (as relevant) for their mastery and teaching, respectively. International in scope and relevance, the textbooks correspond to course syllabi sufficiently to serve as required reading. Their didactic style, comprehensiveness and coverage of fundamental material also make them suitable as introductions or references for scientists entering, or requiring timely knowledge of, a research field.

More information about this series at http://www.springer.com/series/8431

Nhan Phan-Thien · Nam Mai-Duy

Understanding Viscoelasticity

An Introduction to Rheology

Third Edition

 Springer

Nhan Phan-Thien
Department of Mechanical Engineering
National University of Singapore
Singapore
Singapore

Nam Mai-Duy
Faculty of Health, Engineering and Sciences
University of Southern Queensland
Toowoomba, QLD
Australia

ISSN 1868-4513   ISSN 1868-4521   (electronic)
Graduate Texts in Physics
ISBN 978-3-319-87210-0   ISBN 978-3-319-62000-8   (eBook)
DOI 10.1007/978-3-319-62000-8

Printed on acid-free paper

This Springer imprint is published by Springer Nature
The registered company is Springer International Publishing AG
The registered company address is: Gewerbestrasse 11, 6330 Cham, Switzerland

Preface to the Third Edition

In this third edition, we have updated on the Dissipative Particle Dynamics method with recent results and provided some MATLAB programs for students to try out in fluid/flow modelling. In addition, a solution manual is provided, to guide students in their attempts at the questions in the book. The book itself remains compact, but sufficient in content for a first-year graduate module.

We sincerely thank our wives, Kim-Thoa (N. Phan-Thien) and Oanh (N. Mai-Duy) for their encouragement and their unfailing support—this is their achievement as well as ours!

Singapore   Nhan Phan-Thien
Toowoomba, Australia   Nam Mai-Duy

Preface to the Second Edition

In this second edition, typographical errors brought about by the conversion process to LATEX were corrected; my gratitude went to Brittany Bannish (University of Utah) for painstakingly going through the first edition. My main aim in revising this is to produce a still compact book, sufficient at the level of first-year graduate course for those who wish to understand viscoelasticity, and to embark in modelling viscoelastic multiphase fluids. To this end, I have decided to introduce a new chapter on Dissipative Particle Dynamics (DPD), which I believe is relevant in modelling complex-structured fluids. All the basic ideas in DPD are reviewed, with some sample problems to illustrate the methodology. My gratitude goes to A*STAR, the Agency for Science, Technology and Research, for funding Multiphase Modelling Projects, and Prof. Khoo Boo Cheong, a colleague and above all a friend, for his support, which made the writing of Chapter 9 possible. I wish to acknowledge Prof. Mai-Duy Nam and Dr. Pan Dingyi, for their contributions to the DPD research, and Prof. Yu Shaozheng, for his comments on the revised book. Lastly, my humble thanks to the continuing support and constant encouragement of my wife, Kim-Thoa—without her capable hands, normal daily tasks would be impossible, let alone revising this book! It has been good for me to go through this revision, and I sincerely hope that the readers find the book useful in their research works.

Singapore
June 2012

Nhan Phan-Thien

Preface to the First Edition

This book presents an introduction to viscoelasticity, in particular, to the theories of dilute polymer solutions and dilute suspensions of rigid particles in viscous and incompressible fluids. These theories are important, not just because they apply to practical problems of industrial interest, but because they form a solid theoretical base upon which mathematical techniques can be built, from which more complex theories can be constructed, to better mimic material behaviour. The emphasis is not on the voluminous current topical research, but on the necessary tools to understand viscoelasticity at a first-year graduate level.

Viscoelasticity, or Continuum Mechanics, or Rheology[1] (certainly not to be confused with Theology) is *the science of deformation and flow*. This definition was due to Bingham, who, together with Scott-Blair[2] and Reiner,[3] helped form The Society of Rheology in 1929. Rheology has a distinguished history involving high-profile scientists. The idea that everything has a timescale and that if we are prepared to wait long enough, then everything will flow was known to the Greek philosopher Heraclitus, and prior to him, to the Prophetess Deborah—*The Mountains Flowed Before The Lord.*[4] Not surprisingly, the motto of the Society of Rheology is $\pi\alpha\nu\tau\alpha\,\rho\varepsilon\iota$ (everything flows), a saying attributed to Heraclitus.

From the rheological viewpoint, there is no clear distinction between solid and liquid; it is a matter between the relative timescale T of the experiment to the timescale τ of the material concerned. The timescale ratio, $De = \tau/T$ is called the *Deborah number*. If this ratio is negligibly small, then one has a viscous fluid (more precise definition later), and if it is large, a solid, and in-between, a viscoelastic liquid. The timescale of the fluid varies considerably, from 10^{-13} s for water, to a

---

[1]This word was coined by E.C. Bingham (1878–1946), Professor of Chemistry at Lafayette College, Pennsylvania. The Bingham fluid is named after him.

[2]G.W. Scott Blair (1902–1987), Professor of Chemistry at the University of Reading. His main contributions were in biorheology.

[3]M. Reiner (1886–1976), Professor of Mathematics at the Technion University of Haifa, Israel. He is remembered for contributing to the Reiner–Rivlin fluid.

[4]The Book of Judges.

The logo of the Society of Rheology

few milliseconds for automotive oils, to minutes for polymer solutions and to hours for melts and soft solids.

Graduate students of Rheology naturally have the unenviable task of walking the bridge between solid mechanics and fluid mechanics, and at the same time trying to grasp the more significant and relevant concepts. They often find it hard (at least for me, during my graduate days) to piece together useful information from several comprehensive monographs and published articles on this subject. This set of lectures is an attempt to address this problem—it contains the necessary tools to understand viscoelasticity but does not insist on giving the latest piece of information on the topic.

The book starts with an introduction to the basic tools from tensor and dyadic analysis. Some authors prefer Cartesian tensor notation, others, dyadic notation. We use both notations, and they will be summarised here. Chapter 2 is a review of non-Newtonian behaviour in flows; here, the elasticity of the liquid and its ability to support large tension in stretching can be responsible for variety of phenomena, sometimes counter-intuitive. Kinematics and the equations of balance are discussed in detail in Chapter 3, including the finite strain and Rivlin–Ericksen tensors. In Chapter 4, some classical constitutive equations are reviewed, and the general principles governing the constitutive modelling are outlined. In this Chapter, the order fluid models are also discussed, leading to the well-known result that the Newtonian velocity field is admissible to a second-order fluid in plane flow. Chapter 5 describes some of the popular engineering inelastic and the linear elastic models. The inelastic models are very useful in shear-like flows where viscosity/shear rate relation plays a dominant role. The linear viscoelastic model is a limit of the simple fluid at small strain—any model must reduce to this limit when the strain amplitude is small enough. In Chapter 6, we discuss a special class of flows known as viscometric flows in which both the kinematics and the stress are fully determined by the flow, irrespective of the constitutive equations. This class of flows is equivalent to the simple shearing flow. Modelling techniques for polymer solutions are discussed next in Chapter 7. Here one has a set of stochastic differential equations for the motion of the particles; the random excitations come from a white noise model of the collision between the solvent molecules and the particles. It is our belief that a relevant model should come from the microstructure; however, when the microstructure is so complex that a detailed model is not tractable, elements of continuum model should be brought in. Finally, an introduction to suspension mechanics is given in Chapter 8. I have deliberately left out a number of topics: instability, processing flows, electro-rheological fluids, magnetised fluids and viscoelastic computational mechanics. It is hoped that the book forms a good foundation for those who wish to embark on the Rheology path.

This has been tested out in a one-semester course in Viscoelasticity at the National University of Singapore. It is entirely continuous-assessment based, with the assignments graded at different difficulty levels to be attempted—solving problems is an indispensable part of the education process. A good knowledge of fluid mechanics is helpful, but it is more important to have a solid foundation in Mathematics and Physics (Calculus, Linear Algebra, Partial Differential Equations), of a standard that every one gets in the first two years in an undergraduate Engineering curriculum.

I have greatly benefitted from numerous correspondence with my academic brother, Prof. Raj Huilgol, and my mentor, Prof. Roger Tanner. Prof. Jeff Giacomin read the first draft of this; his help is gratefully acknowledged.

Singapore   Nhan Phan-Thien
February 2002

Contents

Chapter 1
Tensor Notation

A Working Knowledge in Tensor Analysis

This chapter is not meant as a replacement for a course in tensor analysis, but it will provide a sufficient working background to tensor notation and algebra.

1.1 Cartesian Frame of Reference

Physical quantities encountered are either scalars (e.g., time, temperature, pressure, volume, density), or vectors (e.g., displacement, velocity, acceleration, force, torque), or tensors (e.g., stress, displacement gradient, velocity gradient, alternating tensors – we deal mostly with second-order tensors). These quantities are distinguished by the following generic notation:

s denotes a scalar (lightface italic)
u denotes a vector (boldface)
F denotes a tensor (boldface)

The distinction between vector and tensor is usually clear from the context. When they are functions of points in a three-dimensional Euclidean space $\mathbb{E}$, they are called **fields**. The set of all vectors (or tensors) form a normed vector space $\mathbb{U}$.

Distances and time are measured in the Cartesian frame of reference, or simply frame of reference, $\mathcal{F} = \{O; \mathbf{e}_1, \mathbf{e}_2, \mathbf{e}_3\}$, which consists of an origin O, a clock, and an orthonormal basis $\{\mathbf{e}_1, \mathbf{e}_2, \mathbf{e}_3\}$, see Fig. 1.1,

$$\mathbf{e}_i \cdot \mathbf{e}_j = \delta_{ij}, \quad i, j = 1, 2, 3 \tag{1.1}$$

where the Kronecker delta is defined as

© Springer International Publishing AG 2017
N. Phan-Thien and N. Mai-Duy, *Understanding Viscoelasticity*,
Graduate Texts in Physics, DOI 10.1007/978-3-319-62000-8_1

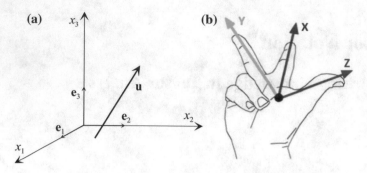

Fig. 1.1 a Cartesian frame of reference and **b** right hand rule

$$\delta_{ij} = \begin{cases} 1, & i = j, \\ 0, & i \neq j. \end{cases} \tag{1.2}$$

We only deal with right-handed frames of reference (applying the right-hand rule, when the thumb is in direction 1, and the forefinger in direction 2, the middle finger lies in direction 3, or an even permutation of this), where $(\mathbf{e}_1 \times \mathbf{e}_2) \cdot \mathbf{e}_3 = 1$.

The Cartesian components of a vector **u** are given by

$$u_i = \mathbf{u} \cdot \mathbf{e}_i \tag{1.3}$$

so that one may write

$$\mathbf{u} = \sum_{i=1}^{3} u_i \mathbf{e}_i = u_i \mathbf{e}_i. \tag{1.4}$$

Here we have employed the *summation convention*, i.e., whenever there are repeated subscripts, a summation is implied over the range of the subscripts, from (1, 2, 3). For example,

$$A_{ij} B_{jk} = \sum_{j=1}^{3} A_{ij} B_{jk}. \tag{1.5}$$

This short-hand notation is due to Einstein (Fig. 1.2), who argued that physical laws must not depend on coordinate systems, and therefore must be expressed in tensorial format. This is the essence of the *Principle of Frame Indifference*, to be discussed later.

The *alternating tensor* is defined as

$$\varepsilon_{ijk} = \begin{cases} +1, & \text{if } (i, j, k) \text{ is an even permutation of } (1, 2, 3) \\ -1, & \text{if } (i, j, k) \text{ is an odd permutation of } (1, 2, 3) \\ 0, & \text{otherwise.} \end{cases} \tag{1.6}$$

Fig. 1.2 Albert Einstein
(1879–1955) got the Nobel
Prize in Physics in 1921 for
his explanation in
photoelectricity. He derived
the effective viscosity of a
dilute suspension of
neutrally buoyant spheres,
$\eta = \eta_s(1 + \frac{5}{2}\phi)$, η_s: the
solvent viscosity, ϕ: the
sphere volume fraction

1.1.1   Position Vector

In the frame $\mathcal{F} = \{O; \mathbf{e}_1, \mathbf{e}_2, \mathbf{e}_3\}$, the position vector is denoted by

$$\mathbf{x} = x_i \mathbf{e}_i, \tag{1.7}$$

where x_i are the components of $\mathbf{x}$.

1.2   Frame Rotation

Consider the two frames of references, $\mathcal{F} = \{O; \mathbf{e}_1, \mathbf{e}_2, \mathbf{e}_3\}$ and $\mathcal{F}' = \{O; \mathbf{e}'_1, \mathbf{e}'_2, \mathbf{e}'_3\}$, as shown in Fig. 1.3, one obtained from the other by a rotation. Hence,

$$\mathbf{e}_i \cdot \mathbf{e}_j = \delta_{ij}, \quad \mathbf{e}'_i \cdot \mathbf{e}'_j = \delta_{ij}.$$

Define the cosine of the angle between $\left(\mathbf{e}_i, \mathbf{e}'_j\right)$ as

$$A_{ij} = \mathbf{e}'_i \cdot \mathbf{e}_j.$$

Thus A_{ij} can be regarded as the components of $\mathbf{e}'_i$ in $\mathcal{F}$, or the components of $\mathbf{e}_j$ in $\mathcal{F}'$. We write

$$\mathbf{e}'_p = A_{pi} \mathbf{e}_i \quad A_{pi} A_{qi} = \delta_{pq}.$$

Fig. 1.3 Two frames of reference sharing a common origin

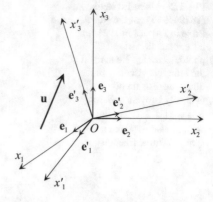

Similarly

$$\mathbf{e}_i = A_{pi}\mathbf{e}'_p \quad A_{pi}A_{pj} = \delta_{ij}.$$

1.2.1 Orthogonal Matrix

A matrix is said to be an orthogonal matrix if its inverse is also its transpose; furthermore, if its determinant is $+1$, then it is a proper orthogonal matrix. Thus $[\mathbf{A}]$ is a proper orthogonal matrix.

We now consider a vector $\mathbf{u}$, expressed in either frame $\mathcal{F}$ or $\mathcal{F}'$,

$$\mathbf{u} = u_i\mathbf{e}_i = u'_j\mathbf{e}'_j.$$

Taking scalar product with either base vector,

$$u'_i = \mathbf{e}'_i \cdot \mathbf{e}_j u_j = A_{ij}u_j,$$
$$u_j = \mathbf{e}_j \cdot \mathbf{e}'_i u'_i = A_{ij}u'_i.$$

In matrix notation,

$$[\mathbf{A}] = \begin{bmatrix} A_{11} & A_{12} & A_{13} \\ A_{21} & A_{22} & A_{23} \\ A_{31} & A_{32} & A_{33} \end{bmatrix}, \quad [\mathbf{u}] = \begin{bmatrix} u_1 \\ u_2 \\ u_3 \end{bmatrix}, \quad [\mathbf{u}'] = \begin{bmatrix} u'_1 \\ u'_2 \\ u'_3 \end{bmatrix},$$

we have

$$[\mathbf{u}'] = [\mathbf{A}] \cdot [\mathbf{u}], \quad [\mathbf{u}] = [\mathbf{A}]^T \cdot [\mathbf{u}'],$$
$$u'_i = A_{ij}u_j, \qquad u_j = A_{ij}u'_i. \tag{1.8}$$

In particular, the position transforms according to this rule

$$\mathbf{x} = x_i' \mathbf{e}_i' = x_j \mathbf{e}_j \qquad x_i' = A_{ij} x_j \text{ or } x_j = A_{ij} x_i'.$$

1.2.2 Rotation Matrix

The matrix A is called a rotation – in fact a proper rotation ($det\mathbf{A} = +1$).

1.3 Tensors

1.3.1 Zero-Order Tensors

Scalars, which are invariant under a frame rotation, are said to be tensors of zero order.

1.3.2 First-Order Tensor

A set of three scalars referred to one frame of reference, written collectively as $\mathbf{v} = (v_1, v_2, v_3)$, is called a tensor of first order, or a vector, if the three components transform according to (1.8) under a frame rotation.
Clearly,

- If $\mathbf{u}$ and $\mathbf{v}$ are vectors, then $\mathbf{u} + \mathbf{v}$ is also a vector.
- If $\mathbf{u}$ is a vector, then $\alpha\mathbf{u}$ is also a vector, where α is a real number.

The set of all vectors form a vector space $\mathcal{U}$ under addition and multiplication. In this space, the usual scalar product can be shown to be an inner product. With the norm induced by this inner product, $|\mathbf{u}|^2 = \mathbf{u} \cdot \mathbf{u}$, $\mathcal{U}$ is a normed vector space. We also refer to a vector $\mathbf{u}$ by its components, u_i.

1.3.3 Outer Products

Consider now two tensors of first order, u_i and v_i. The product $u_i v_j$ represents the outer product of $\mathbf{u}$ and $\mathbf{v}$, and written as (the subscripts are assigned from left to right by convention),

$$[\mathbf{uv}] = \begin{bmatrix} u_1 v_1 & u_1 v_2 & u_1 v_3 \\ u_2 v_1 & u_2 v_2 & u_2 v_3 \\ u_3 v_1 & u_3 v_2 & u_3 v_3 \end{bmatrix}.$$

In a frame rotation, from $\mathcal{F}$ to $\mathcal{F}'$, the components of this change according to

$$u_i' v_j' = A_{im} A_{jn} u_m v_n.$$

1.3.4 Second-Order Tensors

In general, a set of 9 scalars referred to one frame of reference, collectively written as $\mathbf{W} = [W_{ij}]$, transformed to another set under a frame rotation according to

$$W_{ij}' = A_{im} A_{jn} W_{mn}, \tag{1.9}$$

is said to be a second-order tensor, or a two-tensor, or simply a tensor (when the order does not have to be explicit). In matrix notation, we write

$$\left[\mathbf{W}'\right] = [\mathbf{A}][\mathbf{W}][\mathbf{A}]^T \text{ or } \mathbf{W}' = \mathbf{A}\mathbf{W}\mathbf{A}^T \text{ or } W_{ij}' = A_{ik} W_{kl} A_{jl}.$$

In the direct notation, we denote a tensor by a bold face letter (without the square brackets). This direct notation is intimately connected to the concept of a linear operator, e.g., Gurtin [34].

1.3.5 Third-Order Tensors

A set of 27 scalars referred to one frame of reference, collectively written as $\mathbf{W} = [W_{ijk}]$, transformed to another set under a frame rotation according to

$$W_{ijk}' = A_{il} A_{jm} A_{kn} W_{lmn}, \tag{1.10}$$

is said to be a third-order tensor.

Obviously, the definition can be extended to a set of $3n$ scalars, and $\mathbf{W} = [W_{i_1 i_2 \ldots i_n}]$ (n indices) is said to be an n-order tensor if its components transform under a frame rotation according to

$$W_{i_1 i_2 \ldots i_n}' = A_{i_1 j_1} A_{i_2 j_2} \cdots A_{i_n j_n} W_{j_1 j_2 \ldots j_n}. \tag{1.11}$$

We will deal mainly with vectors and tensors of second order. Usually, a higher-order (higher than 2) tensor is formed by taking outer products of tensors of lower orders, for example the outer product of a two-tensor $\mathbf{T}$ and a vector $\mathbf{n}$ is a third-order tensor $\mathbf{T} \otimes \mathbf{n}$. One can verify that the transformation rule (1.11) is obeyed.

1.3.6 Transpose Operation

The components of the transpose of a tensor $\mathbf{W}$ are obtained by swapping the indices:

$$[\mathbf{W}]_{ij} = W_{ij}, \quad [\mathbf{W}]^T_{ij} = W_{ji}.$$

A tensor $\mathbf{S}$ is *symmetric* if it is unaltered by the transpose operation,

$$\mathbf{S} = \mathbf{S}^T, \quad S_{ij} = S_{ji}.$$

It is *anti-symmetric* (or *skew*) if

$$\mathbf{S} = -\mathbf{S}^T, \quad S_{ij} = -S_{ji}.$$

An anti-symmetric tensor must have zero diagonal terms (when $i = j$). Clearly

- If $\mathbf{U}$ and $\mathbf{V}$ are two-tensors, then $\mathbf{U} + \mathbf{V}$ is also a two-tensor.
- If $\mathbf{U}$ is a two-tensor, then $\alpha\mathbf{U}$ is also a two-tensor, where α is a real number. The set of $\mathbf{U}$ form a vector space under addition and multiplication.

1.3.7 Decomposition

Any second-order tensor can be decomposed into symmetric and anti-symmetric parts:

$$\mathbf{W} = \frac{1}{2}\left(\mathbf{W} + \mathbf{W}^T\right) + \frac{1}{2}\left(\mathbf{W} - \mathbf{W}^T\right),$$
$$W_{ij} = \frac{1}{2}\left(W_{ij} + W_{ji}\right) + \frac{1}{2}\left(W_{ij} - W_{ji}\right). \tag{1.12}$$

Returning to (1.9), if we interchange i and j, we get

$$W'_{ji} = A_{jm}A_{in}W_{mn} = A_{jn}A_{im}W_{nm}.$$

The second equality arises because m and n are dummy indices, mere labels in the summation. The left side of this expression is recognised as the components of the transpose of $\mathbf{W}$. The equation asserts that the components of the transpose of $\mathbf{W}$ are also transformed according to (1.9). Thus, if $\mathbf{W}$ is a two-tensor, then its transpose is also a two-tensor, and the Cartesian decomposition (1.12) splits an arbitrary two-tensor into a symmetric and an anti-symmetric tensor (of second order).

We now go through some of the first and second-order tensors that will be encountered in this course.

1.3.8   Some Common Vectors

Position, displacement, velocity, acceleration, linear and angular momentum, linear
and angular impulse, force, torque, are vectors. This is because the position vector
transforms under a frame rotation according to (1.8). Any other quantity linearly
related to the position (including the derivative and integral operation) will also be a
vector.

1.3.9   Gradient of a Scalar

The gradient of a scalar is a vector. Let ϕ be a scalar, its gradient is written as

$$\mathbf{g} = \nabla\phi, \quad g_i = \frac{\partial\phi}{\partial x_i}.$$

Under a frame rotation, the new components of $\nabla\phi$ are

$$\frac{\partial\phi}{\partial x_i'} = \frac{\partial\phi}{\partial x_j}\frac{\partial x_j}{\partial x_i'} = A_{ij}\frac{\partial\phi}{\partial x_j},$$

which qualifies $\nabla\phi$ as a vector.

1.3.10   Some Common Tensors

We have met a second-order tensor formed by the outer product of two vectors,
written compactly as $\mathbf{uv}$, with components (for vectors, the outer products is written
without the symbol $\otimes$)

$$(\mathbf{uv})_{ij} = u_i v_j.$$

In general, the outer product of n vectors is an n-order tensor.

Unit Tensor. The Kronecker delta is a second-order tensor. In fact it is invariant in
any coordinate system, and therefore is an *isotropic* tensor of second-order. To show
that it is a second-order tensor, note that

$$\delta_{ij} = A_{ik}A_{jk} = A_{ik}A_{jl}\delta_{kl},$$

which follows from the orthogonality of the transformation matrix. δ_{ij} are said to
be the components of the second-order unit tensor $\mathbf{I}$. Finding isotropic tensors of
arbitrary orders is not a trivial task.

Gradient of a Vector. The gradient of a vector is a two-tensor: if u_i and u_i' are the components of **u** in $\mathcal{F}$ and $\mathcal{F}'$,

$$\frac{\partial u_i'}{\partial x_j'} = \frac{\partial x_l}{\partial x_j'} \frac{\partial}{\partial x_l} (A_{ik} u_k) = A_{ik} A_{jl} \frac{\partial u_k}{\partial x_l}.$$

This qualifies the gradient of a vector as a two-tensor.

Velocity Gradient. If **u** is the velocity field, then ∇**u** is the gradient of the velocity. Be careful with the notation here. By our convention, the subscripts are assigned from left to right, so

$$(\nabla \mathbf{u})_{ij} = \nabla_i u_j = \frac{\partial u_j}{\partial x_i}.$$

In most books on viscoelasticity including this, the term "velocity gradient" is taken to mean the second-order tensor $\mathbf{L} = (\nabla \mathbf{u})^T$ with components

$$L_{ij} = \frac{\partial u_i}{\partial x_j}. \tag{1.13}$$

Strain Rate and Vorticity Tensors. The velocity gradient tensor can be decomposed into a symmetric part **D**, called the strain rate tensor, and an anti-symmetric part **W**, called the vorticity tensor:

$$\mathbf{D} = \frac{1}{2} \left(\nabla \mathbf{u} + \nabla \mathbf{u}^T \right), \quad \mathbf{W} = \frac{1}{2} \left(\nabla \mathbf{u}^T - \nabla \mathbf{u} \right). \tag{1.14}$$

Stress Tensor and Quotient Rule. We are given that stress $\mathbf{T} = [T_{ij}]$ at a point **x** is defined by, (see Fig. 1.4),

$$\mathbf{t} = \mathbf{Tn}, \quad t_i = T_{ij} n_j, \tag{1.15}$$

where **n** is a normal unit vector on an infinitesimal surface ΔS at point **x**, and **t** is the surface traction (force per unit area) representing the force the material on the positive side of **n** is pulling on the material on the negative side of **n**. Under a frame rotation, since both **t** (force) and **n** are vectors,

$$\mathbf{t}' = \mathbf{At}, \quad \mathbf{t} = \mathbf{A}^T \mathbf{t}' \qquad \mathbf{n}' = \mathbf{An}, \quad \mathbf{n} = \mathbf{A}^T \mathbf{n}',$$
$$\mathbf{A}^T \mathbf{t}' = \mathbf{t} = \mathbf{Tn} = \mathbf{TA}^T \mathbf{n}' \qquad \mathbf{t}' = \mathbf{ATA}^T \mathbf{n}'.$$

From the definition of the stress, $\mathbf{t}' = \mathbf{T}' \mathbf{n}'$, and therefore

$$\mathbf{T}' = \mathbf{ATA}^T.$$

So the stress is a second-order tensor.

In fact, as long as **t** and **n** are vector, the 9 components T_{ij} defined in the manner indicated by (1.15) form a second-order tensor. This is known as the *quotient rule*.

Fig. 1.4 Defining the stress
tensor

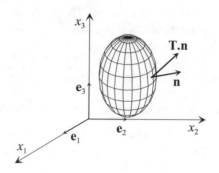

1.4 Tensor and Linear Vector Function

L is a linear vector function on $\mathcal{U}$ if it satisfies

- $L(\mathbf{u}_1 + \mathbf{u}_2) = L(\mathbf{u}_1) + L(\mathbf{u}_2)$,
- $L(\alpha\mathbf{u}) = \alpha L(\mathbf{u})$, $\forall \mathbf{u}, \mathbf{u}_1, \mathbf{u}_2 \in \mathcal{U}$, $\forall \alpha \in \mathbb{R}$

1.4.1 Claim

Let $\mathbf{W}$ be a two-tensor, and define a vector-valued function through

$$\mathbf{v} = L(\mathbf{u}) = \mathbf{W}\mathbf{u},$$

then L is a linear function. Conversely, for any linear function on $\mathcal{U}$, there is a unique two-tensor $\mathbf{W}$ such that

$$L(\mathbf{u}) = \mathbf{W}\mathbf{u}, \quad \forall \mathbf{u} \in \mathcal{U}.$$

The first statement can be easily verified. For the converse part, given the linear function, let define W_{ij} through

$$L(\mathbf{e}_i) = W_{ji}\mathbf{e}_j.$$

Now, $\forall \mathbf{u} \in \mathcal{U}$,

$$\mathbf{v} = L(\mathbf{u}) = L(u_i\mathbf{e}_i) = u_i W_{ji}\mathbf{e}_j$$
$$v_j = W_{ji}u_i.$$

$\mathbf{W}$ is a second-order tensor because $\mathbf{u}$ and $\mathbf{v}$ are vectors. The uniqueness part of $\mathbf{W}$ can be demonstrated by assuming that there is another $\mathbf{W}'$, then

$$\left(\mathbf{W} - \mathbf{W}'\right)\mathbf{u} = 0, \quad \forall \mathbf{u} \in \mathcal{U},$$

which implies that $\mathbf{W}' = \mathbf{W}$.

In this connection, one can define a second-order tensor as a linear function, taking one vector into another. This is the direct approach, e.g., Gurtin [34], emphasising linear algebra. We use whatever notation is convenient for the purpose at hand. The set of all linear vector functions forms a vector space under addition and multiplication. The main result here is that

$$L\left(\mathbf{e}_i\right) = \mathbf{W}\mathbf{e}_i = W_{ji}\mathbf{e}_j \quad W_{ji} = \mathbf{e}_j \cdot \left(\mathbf{W}\mathbf{e}_i\right).$$

1.4.2 Dyadic Notation

Thus, one may write

$$\mathbf{W} = W_{ij}\mathbf{e}_i\mathbf{e}_j. \tag{1.16}$$

This is the basis for the *dyadic* notation, the $\mathbf{e}_i\mathbf{e}_j$ play the role of the basis "vectors" for the tensor $\mathbf{W}$.

1.5 Tensor Operations

1.5.1 Substitution

The operation $\delta_{ij}u_j = u_i$ replaces the subscript j by i – the tensor δ_{ij} is therefore sometimes called the substitution tensor.

1.5.2 Contraction

Given a two-tensor W_{ij}, the operation

$$W_{ii} = \sum_{i=1}^{3} W_{ii} = W_{11} + W_{22} + W_{33}$$

is called a contraction. It produces a scalar. The invariance of this scalar under a frame rotation is seen by noting that

$$W_{ii}' = A_{ik}A_{il}W_{kl} = \delta_{kl}W_{kl} = W_{kk}.$$

This scalar is also called the trace of $\mathbf{W}$, written as

$$\operatorname{tr}\mathbf{W} = W_{ii}. \tag{1.17}$$

It is one of the invariants of $\mathbf{W}$ (i.e., unchanged in a frame rotation). If the trace of $\mathbf{W}$ is zero, then $\mathbf{W}$ is said to be traceless. In general, given an n-order tensor, contracting any two subscripts produces a tensor of $(n-2)$ order.

1.5.3 Transpose

Given a two-tensor $\mathbf{W} = [W_{ij}]$, the transpose operation swaps the two indices

$$\mathbf{W}^T = \left(W_{ij}\mathbf{e}_i\mathbf{e}_j\right)^T = W_{ij}\mathbf{e}_j\mathbf{e}_i, \quad \left[\mathbf{W}^T\right]_{ij} = W_{ji}. \tag{1.18}$$

1.5.4 Products of Two Second-Order Tensors

Given two second-order tensors, $\mathbf{U}$ and $\mathbf{V}$,

$$\mathbf{U} = U_{ij}\mathbf{e}_i\mathbf{e}_j, \quad \mathbf{V} = V_{ij}\mathbf{e}_i\mathbf{e}_j,$$

one can form different products from them, and it is helpful to refer to the dyadic notation here.

- The tensor product $\mathbf{U} \otimes \mathbf{V}$ is a 4th-order tensor, with component $U_{ij}V_{kl}$,

$$\mathbf{U} \otimes \mathbf{V} = U_{ij}V_{kl}\mathbf{e}_i\mathbf{e}_j\mathbf{e}_k\mathbf{e}_l. \tag{1.19}$$

- The single dot product $\mathbf{U}.\mathbf{V}$ is a 2nd-order tensor, sometimes written without the dot (the dot is the contraction operator),

$$\begin{aligned}
\mathbf{U} \cdot \mathbf{V} = \mathbf{U}\mathbf{V} &= \left(U_{ij}\mathbf{e}_i\mathbf{e}_j\right) \cdot (V_{kl}\mathbf{e}_k\mathbf{e}_l) \\
&= U_{ij}\mathbf{e}_i\delta_{jk}V_{kl}\mathbf{e}_l = U_{ij}V_{jl}\mathbf{e}_i\mathbf{e}_l,
\end{aligned} \tag{1.20}$$

with components $U_{ik}V_{kl}$, just like multiplying two matrices U_{ik} and V_{kj}. This single dot product induces a contraction of a pair of subscripts (j and k) in $U_{ij}V_{kl}$, and acts just like a vector dot product.
- The double dot (or scalar, or inner) product produces a scalar,

$$\begin{aligned}
\mathbf{U} : \mathbf{V} = \left(U_{ij}\mathbf{e}_i\mathbf{e}_j\right) : (V_{kl}\mathbf{e}_k\mathbf{e}_l) &= \left(U_{ij}\mathbf{e}_i\right)\delta_{jk} \cdot (V_{kl}\mathbf{e}_l) \\
&= U_{ij}V_{kl}\delta_{jk}\delta_{il} = U_{ij}V_{ji}.
\end{aligned} \tag{1.21}$$

The dot operates on a pair of base vectors until we run out of dots. The end result is a scalar (remember our summation convention). It can be shown that the scalar product is in fact an inner product.

- The norm of a two-tensor is defined from the inner product in the usual manner,

$$\|\mathbf{U}\|^2 = \mathbf{U}^T : \mathbf{U} = U_{ij}U_{ij} = \text{tr}\left(\mathbf{U}^T\mathbf{U}\right). \tag{1.22}$$

The space of all linear vector functions therefore form a normed vector space.

- One writes $\mathbf{U}^2 = \mathbf{UU}$, $\mathbf{U}^3 = \mathbf{U}^2\mathbf{U}$, etc.
- A tensor $\mathbf{U}$ is invertible if there exists a tensor, $\mathbf{U}^{-1}$, called the inverse of $\mathbf{U}$, such that

$$\mathbf{UU}^{-1} = \mathbf{U}^{-1}\mathbf{U} = \mathbf{I} \tag{1.23}$$

One can also define the vector cross product between two second-order tensors (and indeed any combination of dot and cross vector products). However, we refrain from listing all possible combinations here.

1.6 Invariants

1.6.1 Invariant of a Vector

When a quantity is unchanged with a frame rotation, it is said to be invariant. From a vector, a scalar can be formed by taking the scalar product with itself, $v_i v_i = v^2$. This is of course the magnitude of the vector and it is the only independent scalar invariant for a vector.

1.6.2 Invariants of a Tensor

From a second-order tensor $\mathbf{S}$, there are three independent scalar invariants that can be formed, by taking the trace of $\mathbf{S}$, $\mathbf{S}^2$ and $\mathbf{S}^3$,

$$I = \text{tr}\mathbf{S} = S_{ii}, \quad II = \text{tr}\mathbf{S}^2 = S_{ij}S_{ji}, \quad III = \text{tr}\mathbf{S}^3 = S_{ij}S_{jk}S_{ki}.$$

However, it is customary to use the following invariants

$$I_1 = I, \quad I_2 = \frac{1}{2}\left(I^2 - II\right), \quad I_3 = \frac{1}{6}\left(I^3 - 3I\,II + 2III\right) = \det\mathbf{S}.$$

It is also possible to form ten invariants between two tensors (Gurtin [34]).

1.7   Decompositions

We now quote some of the well-known results without proof, some are intuitively obvious, others not.

1.7.1   Eigenvalue and Eigenvector

A scalar ω is an *eigenvalue* of a two-tensor $\mathbf{S}$ if there exists a non-zero vector $\mathbf{e}$, called the *eigenvector*, satisfying

$$\mathbf{Se} = \omega\mathbf{e}. \tag{1.24}$$

The characteristic space for $\mathbf{S}$ corresponding to the eigenvalue ω consists of all vectors in the eigenspace, $\{\mathbf{v} : \mathbf{Sv} = \omega\mathbf{v}\}$. If the dimension of this space is n, then ω is said to have geometric multiplicity of n. The *spectrum* of $\mathbf{S}$ is the ordered list $\{\omega_1, \omega_2, \ldots\}$ of all the eigenvalues of S.

A tensor $\mathbf{S}$ is said to be positive definite if it satisfies

$$\mathbf{S} : \mathbf{vv} > 0, \qquad \forall \mathbf{v} \neq \mathbf{0}. \tag{1.25}$$

We record the following theorems:

- The eigenvalues of a positive definite tensor are strictly positive.
- The characteristic spaces of a symmetric tensor are mutually orthogonal.
- Spectral decomposition theorem: Let $\mathbf{S}$ be a symmetric two-tensor. Then there is a basis consisting entirely of eigenvectors of $\mathbf{S}$. For such a basis, $\{\mathbf{e}_i, i = 1, 2, 3\}$, the corresponding eigenvalues $\{\omega_i, i = 1, 2, 3\}$ form the entire spectrum of $\mathbf{S}$, and $\mathbf{S}$ can be represented by the *spectral representation*, where

$$
\begin{aligned}
\mathbf{S} &= \sum_{i=1}^{3} \omega_i \mathbf{e}_i \mathbf{e}_i, \text{ when } \mathbf{S} \text{ has three distinct eigenvalues,} \\
\mathbf{S} &= \omega_1 \mathbf{ee} + \omega_2 (\mathbf{I} - \mathbf{ee}), \text{ when } \mathbf{S} \text{ has two distinct eigenvalues,} \\
\mathbf{S} &= \omega\mathbf{I}, \text{ when } \mathbf{S} \text{ has only one eigenvalue.}
\end{aligned}
\tag{1.26}
$$

1.7.2   Square Root Theorem

Let $\mathbf{S}$ be a symmetric positive definite tensor. Then there is a unique positive definite tensor $\mathbf{U}$ such that $\mathbf{U}^2 = \mathbf{S}$. We write

$$\mathbf{U} = \mathbf{S}^{1/2}.$$

The proof of this follows from the spectral representation of $\mathbf{S}$.

1.7.3 Polar Decomposition Theorem

For any given tensor $\mathbf{F}$, there exist positive definite tensors $\mathbf{U}$ and $\mathbf{V}$, and a rotation tensor $\mathbf{R}$, such that

$$\mathbf{F} = \mathbf{RU} = \mathbf{VR}. \tag{1.27}$$

Each of these representations is unique, and

$$\mathbf{U} = \left(\mathbf{F}^T\mathbf{F}\right)^{1/2}, \quad \mathbf{V} = \left(\mathbf{F}\mathbf{F}^T\right)^{1/2}. \tag{1.28}$$

The first representation ($\mathbf{RU}$) is called the right, and the second ($\mathbf{VR}$) is called the left polar decomposition.

1.7.4 Cayley–Hamilton Theorem

The most important theorem is the Cayley–Hamilton theorem: Every tensor $\mathbf{S}$ satisfies its own characteristic equation

$$\mathbf{S}^3 - I_1\mathbf{S}^2 + I_2\mathbf{S} - I_3\mathbf{I} = \mathbf{0}, \tag{1.29}$$

where $I_1 = \text{tr}\mathbf{S}$, $I_2 = \frac{1}{2}\left((\text{tr}\mathbf{S})^2 - \text{tr}\mathbf{S}^2\right)$, and $I_3 = \det\mathbf{S}$ are the three scalar invariants for $\mathbf{S}$, and $\mathbf{I}$ is the unit tensor in three dimensions.

In two dimensions, this equation reads

$$\mathbf{S}^2 - I_1\mathbf{S} + I_2\mathbf{I} = \mathbf{0}, \tag{1.30}$$

where $I_1 = \text{tr}\mathbf{S}$, $I_2 = \det\mathbf{S}$ are the two scalar invariants for $\mathbf{S}$, and $\mathbf{I}$ is the unit tensor in two dimensions.

Cayley–Hamilton theorem is used to reduce the number of independent tensorial groups in tensor-valued functions. We record here one possible use of the Cayley–Hamilton theorem in two dimensions. The three-dimensional case is reserved as an exercise.

Suppose $\mathbf{C}$ is a given symmetric positive definite tensor in 2-D,

$$[\mathbf{C}] = \begin{bmatrix} C_{11} & C_{12} \\ C_{12} & C_{22} \end{bmatrix},$$

and its square root $\mathbf{U} = \mathbf{C}^{1/2}$ is desired. From the characteristic equation for $\mathbf{U}$,

$$\mathbf{U} = I_1^{-1}\left(\mathbf{U}\right)\left[\mathbf{C} + I_3\left(\mathbf{U}\right)\mathbf{I}\right],$$

so if we can express the invariants of $\mathbf{U}$ in terms of the invariant of $\mathbf{C}$, we're done. Now, if the eigenvalues of $\mathbf{U}$ are λ_1 and λ_2, then

$$I_1\,(\mathbf{U}) = \lambda_1 + \lambda_2,\ I_2\,(\mathbf{U}) = \lambda_1\lambda_2,$$
$$I_1\,(\mathbf{C}) = \lambda_1^2 + \lambda_2^2,\ I_2\,(\mathbf{C}) = \lambda_1^2\lambda_2^2.$$

Thus

$$I_2\,(\mathbf{U}) = \sqrt{I_2\,(\mathbf{C})},$$
$$I_1^2\,(\mathbf{U}) = I_1\,(\mathbf{C}) + 2\sqrt{I_2\,(\mathbf{C})}.$$

Therefore

$$\mathbf{U} = \frac{\mathbf{C} + \sqrt{I_2\,(\mathbf{C})}\mathbf{I}}{\sqrt{I_1\,(\mathbf{C}) + 2\sqrt{I_2\,(\mathbf{C})}}}.$$

1.8 Derivative Operations

Suppose $\varphi\,(\mathbf{u})$ is a scalar-valued function of a vector $\mathbf{u}$. The derivative of $\varphi(\mathbf{u})$ with respect to $\mathbf{u}$ in the direction $\mathbf{v}$ is defined as the linear operator $D\varphi\,(\mathbf{u})\,[\mathbf{v}]$:

$$\varphi\,(\mathbf{u} + \alpha\mathbf{v}) = \varphi\,(\mathbf{u}) + \alpha D\varphi\,(\mathbf{u})\,[\mathbf{v}] + HOT,$$

where HOT are terms of higher order, which vanish faster than α. Also, the square brackets enclosing $\mathbf{v}$ are used to emphasise the linearity of in $\mathbf{v}$. An operational definition for the derivative of $\varphi(\mathbf{u})$ in the direction $\mathbf{v}$ is therefore,

$$D\varphi\,(\mathbf{u})\,[\mathbf{v}] = \frac{d}{d\alpha}\,[\varphi\,(\mathbf{u} + \alpha\mathbf{v})]_{\alpha=0}. \tag{1.31}$$

This definition can be extended verbatim to derivatives of a tensor-valued (of any order) function of a tensor (of any order). The argument $\mathbf{v}$ is a part of the definition. We illustrate this with a few examples.

Example 1 Consider the scalar-valued function of a vector, $\varphi\,(\mathbf{u}) = u^2 = \mathbf{u} \cdot \mathbf{u}$. Its derivative in the direction of $\mathbf{v}$ is

$$D\varphi\,(\mathbf{u})\,[\mathbf{v}] = \frac{d}{d\alpha}\varphi\,(\mathbf{u} + \alpha\mathbf{v})_{\alpha=0} = \frac{d}{d\alpha}\left[u^2 + 2\alpha\mathbf{u} \cdot \mathbf{v} + \alpha^2 v^2\right]_{\alpha=0}$$
$$= 2\mathbf{u} \cdot \mathbf{v}.$$

Example 2 Consider the tensor-valued function of a tensor, $\mathbf{G}\,(\mathbf{A}) = \mathbf{A}^2 = \mathbf{A}\mathbf{A}$. Its derivative in the direction of $\mathbf{B}$ is

$$DG\left(\mathbf{A}\right)\left[\mathbf{B}\right] = \frac{d}{d\alpha}\left[\mathbf{G}\left(\mathbf{A} + \alpha\mathbf{B}\right)\right]_{\alpha=0}$$

$$= \frac{d}{d\alpha}\left[\mathbf{A}^2 + \alpha\left(\mathbf{AB} + \mathbf{BA}\right) + O\left(\alpha^2\right)\right]_{\alpha=0}$$

$$= \mathbf{AB} + \mathbf{BA}.$$

1.8.1 Derivative of det(A)

Consider the scalar-valued function of a tensor, $\varphi\left(\mathbf{A}\right) = \det\mathbf{A}$. Its derivative in the direction of $\mathbf{B}$ can be calculated using

$$\det(\mathbf{A} + \alpha\mathbf{B}) = \det\alpha\mathbf{A}\left(\mathbf{A}^{-1}\mathbf{B} + \alpha^{-1}\mathbf{I}\right) = \alpha^3\det\mathbf{A}\det\left(\mathbf{A}^{-1}\mathbf{B} + \alpha^{-1}\mathbf{I}\right)$$

$$= \alpha^3\det\mathbf{A}\left(\alpha^{-3} + \alpha^{-2}I_1\left(\mathbf{A}^{-1}\mathbf{B}\right) + \alpha^{-1}I_2\left(\mathbf{A}^{-1}\cdot\mathbf{B}\right) + I_3\left(\mathbf{A}^{-1}\mathbf{B}\right)\right)$$

$$= \det\mathbf{A}\left(1 + \alpha I_1\left(\mathbf{A}^{-1}\mathbf{B}\right) + O\left(\alpha^2\right)\right).$$

Thus

$$D\varphi\left(\mathbf{A}\right)\left[\mathbf{B}\right] = \frac{d}{d\alpha}\left[\varphi\left(\mathbf{A} + \alpha\mathbf{B}\right)\right]_{\alpha=0} = \det\mathbf{A}\mathrm{tr}\left(\mathbf{A}^{-1}\mathbf{B}\right).$$

1.8.2 Derivative of tr(A)

Consider the first invariant $I\left(\mathbf{A}\right) = \mathrm{tr}\mathbf{A}$. Its derivative in the direction of $\mathbf{B}$ is

$$DI\left(\mathbf{A}\right)\left[\mathbf{B}\right] = \frac{d}{d\alpha}\left[I\left(\mathbf{A} + \alpha\mathbf{B}\right)\right]_{\alpha=0}$$

$$= \frac{d}{d\alpha}\left[\mathrm{tr}\mathbf{A} + \alpha\mathrm{tr}\mathbf{B}\right]_{\alpha=0} = \mathrm{tr}\mathbf{B} = \mathbf{I} : \mathbf{B}.$$

1.8.3 Derivative of tr(A²)

Consider the second invariant $II\left(\mathbf{A}\right) = \mathrm{tr}\mathbf{A}^2$. Its derivative in the direction of $\mathbf{B}$ is

$$DII\left(\mathbf{A}\right)\left[\mathbf{B}\right] = \frac{d}{d\alpha}\left[II\left(\mathbf{A} + \alpha\mathbf{B}\right)\right]_{\alpha=0}$$

$$= \frac{d}{d\alpha}\left[\mathbf{A} : \mathbf{A} + \alpha\left(\mathbf{A} : \mathbf{B} + \mathbf{B} : \mathbf{A}\right) + O\left(\alpha^2\right)\right]_{\alpha=0}$$

$$= 2\mathbf{A} : \mathbf{B}.$$

1.9 Gradient of a Field

1.9.1 Field

A function of the position vector $\mathbf{x}$ is called a field. One has a scalar field, for example the temperature field $T(\mathbf{x})$, a vector field, for example the velocity field $\mathbf{u}(\mathbf{x})$, or a tensor field, for example the stress field $\mathbf{S}(\mathbf{x})$. Higher-order tensor fields are rarely encountered, as in the many-point correlation fields. Conservation equations in continuum mechanics involve derivatives (derivatives with respect to position vectors are called *gradients*) of different fields, and it is absolutely essential to know how to calculate the gradients of fields in different coordinate systems. We also find it more convenient to employ the dyadic notation at this point.

1.9.2 Cartesian Frame

We consider first a scalar field, $\varphi(\mathbf{x})$. The Taylor expansion of this about point x is

$$\varphi(\mathbf{x} + \alpha\mathbf{r}) = \varphi(\mathbf{x}) + \alpha r_j \frac{\partial}{\partial x_j}\varphi(\mathbf{x}) + O(\alpha^2).$$

Thus the gradient of $\varphi(\mathbf{x})$ at point $\mathbf{x}$, now written as $\nabla\varphi$, defined in (1.31), is given by

$$\nabla\varphi[\mathbf{r}] = \mathbf{r} \cdot \frac{\partial\varphi}{\partial\mathbf{x}}. \tag{1.32}$$

This remains unchanged for a vector or a tensor field.

Gradient Operator. This leads us to define the gradient operator as

$$\nabla = \mathbf{e}_j \frac{\partial}{\partial x_j} = \mathbf{e}_1 \frac{\partial}{\partial x_1} + \mathbf{e}_2 \frac{\partial}{\partial x_2} + \mathbf{e}_3 \frac{\partial}{\partial x_3}. \tag{1.33}$$

This operator can be treated as a vector, operating on its arguments. By itself, it has no meaning; it must operate on a scalar, a vector or a tensor.

Gradient of a Scalar. For example, the gradient of a scalar is

$$\nabla\varphi = \mathbf{e}_j \frac{\partial\varphi}{\partial x_j} = \mathbf{e}_1 \frac{\partial\varphi}{\partial x_1} + \mathbf{e}_2 \frac{\partial\varphi}{\partial x_2} + \mathbf{e}_3 \frac{\partial\varphi}{\partial x_3}. \tag{1.34}$$

Gradient of a Vector. The gradient of a vector can be likewise calculated

$$\nabla\mathbf{u} = \left(\mathbf{e}_i \frac{\partial}{\partial x_i}\right)(u_j\mathbf{e}_j) = \mathbf{e}_i\mathbf{e}_j \frac{\partial u_j}{\partial x_i}. \tag{1.35}$$

In matrix notation,

$$[\nabla \mathbf{u}] = \begin{bmatrix} \dfrac{\partial u_1}{\partial x_1} & \dfrac{\partial u_2}{\partial x_1} & \dfrac{\partial u_3}{\partial x_1} \\[2ex] \dfrac{\partial u_1}{\partial x_2} & \dfrac{\partial u_2}{\partial x_2} & \dfrac{\partial u_3}{\partial x_2} \\[2ex] \dfrac{\partial u_1}{\partial x_3} & \dfrac{\partial u_2}{\partial x_3} & \dfrac{\partial u_3}{\partial x_3} \end{bmatrix}.$$

The component $(\nabla \mathbf{u})_{ij}$ is $\partial u_j / \partial x_i$; some books define this differently.

Transpose of a Gradient. The transpose of a gradient of a vector is therefore

$$\nabla \mathbf{u}^T = \mathbf{e}_i \mathbf{e}_j \frac{\partial u_i}{\partial x_j}. \tag{1.36}$$

In matrix notation,

$$[\nabla \mathbf{u}]^T = \begin{bmatrix} \dfrac{\partial u_1}{\partial x_1} & \dfrac{\partial u_1}{\partial x_2} & \dfrac{\partial u_1}{\partial x_3} \\[2ex] \dfrac{\partial u_2}{\partial x_1} & \dfrac{\partial u_2}{\partial x_2} & \dfrac{\partial u_2}{\partial x_3} \\[2ex] \dfrac{\partial u_3}{\partial x_1} & \dfrac{\partial u_3}{\partial x_2} & \dfrac{\partial u_3}{\partial x_3} \end{bmatrix}.$$

Divergence of a Vector. The divergence of a vector is a scalar defined by

$$\nabla \cdot \mathbf{u} = \left(\mathbf{e}_i \frac{\partial}{\partial x_i} \right) \cdot (u_j \mathbf{e}_j) = \mathbf{e}_i \cdot \mathbf{e}_j \frac{\partial u_j}{\partial x_i} = \delta_{ij} \frac{\partial u_j}{\partial x_i}$$

$$\nabla \cdot \mathbf{u} = \frac{\partial u_i}{\partial x_i} = \frac{\partial u_1}{\partial x_1} + \frac{\partial u_2}{\partial x_2} + \frac{\partial u_3}{\partial x_3}. \tag{1.37}$$

The divergence of a vector is also an invariant, being the trace of a tensor.

Curl of a Vector. The curl of a vector is a vector defined by

$$\nabla \times \mathbf{u} = \left(\mathbf{e}_i \frac{\partial}{\partial x_i} \right) \times (u_j \mathbf{e}_j) = \mathbf{e}_i \times \mathbf{e}_j \frac{\partial u_j}{\partial x_i} = \varepsilon_{kij} \mathbf{e}_k \frac{\partial u_j}{\partial x_i}$$

$$= \mathbf{e}_1 \left(\frac{\partial u_3}{\partial x_2} - \frac{\partial u_2}{\partial x_3} \right) + \mathbf{e}_2 \left(\frac{\partial u_1}{\partial x_3} - \frac{\partial u_3}{\partial x_1} \right) + \mathbf{e}_3 \left(\frac{\partial u_2}{\partial x_1} - \frac{\partial u_1}{\partial x_2} \right). \tag{1.38}$$

The curl of a vector is sometimes denoted by rot.

Divergence of a Tensor. The divergence of a tensor is a vector field defined by

$$\nabla \cdot \mathbf{S} = \left(\mathbf{e}_k \frac{\partial}{\partial x_k} \right) \cdot (S_{ij} \mathbf{e}_i \mathbf{e}_j) = \mathbf{e}_j \frac{\partial S_{ij}}{\partial x_i}. \tag{1.39}$$

Fig. 1.5 Cylindrical and
spherical frame of references

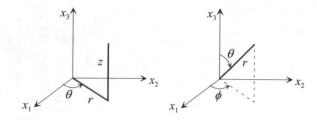

1.9.3 Non-Cartesian Frames

All the above definitions for gradient and divergence of a tensor remain valid in a
non-Cartesian frame, provided that the derivative operation is also applied to the
basis vectors as well. We illustrate this process in two important frames, cylindrical
and spherical coordinate systems (Fig. 1.5); for other systems, consult Bird et al. [6].

Cylindrical Coordinates. In a cylindrical coordinate system (Fig. 1.5, left), points
are located by giving them values to $\{r, \theta, z\}$, which are related to
$\{x = x_1, y = x_2, z = x_3\}$ by

$$x = r\cos\theta, \quad y = r\sin\theta, \quad z = z$$
$$r = \sqrt{x^2 + y^2}, \quad \theta = \tan^{-1}\left(\frac{y}{x}\right), \quad z = z$$

The basis vectors in this frame are related to the Cartesian ones by

$$\mathbf{e}_r = \cos\theta\,\mathbf{e}_x + \sin\theta\,\mathbf{e}_y, \qquad \mathbf{e}_x = \cos\theta\,\mathbf{e}_r - \sin\theta\,\mathbf{e}_\theta$$
$$\mathbf{e}_\theta = -\sin\theta\,\mathbf{e}_x + \cos\theta\,\mathbf{e}_y, \qquad \mathbf{e}_y = \sin\theta\,\mathbf{e}_r + \cos\theta\,\mathbf{e}_\theta$$

Physical components. In this system, a vector $\mathbf{u}$, or a tensor $\mathbf{S}$, are represented by,
respectively,

$$\mathbf{u} = u_r\mathbf{e}_r + u_\theta\mathbf{e}_\theta + u_z\mathbf{e}_z,$$
$$\mathbf{S} = S_{rr}\mathbf{e}_r\mathbf{e}_r + S_{r\theta}\mathbf{e}_r\mathbf{e}_\theta + S_{rz}\mathbf{e}_r\mathbf{e}_z + S_{\theta r}\mathbf{e}_\theta\mathbf{e}_r$$
$$+ S_{\theta\theta}\mathbf{e}_\theta\mathbf{e}_\theta + S_{\theta z}\mathbf{e}_\theta\mathbf{e}_z + S_{zr}\mathbf{e}_z\mathbf{e}_r + S_{z\theta}\mathbf{e}_z\mathbf{e}_\theta + S_{zz}\mathbf{e}_z\mathbf{e}_z.$$

Gradient operator. The components expressed this way are called physical components. The gradient operator is converted from one system to another by the chain
rule,

$$\nabla = \mathbf{e}_x \frac{\partial}{\partial x} + \mathbf{e}_y \frac{\partial}{\partial y} + \mathbf{e}_z \frac{\partial}{\partial z} = (\cos\theta\mathbf{e}_r - \sin\theta\mathbf{e}_\theta)\left(\cos\theta\frac{\partial}{\partial r} - \frac{\sin\theta}{r}\frac{\partial}{\partial\theta}\right)$$

$$+ (\sin\theta\mathbf{e}_r + \cos\theta\mathbf{e}_\theta)\left(\sin\theta\frac{\partial}{\partial r} + \frac{\cos\theta}{r}\frac{\partial}{\partial\theta}\right) + \mathbf{e}_z\frac{\partial}{\partial z}$$

$$= \mathbf{e}_r\frac{\partial}{\partial r} + \mathbf{e}_\theta\frac{1}{r}\frac{\partial}{\partial\theta} + \mathbf{e}_z\frac{\partial}{\partial z}. \tag{1.40}$$

When carrying out derivative operations, remember that

$$\frac{\partial}{\partial r}\mathbf{e}_r = 0, \quad \frac{\partial}{\partial r}\mathbf{e}_\theta = 0, \quad \frac{\partial}{\partial r}\mathbf{e}_z = 0$$

$$\frac{\partial}{\partial\theta}\mathbf{e}_r = \mathbf{e}_\theta, \quad \frac{\partial}{\partial\theta}\mathbf{e}_\theta = -\mathbf{e}_r, \quad \frac{\partial}{\partial\theta}\mathbf{e}_z = 0$$

$$\frac{\partial}{\partial z}\mathbf{e}_r = 0, \quad \frac{\partial}{\partial z}\mathbf{e}_\theta = 0, \quad \frac{\partial}{\partial z}\mathbf{e}_z = 0 \tag{1.41}$$

Gradient of a vector. The gradient of any vector is

$$\nabla\mathbf{u} = \left(\mathbf{e}_r\frac{\partial}{\partial r} + \mathbf{e}_\theta\frac{1}{r}\frac{\partial}{\partial\theta} + \mathbf{e}_z\frac{\partial}{\partial z}\right)(u_r\mathbf{e}_r + u_\theta\mathbf{e}_\theta + u_z\mathbf{e}_z)$$

$$= \mathbf{e}_r\mathbf{e}_r\frac{\partial u_r}{\partial r} + \mathbf{e}_r\mathbf{e}_\theta\frac{\partial u_\theta}{\partial r} + \mathbf{e}_r\mathbf{e}_z\frac{\partial u_z}{\partial r} + \mathbf{e}_\theta\mathbf{e}_r\frac{1}{r}\frac{\partial u_r}{\partial\theta} + \mathbf{e}_\theta\mathbf{e}_\theta\frac{u_r}{r}$$

$$+ \mathbf{e}_\theta\mathbf{e}_\theta\frac{1}{r}\frac{\partial u_\theta}{\partial\theta} - \mathbf{e}_\theta\mathbf{e}_r\frac{u_\theta}{r} + \mathbf{e}_\theta\mathbf{e}_z\frac{1}{r}\frac{\partial u_z}{\partial\theta} + \mathbf{e}_z\mathbf{e}_r\frac{\partial u_r}{\partial z} + \mathbf{e}_z\mathbf{e}_\theta\frac{\partial u_\theta}{\partial z}$$

$$+ \mathbf{e}_z\mathbf{e}_z\frac{\partial u_z}{\partial z}$$

$$\nabla\mathbf{u} = \mathbf{e}_r\mathbf{e}_r\frac{\partial u_r}{\partial r} + \mathbf{e}_r\mathbf{e}_\theta\frac{\partial u_\theta}{\partial r} + \mathbf{e}_r\mathbf{e}_z\frac{\partial u_z}{\partial r} + \mathbf{e}_\theta\mathbf{e}_r\left(\frac{1}{r}\frac{\partial u_r}{\partial\theta} - \frac{u_\theta}{r}\right)$$

$$+ \mathbf{e}_\theta\mathbf{e}_\theta\left(\frac{1}{r}\frac{\partial u_\theta}{\partial\theta} + \frac{u_r}{r}\right) + \mathbf{e}_\theta\mathbf{e}_z\frac{1}{r}\frac{\partial u_z}{\partial\theta} + \mathbf{e}_z\mathbf{e}_r\frac{\partial u_r}{\partial z} + \mathbf{e}_z\mathbf{e}_\theta\frac{\partial u_\theta}{\partial z}$$

$$+ \mathbf{e}_z\mathbf{e}_z\frac{\partial u_z}{\partial z}. \tag{1.42}$$

Divergence of a vector. The divergence of a vector is obtained by a contraction of the above equation:

$$\nabla\cdot\mathbf{u} = \frac{\partial u_r}{\partial r} + \frac{1}{r}\frac{\partial u_\theta}{\partial\theta} + \frac{u_r}{r} + \frac{\partial u_z}{\partial z}. \tag{1.43}$$

1.9.4   Spherical Coordinates

In a spherical coordinate system (Fig. 1.5, right), points are located by giving them values to $\{r, \theta, \phi\}$, which are related to $\{x = x_1, y = x_2, z = x_3\}$ by

$$x = r \sin \theta \cos \phi, \quad y = r \sin \theta \sin \phi, \quad z = r \cos \theta,$$

$$r = \sqrt{x^2 + y^2 + z^2}, \quad \theta = \tan^{-1}\left(\frac{\sqrt{x^2 + y^2}}{z}\right), \quad \phi = \tan^{-1}\left(\frac{y}{x}\right).$$

The basis vectors are related by

$$\mathbf{e}_r = \mathbf{e}_1 \sin \theta \cos \phi + \mathbf{e}_2 \sin \theta \sin \phi + \mathbf{e}_3 \cos \theta,$$

$$\mathbf{e}_\theta = \mathbf{e}_1 \cos \theta \cos \phi + \mathbf{e}_2 \cos \theta \sin \phi - \mathbf{e}_3 \sin \theta,$$

$$\mathbf{e}_\phi = -\mathbf{e}_1 \sin \phi + \mathbf{e}_2 \cos \phi,$$

and

$$\mathbf{e}_1 = \mathbf{e}_r \sin \theta \cos \phi + \mathbf{e}_\theta \cos \theta \cos \phi - \mathbf{e}_\phi \sin \phi,$$

$$\mathbf{e}_2 = \mathbf{e}_r \sin \theta \sin \phi + \mathbf{e}_\theta \cos \theta \sin \phi + \mathbf{e}_\phi \cos \phi,$$

$$\mathbf{e}_3 = \mathbf{e}_r \cos \theta - \mathbf{e}_\theta \sin \theta.$$

Gradient operator. Using the chain rule, it can be shown that the gradient operator in spherical coordinates is

$$\nabla = \mathbf{e}_r \frac{\partial}{\partial r} + \mathbf{e}_\theta \frac{1}{r} \frac{\partial}{\partial \theta} + \mathbf{e}_\phi \frac{1}{r \sin \theta} \frac{\partial}{\partial \phi}. \tag{1.44}$$

We list below a few results of interest.

Gradient of a scalar. The gradient of a scalar is given by

$$\nabla \varphi = \mathbf{e}_r \frac{\partial \varphi}{\partial r} + \mathbf{e}_\theta \frac{1}{r} \frac{\partial \varphi}{\partial \theta} + \mathbf{e}_\phi \frac{1}{r \sin \theta} \frac{\partial \varphi}{\partial \phi}. \tag{1.45}$$

Gradient of a vector. The gradient of a vector is given by

$$\nabla \mathbf{u} = \mathbf{e}_r \mathbf{e}_r \frac{\partial u_r}{\partial r} + \mathbf{e}_r \mathbf{e}_\theta \frac{\partial u_\theta}{\partial r} + \mathbf{e}_r \mathbf{e}_\phi \frac{\partial u_\phi}{\partial r} + \mathbf{e}_\theta \mathbf{e}_r \left(\frac{1}{r} \frac{\partial u_r}{\partial \theta} - \frac{u_\theta}{r} \right)$$

$$+ \mathbf{e}_\theta \mathbf{e}_\theta \left(\frac{1}{r} \frac{\partial u_\theta}{\partial \theta} + \frac{u_r}{r} \right) + \mathbf{e}_\phi \mathbf{e}_r \left(\frac{1}{r \sin \theta} \frac{\partial u_r}{\partial \phi} - \frac{u_\phi}{r} \right)$$

$$+ \mathbf{e}_\theta \mathbf{e}_\phi \frac{1}{r} \frac{\partial u_\phi}{\partial \theta} + \mathbf{e}_\phi \mathbf{e}_\theta \left(\frac{1}{r \sin \theta} \frac{\partial u_\theta}{\partial \phi} - \frac{u_\phi}{r} \cot \theta \right)$$

$$+ \mathbf{e}_\phi \mathbf{e}_\phi \left(\frac{1}{r \sin \theta} \frac{\partial u_\phi}{\partial \phi} + \frac{u_r}{r} + \frac{u_\theta}{r} \cot \theta \right). \tag{1.46}$$

Divergence of a vector. The divergence of a vector is given by

$$\nabla \cdot \mathbf{u} = \frac{1}{r^2} \frac{\partial}{\partial r} \left(r^2 u_r \right) + \frac{1}{r} \frac{\partial}{\partial \theta} \left(u_\theta \sin \theta \right) + \frac{1}{r \sin \theta} \frac{\partial u_\phi}{\partial \phi}. \tag{1.47}$$

Divergence of a tensor. The divergence of a tensor is given by

$$\nabla \cdot \mathbf{S} = \mathbf{e}_r \left[\frac{1}{r^2} \frac{\partial}{\partial r} \left(r^2 S_{rr} \right) + \frac{1}{r \sin \theta} \frac{\partial}{\partial \theta} \left(S_{\theta r} \sin \theta \right) + \frac{1}{r \sin \theta} \frac{\partial S_{\phi r}}{\partial \phi} \right. \tag{1.48}$$
$$\left. - \frac{S_{\theta\theta} + S_{\phi\phi}}{r} \right] + \mathbf{e}_\theta \left[\frac{1}{r^3} \frac{\partial}{\partial r} \left(r^3 S_{r\theta} \right) + \frac{1}{r \sin \theta} \frac{\partial}{\partial \theta} \left(S_{\theta\theta} \sin \theta \right) \right.$$
$$\left. + \frac{1}{r \sin \theta} \frac{\partial S_{\phi\theta}}{\partial \phi} + \frac{S_{\theta r} - S_{r\theta} - S_{\phi\phi} \cot \theta}{r} \right] + \mathbf{e}_\phi \left[\frac{1}{r^3} \frac{\partial}{\partial r} \left(r^3 S_{r\phi} \right) \right.$$
$$\left. + \frac{1}{r \sin \theta} \frac{\partial}{\partial \theta} \left(S_{\theta\phi} \sin \theta \right) + \frac{1}{r \sin \theta} \frac{\partial S_{\phi\phi}}{\partial \phi} + \frac{S_{\phi r} - S_{r\phi} + S_{\phi\theta} \cot \theta}{r} \right].$$

1.10 Integral Theorems

1.10.1 Gauss Divergence Theorem

Various volume integrals can be converted to surface integrals by the following theorems, due to Gauss (Fig. 1.6):

$$\int_V \nabla \varphi \, dV = \int_S \varphi \mathbf{n} \, dS, \tag{1.49}$$

Fig. 1.6 Carl Friedrich Gauss (1777–1855) was a Professor of Mathematics at the University of Göttingen. He made several contributions to Number Theory, Geodesy, Statistics, Geometry, Physics. His motto was "few, but ripe" (*Pauca, Sed Matura*). He did not publish several important papers because they did not satisfy these requirements

Fig. 1.7 A region enclosed
by a closed surface with
outward unit vector field

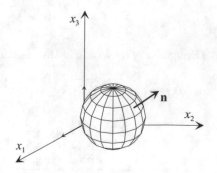

$$\int_V \nabla \cdot \mathbf{u} \, dV = \int_S \mathbf{n} \cdot \mathbf{u} \, dS, \tag{1.50}$$

$$\int_V \nabla \cdot \mathbf{S} \, dV = \int_S \mathbf{n} \cdot \mathbf{S} \, dS. \tag{1.51}$$

The proofs may be found in Kellogg [43]. In these, V is a bounded regular region, with bounding surface S and outward unit vector $\mathbf{n}$ (Fig. 1.7), φ, $\mathbf{u}$, and $\mathbf{S}$ are differentiable scalar, vector, and tensor fields with continuous gradients. Indeed the indicial version of (1.50) is valid even if u_i are merely three scalar fields of the required smoothness (rather than three components of a vector field).

1.10.2 Stokes Curl Theorem

Various surfaces integrals can be converted into contour integrals using the following theorems:

$$\int_S \mathbf{n} \cdot (\nabla \times \mathbf{u}) \, dS = \oint_C \mathbf{t} \cdot \mathbf{u} \, dC, \tag{1.52}$$

$$\int_S \mathbf{n} \cdot (\nabla \times \mathbf{S}) \, dS = \oint_C \mathbf{t} \cdot \mathbf{S} \, dC. \tag{1.53}$$

In these, $\mathbf{t}$ is a tangential unit vector along the contour C. The direction of integration is determined by the right-hand rule: thumb pointing in the direction of $\mathbf{n}$, fingers curling in the direction of C.

Fig. 1.8 Gottfried W.
Leibniz (1646–1716) was a
German philosopher and
mathematician, who
independently with Newton,
laid the foundation for
integral and differential
calculus in 1675

1.10.3 Leibniz Formula

If φ is a field (a scalar, a vector, or a tensor) define on a region $V(t)$, which is changing
in time, with bounding surface $S(t)$, also changing in time with velocity $\mathbf{u}_S$, then
(Leibniz formula, Fig. 1.8)

$$\frac{d}{dt} \int_V \varphi dV = \int_V \frac{\partial \varphi}{\partial t} dV + \int_S \varphi \mathbf{u}_S \cdot \mathbf{n} dS. \tag{1.54}$$

Problems

Problem 1.1 The components of vectors $\mathbf{u}$, $\mathbf{v}$, and $\mathbf{w}$ are given by u_i, v_i, w_i. Verify
that

$$\mathbf{u} \cdot \mathbf{v} = u_i v_i,$$
$$\mathbf{u} \times \mathbf{v} = \varepsilon_{ijk} \mathbf{e}_i u_j v_k,$$
$$(\mathbf{u} \times \mathbf{v}) \cdot \mathbf{w} = \varepsilon_{ijk} u_i v_j w_k,$$
$$(\mathbf{u} \times \mathbf{v}) \cdot \mathbf{w} = \mathbf{u} \cdot (\mathbf{v} \times \mathbf{w}),$$
$$(\mathbf{u} \times \mathbf{v}) \times \mathbf{w} = (\mathbf{u} \cdot \mathbf{w}) \mathbf{v} - (\mathbf{v} \cdot \mathbf{w}) \mathbf{u},$$
$$(\mathbf{u} \times \mathbf{v})^2 = u^2 v^2 - (\mathbf{u} \cdot \mathbf{v})^2,$$

where $u^2 = |\mathbf{u}|^2$ and $v^2 = |\mathbf{v}|^2$.

Problem 1.2 Let **A** be a 3×3 matrix with entries A_{ij},

$$[\mathbf{A}] = \begin{bmatrix} A_{11} & A_{12} & A_{13} \\ A_{21} & A_{22} & A_{23} \\ A_{31} & A_{32} & A_{33} \end{bmatrix}.$$

Verify that

$$\det[\mathbf{A}] = \varepsilon_{ijk} A_{1i} A_{2j} A_{3k} = \varepsilon_{ijk} A_{i1} A_{j2} A_{k3},$$
$$\varepsilon_{lmn} \det[\mathbf{A}] = \varepsilon_{ijk} A_{il} A_{jm} A_{kn} = \varepsilon_{ijk} A_{li} A_{mj} A_{nk},$$
$$\det[\mathbf{A}] = \frac{1}{6} \varepsilon_{ijk} \varepsilon_{lmn} A_{il} A_{jm} A_{kn}.$$

Problem 1.3 Verify that

$$\varepsilon_{ijk} \varepsilon_{imn} = \delta_{jm} \delta_{kn} - \delta_{jn} \delta_{km}.$$

Given that two 3×3 matrices of components

$$[\mathbf{A}] = \begin{bmatrix} A_{11} & A_{12} & A_{13} \\ A_{21} & A_{22} & A_{23} \\ A_{31} & A_{32} & A_{33} \end{bmatrix}, \quad [\mathbf{B}] = \begin{bmatrix} B_{11} & B_{12} & B_{13} \\ B_{21} & B_{22} & B_{23} \\ B_{31} & B_{32} & B_{33} \end{bmatrix}$$

verify that if $[\mathbf{C}] = [\mathbf{A}] \cdot [\mathbf{B}]$, then the components of **C** are $C_{ij} = A_{ik} B_{kj}$. Thus if $[\mathbf{D}] = [\mathbf{A}]^T [\mathbf{B}]$, then $D_{ij} = A_{ki} B_{kj}$.

Problem 1.4 Show that, if $\left[A_{ij} \right]$ is a frame rotation matrix,

$$\det \left[A_{ij} \right] = \left(\mathbf{e}'_1 \times \mathbf{e}'_2 \right) \cdot \mathbf{e}'_3 = 1.$$
$$[\mathbf{A}]^T [\mathbf{A}] = [\mathbf{A}][\mathbf{A}]^T = [\mathbf{I}], \quad [\mathbf{A}]^{-1} = [\mathbf{A}]^T, \quad \det[\mathbf{A}] = 1.$$

Problem 1.5 Verify that

$$\varepsilon_{ijk} u_i v_j w_k = \det \begin{bmatrix} u_1 & u_2 & u_3 \\ v_1 & v_2 & v_3 \\ w_1 & w_2 & w_3 \end{bmatrix}.$$

Consider a second-order tensor W_{ij} and a vector $u_i = \varepsilon_{ijk} W_{jk}$. Show that if **W** is symmetric, **u** is zero, and if **W** is anti-symmetric the components of **u** are twice those of **W** in magnitude. This vector is said to be the axial vector of **W**.

Hence, show that the axial vector associated with the vorticity tensor of (1.14) is $-\nabla \times \mathbf{u}$.

Problem 1.6 If $\mathbf{D}$, $\mathbf{S}$ and $\mathbf{W}$ are second-order tensors, $\mathbf{D}$ symmetric and $\mathbf{W}$ anti-symmetric, show that

$$\mathbf{D} : \mathbf{S} = \mathbf{D} : \mathbf{S}^T = \mathbf{D} : \tfrac{1}{2} \left(\mathbf{S} + \mathbf{S}^T \right),$$
$$\mathbf{W} : \mathbf{S} = -\mathbf{W} : \mathbf{S}^T = \mathbf{W} : \tfrac{1}{2} \left(\mathbf{W} - \mathbf{W}^T \right),$$
$$\mathbf{D} : \mathbf{W} = 0.$$

Further, show that

if $\mathbf{T} : \mathbf{S} = 0$ $\forall \mathbf{S}$ then $\mathbf{T} = 0$,
if $\mathbf{T} : \mathbf{S} = 0$ $\forall$ symmetric $\mathbf{S}$ then $\mathbf{T}$ is anti-symmetric,
if $\mathbf{T} : \mathbf{S} = 0$ $\forall$ anti-symmetric $\mathbf{S}$ then $\mathbf{T}$ is symmetric.

Problem 1.7 Show that $\mathbf{Q}$ is orthogonal if and only if $\mathbf{H} = \mathbf{Q} - \mathbf{I}$ satisfies

$$\mathbf{H} + \mathbf{H}^T + \mathbf{H}\mathbf{H}^T = 0, \qquad \mathbf{H}\mathbf{H}^T = \mathbf{H}^T\mathbf{H}.$$

Problem 1.8 Show that, if $\mathbf{S}$ is a second-order tensor, then $I = \operatorname{tr} \mathbf{S}$, $II = \operatorname{tr} \mathbf{S}^2$, $III = \det \mathbf{S}$ are indeed invariants. In addition, show that

$$\det \left(\mathbf{S} - \omega\mathbf{I} \right) = -\omega^3 + I_1\omega^2 - I_2\omega + I_3.$$

If ω is an eigenvalue of $\mathbf{S}$ then $\det \left(\mathbf{S} - \omega\mathbf{I} \right) = 0$. This is said to be the characteristic equation for $\mathbf{S}$.

Problem 1.9 Apply the result above to find the square root of the Cauchy-Green tensor in a two-dimensional shear deformation

$$[\mathbf{C}] = \begin{bmatrix} 1 + \gamma^2 & \gamma \\ \gamma & 1 \end{bmatrix}.$$

Investigate the corresponding formula for the square root of a symmetric positive definite tensor $\mathbf{S}$ in three dimensions.

Problem 1.10 Write down all the components of the strain rate tensor and the vorticity tensor in a Cartesian frame.

Problem 1.11 Given that $\mathbf{r} = x_i\mathbf{e}_i$ is the position vector, $\mathbf{a}$ is a constant vector, and $f(r)$ is a function of $r = |\mathbf{r}|$, show that

$$\nabla \cdot \mathbf{r} = 3, \quad \nabla \times \mathbf{r} = \mathbf{0}, \quad \nabla (\mathbf{a} \cdot \mathbf{r}) = \mathbf{a}, \quad \nabla f = \frac{1}{r}\frac{df}{dr}\mathbf{r}.$$

Problem 1.12 Show that the divergence of a second-order tensor $\mathbf{S}$ in cylindrical coordinates is given by

$$\nabla \cdot \mathbf{S} = \mathbf{e}_r \left(\frac{\partial S_{rr}}{\partial r} + \frac{S_{rr} - S_{\theta\theta}}{r} + \frac{1}{r} \frac{\partial S_{\theta r}}{\partial \theta} + \frac{\partial S_{zr}}{\partial z} \right)$$

$$+ \mathbf{e}_\theta \left(\frac{\partial S_{r\theta}}{\partial r} + \frac{2S_{r\theta}}{r} + \frac{1}{r} \frac{\partial S_{\theta\theta}}{\partial \theta} + \frac{\partial S_{z\theta}}{\partial z} + \frac{S_{\theta r} - S_{r\theta}}{r} \right)$$

$$+ \mathbf{e}_z \left(\frac{\partial S_{rz}}{\partial r} + \frac{S_{rz}}{r} + \frac{1}{r} \frac{\partial S_{\theta z}}{\partial \theta} + \frac{\partial S_{zz}}{\partial z} \right). \tag{1.55}$$

Problem 1.13 Show that, in cylindrical coordinates, the Laplacian of a vector $\mathbf{u}$ is given by

$$\nabla^2 \mathbf{u} = \mathbf{e}_r \left[\frac{\partial}{\partial r} \left(\frac{1}{r} \frac{\partial}{\partial r} (ru_r) \right) + \frac{1}{r^2} \frac{\partial^2 u_r}{\partial \theta^2} + \frac{\partial^2 u_r}{\partial z^2} - \frac{2}{r^2} \frac{\partial u_\theta}{\partial \theta} \right]$$

$$+ \mathbf{e}_\theta \left[\frac{\partial}{\partial r} \left(\frac{1}{r} \frac{\partial}{\partial r} (ru_\theta) \right) + \frac{1}{r^2} \frac{\partial^2 u_\theta}{\partial \theta^2} + \frac{\partial^2 u_\theta}{\partial z^2} + \frac{2}{r^2} \frac{\partial u_r}{\partial \theta} \right]$$

$$+ \mathbf{e}_z \left[\frac{1}{r} \frac{\partial}{\partial r} \left(r \frac{\partial u_z}{\partial r} \right) + \frac{1}{r^2} \frac{\partial^2 u_z}{\partial \theta^2} + \frac{\partial^2 u_z}{\partial z^2} \right]. \tag{1.56}$$

Problem 1.14 Show that, in cylindrical coordinates,

$$\mathbf{u} \cdot \nabla \mathbf{u} = \mathbf{e}_r \left[u_r \frac{\partial u_r}{\partial r} + \frac{u_\theta}{r} \frac{\partial u_r}{\partial \theta} + u_z \frac{\partial u_r}{\partial z} - \frac{u_\theta u_\theta}{r} \right]$$

$$+ \mathbf{e}_\theta \left[u_r \frac{\partial u_\theta}{\partial r} + \frac{u_\theta}{r} \frac{\partial u_\theta}{\partial \theta} + u_z \frac{\partial u_\theta}{\partial z} + \frac{u_\theta u_r}{r} \right]$$

$$+ \mathbf{e}_z \left[u_r \frac{\partial u_z}{\partial r} + \frac{u_\theta}{r} \frac{\partial u_z}{\partial \theta} + u_z \frac{\partial u_z}{\partial z} \right]. \tag{1.57}$$

Problem 1.15 The stress tensor in a material satisfies $\nabla \cdot \mathbf{S} = \mathbf{0}$. Show that the volume-average stress in a region V occupied by the material is

$$\langle \mathbf{S} \rangle = \frac{1}{2V} \int_S (\mathbf{x}\mathbf{t} + \mathbf{t}\mathbf{x}) \, dS, \tag{1.58}$$

where $\mathbf{t} = \mathbf{n} \cdot \mathbf{S}$ is the surface traction. The quantity on the left side of (1.58) is called the *stresslet* (Batchelor [4]).

Problem 1.16 Calculate the following integrals on the surface of the unit sphere

$$\langle \mathbf{n}\mathbf{n} \rangle = \frac{1}{S} \int_S \mathbf{n}\mathbf{n} \, dS \tag{1.59}$$

$$\langle \mathbf{n}\mathbf{n}\mathbf{n}\mathbf{n} \rangle = \frac{1}{S} \int_S \mathbf{n}\mathbf{n}\mathbf{n}\mathbf{n} \, dS. \tag{1.60}$$

These are the averages of various moments of a uniformly distributed unit vector on a sphere surface.

Chapter 2
Rheological Properties

Overall Material Properties and Flow Behaviour

Fluids with featureless microstructures are well described by the Newtonian constitutive equation, which states that the stress tensor is proportional to the shear rate tensor (these concepts will be made precise later). Fluids with complex microstructures, for example suspensions of particles or droplets (blood, paint, ink, asphalt, bitumen, foodstuffs, etc.), polymer melts and solutions (molten plastics, fibre-reinforced or particulate-filled plastics), exhibit a wide variety of behaviours. These are summarised here. For more information, consult Bird et al. [6] and Tanner [85].

2.1 Viscosity

2.1.1 Shear-Rate Dependent Viscosity

One of the most important fluid properties for engineering calculations is its viscosity. This quantity is defined as the ratio of the shear stress to the shear rate in a simple shear flow.

Here, as shown in Fig. 2.1, the flow is generated by sliding one plate atop another, with the fluid in-between. The quantities of interest are the shear rate, $\dot{\gamma} = U/h$ (U is the velocity of the top plate in the x-direction, the bottom plate is fixed, h is the distance between the plates), and the shear stress, $S = F/A$ (F is the shear force on the top plate, A the fluid contact area). The shear stress is an odd function of the shear rate. In addition, with viscoelastic fluids, there may be a normal force on the plates. When a steady flow is established, the viscosity is defined as

$$\eta = \frac{S}{\dot{\gamma}}.$$

(2.1)

© Springer International Publishing AG 2017
N. Phan-Thien and N. Mai-Duy, *Understanding Viscoelasticity*,
Graduate Texts in Physics, DOI 10.1007/978-3-319-62000-8_2

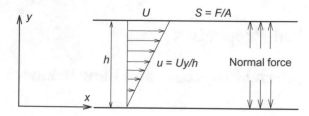

Fig. 2.1 Shear flow generated by sliding one plate on *top* of another. Shear force as well as normal force may be required to keep the plates at a fixed distance

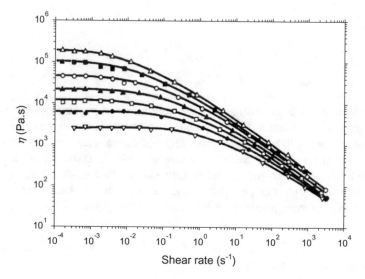

Fig. 2.2 Viscosity of a low density polyethylene melt at different temperature. *Top curve,* $T = 388$ K, and the *bottom curve* $T = 513$ K

For a Newtonian fluid, its viscosity is a constant (having units Pa.s) depending only on the temperature. For most fluids with long chain microstructure (polymer melts and solutions), the viscosity is a decreasing function of the shear rate, sometimes reaching $10^{-3} - 10^{-4}$ of the zero-shear rate viscosity. This type of behaviour is called *shear thinning*. The opposite behaviour, *shear thickening*, is sometimes observed with some suspensions, primarily due to the formation of clusters.

A typical viscosity-shear rate curve is shown in Fig. 2.2 for a low-density polyethylene (LDPE) at different temperatures. It can be seen from this figure that this fluid viscosity is a strong function of its temperature: a 60 °C increase in the temperature induces a ten-fold decrease in its viscosity. In addition, it may be observed that LDPE viscosity decreases by an order of magnitude (compared to its zero-shear-rate value) at a moderate value of shear rate. Note that a constant viscosity does not qualify a fluid to be Newtonian – the term Newtonian is much more restrictive in its

Fig. 2.3 Jean-Louis-Marie
Poiseuille (1797–1869) was
a French physician who
established experimentally
the pressure-drop/flow rate
relationship of laminar flow
in tubes

meaning, implying a whole class of constitutive behaviour in which the stress tensor
is proportional to the strain rate tensor.

For some materials with a solid-like behaviour (for example, bread dough, biological tissues), viscosity measurement makes no sense, since the shear stress just
keeps increasing with time until the sample breaks or flows out of the test cell, and
what has been measured is not a material property, but an indication of the friction
between the sample and the test apparatus (which depends on the flow process).

With suspensions of particles with surface charges, one can get the viscosity to
behave in many different ways; even a discontinuity at a particular shear rate may
be induced.

Incidentally, the CGS units for viscosity is Poise (P), in honour of Poiseuille
(Fig. 2.3), who provided flow rate/pressure drop experimental relationship for pipe
flow in 1846; (1 Pa.s = 10 P). Viscosity of water is about 1 cP (1 centi-Poise = 10^{-2} P).

2.2 Normal Stress Differences

Normal stress differences refer to the differences between the unequal normal stresses
in shear flow (for a Newtonian fluid in shear flow, the normal stresses are always equal,
a consequence of its constitutive equation). With three normal stress components,
we can form two independent quantities, the *first* and the *second normal stress
differences*:

$$N_1 = S_{xx} - S_{yy}, \quad N_2 = S_{yy} - S_{zz}. \tag{2.2}$$

Fig. 2.4 Viscometric functions of 6.8% of polyisobutylene in cetane at 24 °C. Here σ is the shear stress

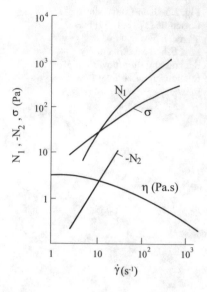

These normal stress differences are even functions of the shear rate, and therefore one defines the normal stress coefficients as

$$\nu_1 = \frac{N_1}{\dot{\gamma}^2}, \qquad \nu_2 = \frac{N_2}{\dot{\gamma}^2}. \tag{2.3}$$

These normal stress coefficients are even functions of the shear rate. The normal stress differences and the shear viscosity are collectively called *viscometric functions*; they are the material properties of the fluid in shear (viscometric) flow.

Figure 2.4 shows some typical measurements of viscometric properties of a polyisobutylene solution, the first (N_1) and the second (N_2) normal stress differences, the shear stress (σ) and the viscosity ($\eta = \sigma/\dot{\gamma}$).

The second normal stress difference is not usually measured. In general, for polymer melts and solutions, it is negative, and about 10% of N_1 in magnitude.

Suspensions have non-zero normal stress differences as well; however, our knowledge of them is still incomplete. Non-equal normal stresses are responsible for some visually striking differences between Newtonian and non-Newtonian fluid. We summarise the key features here.[1]

---

[1] The film *Rheological Behavior of Fluids*, presented by Prof. Hershel Markovitz, should be watched at this point. It contains the main important non-Newtonian flow phenomena and can be found at the site www.web.mit.edu/hml/ncfmf.html. This site is a depository of a large number of other interesting fluid mechanics films. The book by Boger and Walters [10] should also be consulted - it contains a large number of interesting photographs detailing non-Newtonian behaviours.

Fig. 2.5 Weissenberg rod
climbing effect

2.2.1 Weissenberg Rod-Climbing Effect

When a rod rotates in viscoelastic fluid, the fluid climbs the rod. This phenomenon
is called the Weissenberg[2] rod-climbing effect (Fig. 2.5). Rod climbing is due to the
fluid element being able to support a tension along a streamline (due to non-zero
normal stress differences), which forces the fluid up the rod. This effect can occur
without the rod: if one rotates a disk at the bottom of the beaker, then the free surface
bulges up at the middle.

2.2.2 Die Swell

When a viscoelastic fluid exits from a capillary of diameter D, it tends to swell
considerably more than a Newtonian fluid. For a viscous Newtonian fluid, the swell
ratio, D_E/D, where D_E is the extrudate diameter, is a function of the Reynolds
number and is at most 13%. For a polymer melt, the extrudate diameter could be
a few times the capillary diameter. This phenomenon is called *die swell* (Fig. 2.6),
and the dominant mechanism causing this is the first normal stress difference. In
fact, Tanner [84] proposed the simple rule for capillary die swell, based on a simple
analysis,

---

[2]Karl Weissenberg (1893–1976) contributed significantly to Rheology in the early years, and has
several phenomena named after him.

Fig. 2.6 Die swell – *top* a
Newtonian fluid, *bottom* a
viscoelastic fluid

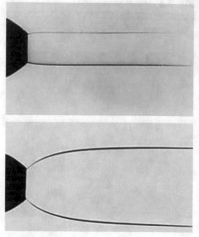

Fig. 2.7 Delay die swell –
increasing Reynolds number
from *left* to *right*

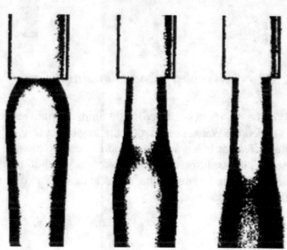

$$\frac{D_E}{D} = 0.13 + \left[1 + \frac{1}{2} \left(\frac{N_1}{2S} \right)^2_w \right]^{1/6}, \tag{2.4}$$

where N_1 and S are the first normal stress difference and the shear stress, both
evaluated at the wall (subscript w). Die swell is mainly due to the fluid elasticity
(normal stress effects), but it can also occur with the shear thinning induced by
viscous heating. Inertia (i.e., Reynolds number) tends to reduce the amount of swell,
and to delay it, see Fig. 2.7.

Fig. 2.8 In a flow down an
inclined channel, the free
surface will bulge up if N_2 is
negative

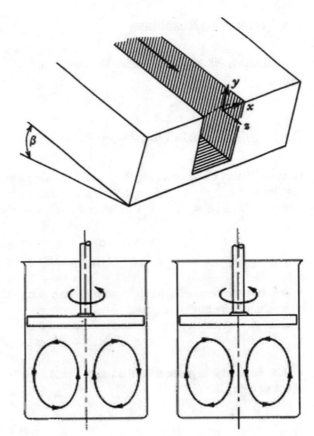

Fig. 2.9 Reversed
secondary flow – *left*
Newtonian fluid, *right*
viscoelastic fluid

2.2.3   Flow down an Inclined Channel

The second normal stress difference, although small in magnitude compared to the
first normal stress difference, is important in some cases. In the flow down an inclined
channel, a Newtonian fluid is seen to have a nearly flat free surface, whereas a convex
surface is seen for a viscoelastic fluid with a negative second normal stress difference
(Fig. 2.8).

Viscoelasticity is also responsible for the reversal of the secondary flow pattern;
one such case is sketched in Fig. 2.9.

2.3 Transient Responses

Viscoelastic fluids have a relaxation time scale, and this can be quantified in several ways.

2.3.1 Small Strain Oscillatory Flow

In an oscillatory shear flow (Fig. 2.1), where the top plate oscillates sinusoidally with angular frequency ω, $x = \delta \sin \omega t$, the plate velocity is $U = \omega \delta \cos \omega t$. The shear rate ($\dot{\gamma}$) and the shear strain ($\gamma$) are given by, respectively,

$$\dot{\gamma} = \dot{\gamma}_0 \cos \omega t, \quad \gamma = \gamma_0 \sin \omega t, \tag{2.5}$$
$$\dot{\gamma}_0 = \delta \omega / h, \quad \gamma_0 = \delta / h.$$

When the strain is small, the shear stress is also sinusoidal, but is not in phase with either the strain or the strain rate,

$$S = S' \sin \omega t + S'' \cos \omega t.$$

The part that is in phase with the strain is used to define the storage modulus (G'), or the storage viscosity (η'')

$$G' = \frac{S'}{\gamma_0}, \quad \eta'' = \frac{S'}{\dot{\gamma}_0}, \quad G' = \omega \eta'', \tag{2.6}$$

and the part that is in phase with the strain rate is used to define the loss modulus (G''), or the dynamic viscosity (η'),

$$G'' = \frac{S''}{\gamma_0}, \quad \eta' = \frac{S''}{\dot{\gamma}_0}, \quad G'' = \omega \eta'. \tag{2.7}$$

These are functions of the frequency, and they are collectively referred to as the *dynamic properties* of the fluid.

Figure 2.10 show the storage and loss moduli of LDPE at different temperatures. The data have been collapsed into a master curve through the use of the time-temperature superposition principle, which involves scaling the frequency by an empirical *shift factor* a_T. The dynamic properties contain time scale information on the fluid expressed in the frequency domain. Large-amplitude oscillatory tests have also been done, but their interpretation is less straightforward.

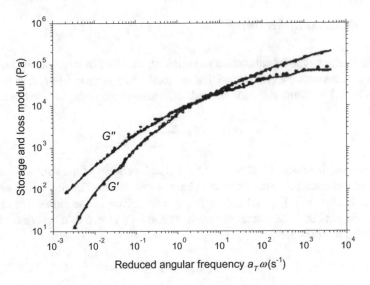

Fig. 2.10 Dynamic properties of a low density polyethylene (LDPE) melt

2.3.2 Stress Overshoot

In a start-up of a shear flow, i.e., $U = U_0 H(t)$, or $\dot\gamma = \dot\gamma_0 H(t)$, where $H(t)$ is the Heaviside function, the shear stress increases with time, then overshoots before approaching its steady value, sometimes with a few oscillations about the steady value. This is seen with the normal stress differences as well. One can define the unsteady viscosity, in a similar manner to the definition of the (steady-state) viscosity:

$$\eta^+(t, \dot\gamma_0) = \frac{S(t)}{\dot\gamma_0}. \tag{2.8}$$

The amount of stress overshoot can be significant at high shear rates, and it may be the main reason why certain biological fluid (e.g., synovial fluid) is a good lubricant.

2.3.3 Stress Relaxation

Corresponding to the start-up of shear flow is stress relaxation, where the fluid motion that has been undergoing a steady-state shear flow at a shear rate is suddenly stopped. The shear stress (and the normal stress differences) is monitored as it relaxes. Again, one can define the stress-relaxation viscosity

$$\eta^-(t, \dot\gamma_0) = \frac{S(t)}{\dot\gamma_0}. \tag{2.9}$$

2.3.4   Relaxation Modulus

There is another type of relaxation experiment, in which a large strain rate $\dot{\gamma}_0$ is applied over a small interval Δt, so that the total strain is $\gamma_0 = \dot{\gamma}_0 \Delta t$, and the shear stress $S(t, \gamma_0)$ is monitored as it relaxes. This allows the *relaxation modulus* to be defined:

$$G(t, \gamma_0) = \frac{S(t, \gamma_0)}{\gamma_0}. \tag{2.10}$$

At small enough strains, $G(t, \gamma_0) = G_0(t)$ is independent of the strain, because of the linearity between the stress and the strain at low strains. The relaxation modulus of LDPE is shown in Fig. 2.11. The nearly parallelism of the curves (at different strains) suggests that the relaxation modulus can be factored in a function of strain and a function of time,

$$G(t, \gamma_0) = h(\gamma_0)G_0(t), \quad h(0) = 1. \tag{2.11}$$

This is known as *strain-time separability*.

2.3.5   Recoil

If the loading is suddenly removed by cutting the liquid column, as seen in Fig. 2.12, the liquid retracts to some previous shape. The liquid is said to have memory (it remembers its original configuration). However, its memory is imperfect, as it can

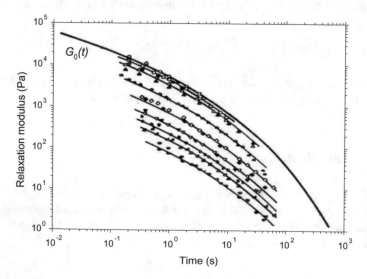

Fig. 2.11 Relaxation modulus of low density polyethylene (LDPE) – increasing strain from *top* to *bottom*

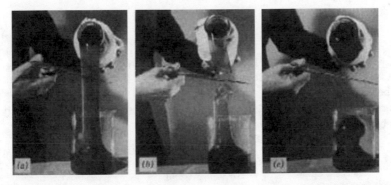

Fig. 2.12 The liquid recoils back into the beaker after Prof. A.S. Lodge cut the liquid column (University of Wisconsin, Madison)

only retract partially. In that sense it has fading memory. A Newtonian liquid has a catastrophic memory: the moment the loading is removed, the motion ceases immediately. An elastic solid has a perfect memory: upon removal of the loads, the solid particles return to exactly the positions they occupied previously.

2.4 Elongational Flows

2.4.1 Elongational Viscosity

Elongational flows refer to flow where the velocity gradient is diagonal, i.e.,

$$u = ax, \quad v = by, \quad w = cz, \tag{2.12}$$

where $a + b + c = 0$ for incompressibility (more about this later). These flows correspond to stretching or elongating a sample fluid specimen. When $b = -a$ and $c = 0$ one has a planar elongational flow, and a uni-axial elongational flow when $a = b = -c/2$. This latter flow occurs in many processes; here a is termed the elongational rate. The elongational viscosity is defined as

$$\eta_E = \frac{S_{xx} - S_{yy}}{a}. \tag{2.13}$$

Except at very low elongational rates, elongational viscosity does not usually reach a steady state (the sample elongates and fails). For a Newtonian fluid its elongational viscosity is thrice its shear viscosity; but for a polymer solution, the elongational viscosity can be orders of magnitude greater. The Trouton ratio is defined as the ratio of the elongation viscosity to the shear viscosity of the fluid

$$\text{Trouton Ratio} = \frac{\eta_E}{\eta}. \tag{2.14}$$

A typical plot of the Trouton ratio for a polybutene solution is shown in Fig. 2.13. For a Newtonian fluid, the Trouton ratio is three, for a viscoelastic fluid, this ratio may be very large.

The ability of a liquid filament to support a significant tensile stress is mainly why the tubeless siphon experiment (Fig. 2.14) works.

Fig. 2.13 Transient Trouton ratio for a high molecular weight polyisobutylene solution – the extensional rate is $a = 2\,\text{s}^{-1}$

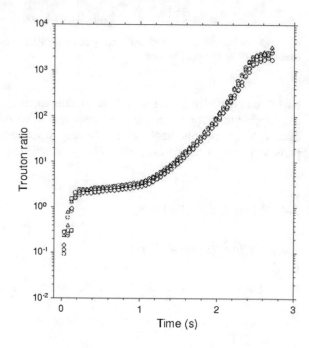

Fig. 2.14 Tubeless siphon

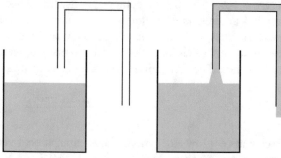

2.5 Viscoelastic Instabilities

Because of the non-linearity in the constitutive equations, viscoelastic flows are full of instabilities. These instabilities may not depend on inertia; they are mainly driven by the fluid normal stresses (elasticity), or by the nature of the boundary conditions. To name a few, we have instability in Taylor-Couette flow, in the torsional flow between two parallel disks, in the shear flow between cone-and-plate, in curved pipe flow, in contraction flows, in the flows from extrusion dies, etc. The extrudate distortion, commonly called melt fracture, is an example of instability due to the interplay between viscoelasticity and the nature of the boundary conditions. The reference [10] contains several photographs of this phenomenon.

Chapter 3
Kinematics and Equations of Balance

A Quick Review of Continuum Mechanics

In this chapter, we review the kinematics and the equations of balance (conservation equations), leaving the question of constitutive description to the next chapter. Bird et al. [6], Tanner [85], Huilgol and Phan-Thien [39], Morrison [59] provide additional reading materials.

3.1 Kinematics

3.1.1 Reference Configuration

We deal with a continuous body $\mathcal{B}$, which occupies a region consisting of points in $\mathbb{E}_3$. We refer to one particular configuration, $\mathcal{B}_R$, for example the configuration at time $t = 0$, as the reference configuration. The particle position in the reference configuration is denoted by a capital letter $\mathbf{X}$. This particle traces out a path in the three-dimensional space $\mathbb{E}_3$ and its current position is denoted by the small letter $\mathbf{x}$; $\mathbf{x}$ is a function of time. The particle is referred to by its position in the reference configuration, $\mathbf{X}$.

A motion is defined to be a twice-differentiable and invertible map (so that acceleration field can be defined, and that every position $\mathbf{x}$ must correspond to a particle $\mathbf{X}$)

$$\mathbf{x} = \mathbf{M}(\mathbf{X}, t), \quad x_i = M_i(X_j, t), \tag{3.1}$$

where t is the time. This is also called the *Lagrangian description* of the motion (after Lagrange, Fig. 3.1). Note that $\mathbf{X} = \mathbf{M}(\mathbf{X}, 0)$, by definition. Collectively $\mathbf{M}(\mathbf{X}, t)$, $\mathbf{X} \in \mathcal{B}_R$ gives us the spatial description of the motion, called the *current configuration*. Since $\mathbf{M}$ is invertible,

$$\mathbf{X} = \mathbf{M}^{-1}(\mathbf{x}, t), \quad X_i = M_i^{-1}(x_j, t) \tag{3.2}$$

gives us the reference in terms of the current configuration.

© Springer International Publishing AG 2017
N. Phan-Thien and N. Mai-Duy, *Understanding Viscoelasticity*,
Graduate Texts in Physics, DOI 10.1007/978-3-319-62000-8_3

Fig. 3.1 Joseph-Louis
Lagrange (1736–1813) was a
Italian/French
mathematician, who made
important contributions to
analytic mechanics, calculus
of variations, and number
theory. His book *Mécanique
Analytique* was an approved
publication by a committee
with members including
Laplace and Legendre in
1788

3.1.2　Velocity and Acceleration Fields

The velocity and the acceleration fields are defined as

$$\hat{\mathbf{u}} = \frac{\partial}{\partial t}\mathbf{M}(\mathbf{X}, t), \quad \hat{u}_i = \frac{\partial}{\partial t}M_i(X_j, t),　　　　(3.3)$$

and

$$\hat{\mathbf{a}} = \frac{\partial^2}{\partial t^2}\mathbf{M}(\mathbf{X}, t), \quad \hat{a}_i = \frac{\partial^2}{\partial t^2}M_i(X_j, t),　　　　(3.4)$$

respectively.

It is customary to refer to velocity and acceleration fields as functions of the current position, the so-called *Eulerian description* (after Euler, Fig. 3.2). This is accomplished using (3.2).

Denoting $\mathbf{u} = \mathbf{u}(\mathbf{x}, t)$ as the Eulerian velocity field, we find that the Eulerian acceleration field is given by

$$\mathbf{a} = \frac{\partial}{\partial t}\mathbf{u}(\mathbf{x}, t)\bigg|_{\mathbf{X}} = \frac{\partial \mathbf{u}}{\partial t} + \frac{\partial \mathbf{u}}{\partial \mathbf{x}} \cdot \frac{\partial \mathbf{x}}{\partial t}\bigg|_{\mathbf{X}}$$

$$= \frac{\partial \mathbf{u}}{\partial t} + \mathbf{u} \cdot \nabla \mathbf{u} = \frac{\partial \mathbf{u}}{\partial t} + \mathbf{L} \cdot \mathbf{u},　　　　(3.5)$$

where we have introduced the velocity gradient tensor (note the transpose operation in the definition)

$$\mathbf{L} = \left(\frac{\partial \mathbf{u}}{\partial \mathbf{x}}\right)^T = (\nabla \mathbf{u})^T, \quad L_{ij} = \frac{\partial u_i}{\partial x_j}.　　　　(3.6)$$

Fig. 3.2 Leonhard Euler (1707–1783) was one of the greatest mathematicians of the 18th century. He occupied Daniel Bernouilli's chair of mathematics at St. Petersburg. He perfected the integral calculus, worked in analytic geometry, theory of lunar motion, introduced the Euler's identity and many other mathematical symbols that we are familiar with, π, $f(),\sum$. He published more than 856 articles and books

An Eulerian velocity is called steady if it does not depend on time, i.e., $\mathbf{u} = \mathbf{u}(\mathbf{x})$. A steady Eulerian velocity field is thus not necessarily Lagrangian steady.

3.1.3 Material Derivative

The derivative on $\mathbf{u}$, as implied in (3.5) is called the material derivative, or total time derivative, or simply time derivative (if there is no confusion),

$$\frac{d}{dt}(.) = \frac{\partial}{\partial t}(.) + \mathbf{u} \cdot \nabla(.).\tag{3.7}$$

The symmetric part of the velocity gradient is called the strain rate tensor $\mathbf{D}$, and its anti-symmetric part is called the vorticity tensor,

$$\mathbf{L} = \mathbf{D} + \mathbf{W}, \quad \mathbf{D} = \frac{1}{2}\left(\mathbf{L} + \mathbf{L}^T\right), \quad \mathbf{W} = \frac{1}{2}\left(\mathbf{L} - \mathbf{L}^T\right).\tag{3.8}$$

3.2 Deformation Gradient and Strain Tensors

3.2.1 Deformation Gradient

The gradient of $\mathbf{x}$ with respect to $\mathbf{X}$ is called the deformation gradient,

$$\mathbf{F} = \left(\frac{\partial \mathbf{x}}{\partial \mathbf{X}}\right)^T, \quad F_{ij} = \frac{\partial x_i}{\partial X_j}.\tag{3.9}$$

Note again our subscript convention. At time $t = 0$ the initial value of $\mathbf{F}$ is

$$\mathbf{F}(0) = \mathbf{I}, \tag{3.10}$$

the identity tensor. The mass in a region V is

$$\int_V \rho d\mathbf{x} = \int_{V_0} \rho \, |J| \, d\mathbf{X},$$

where $J = \det \mathbf{F}$ and V_0 is the region occupied by the reference configuration. Thus, we demand that $\det \mathbf{F} > 0$, so that the mapping is not degenerate. For an incompressible fluid, the kinematic constraint is of course

$$\det \mathbf{F} = 1. \tag{3.11}$$

Because of the chain rule,

$$\left(\frac{\partial \mathbf{X}}{\partial \mathbf{x}}\right)^T = \mathbf{F}^{-1}, \quad F_{ij}^{-1} = \frac{\partial X_i}{\partial x_j}. \tag{3.12}$$

The connection between the deformation and the velocity gradients arises from the equality

$$\frac{\partial}{\partial t} F_{ij} = \frac{\partial}{\partial t}\left(\frac{\partial M_i}{\partial X_j}\right) = \frac{\partial}{\partial X_j}\left(\frac{\partial M_i}{\partial t}\right) = \frac{\partial \hat{u}_i}{\partial X_j}.$$

Using the Eulerian description for the velocity,

$$\frac{\partial \hat{u}_i}{\partial X_j} = \frac{\partial u_i}{\partial x_k}\frac{\partial x_k}{\partial X_j} = L_{ik}F_{kj},$$

one has,

$$\dot{\mathbf{F}} = \mathbf{LF}, \quad \mathbf{F}(0) = \mathbf{I}, \tag{3.13}$$

where the super dot denotes the (total) time derivative. This equation provides an initial-value problem for $\mathbf{F}$.

3.2.2   Cauchy–Green Strain Tensor

The concept of strain is introduced by comparing the length of a fluid element at the current time to that in the reference configuration. We have from the definition of the deformation gradient,

$$d\mathbf{x} = \mathbf{F}d\mathbf{X}.$$

Fig. 3.3 An element $d\mathbf{X}$ at $\mathbf{X}$ in the reference configuration at time $t = 0$ is mapped to $d\mathbf{x}$ at $\mathbf{x}$ at time t

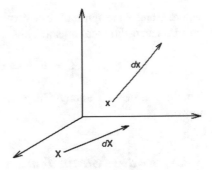

Fig. 3.4 Augustin Cauchy (1789–1857) was a prolific French mathematician. He contributed more than 16 fundamental concepts and theorems, and published more than 800 papers. His name is one of the 72 inscribed on the Eiffel tower

Here $d\mathbf{X}$ is a fluid element at point $\mathbf{X}$, which at time t is mapped to $d\mathbf{x}$ at point $\mathbf{x}$, see Fig. 3.3. Its current length is

$$dx^2 = d\mathbf{x} \cdot d\mathbf{x} = F_{ij}dX_j F_{ik}dX_k = d\mathbf{X}\mathbf{F}^T \cdot \mathbf{F}d\mathbf{X} = \mathbf{F}^T\mathbf{F} : d\mathbf{X}d\mathbf{X}.$$

The tensor

$$\mathbf{C} = \mathbf{F}^T\mathbf{F} \tag{3.14}$$

is therefore a measure of the strain the fluid experiences. It is called the *right Cauchy–Green* tensor (for a portray of Cauchy, see Fig. 3.4). The *left Cauchy–Green* tensor is defined as

$$\mathbf{B} = \mathbf{F}\mathbf{F}^T. \tag{3.15}$$

The name refers to the right or left polar decompositions of $\mathbf{F}$,

$$\mathbf{F} = \mathbf{R}\mathbf{U} \ (\text{right}) \ = \mathbf{V}\mathbf{R} \ (\text{left}),$$

where $\mathbf{U}$ and $\mathbf{V}$ are symmetric positive-definite tensors (right and left stretch tensors), and $\mathbf{R}$ is an orthogonal tensor. Thus

$$\mathbf{C} = \mathbf{F}^T\mathbf{F} = \mathbf{U}^2, \quad \mathbf{B} = \mathbf{F}\mathbf{F}^T = \mathbf{V}^2.$$

The inverse of the Cauchy–Green tensor is also used; it is called the Finger strain tensor.

3.2.3   Relative Strain Tensors

The reference configuration enjoys no particular mathematical status, although it may have a physical significance (e.g., the stress-free state). Suppose the particle X at time τ occupies the position $\boldsymbol{\xi}$, then the relative deformation gradient is defined as

$$\mathbf{F}_t\left(\tau\right) = \left(\frac{\partial \boldsymbol{\xi}}{\partial \mathbf{x}}\right)^T, \quad (\mathbf{F}_t(\tau))_{ij} = \frac{\partial \xi_i}{\partial x_j}. \tag{3.16}$$

Correspondingly, the right relative Cauchy–Green tensor is given by

$$\mathbf{C}_t\left(\tau\right) = \mathbf{F}_t\left(\tau\right)^T \mathbf{F}_t\left(\tau\right), \quad (\mathbf{C}_t(\tau))_{ij} = \frac{\partial \xi_k}{\partial x_i}\frac{\partial \xi_k}{\partial x_j}. \tag{3.17}$$

Similarly, the left Cauchy-Green tensor is given by

$$\mathbf{B}_t\left(\tau\right) = \mathbf{F}_t\left(\tau\right) \mathbf{F}_t\left(\tau\right)^T, \quad (\mathbf{B}_t(\tau))_{ij} = \frac{\partial \xi_i}{\partial x_k}\frac{\partial \xi_j}{\partial x_k}. \tag{3.18}$$

Because of the chain rule, the relative deformation gradient satisfies

$$\mathbf{F}_t\left(\tau\right) = \left(\frac{\partial \boldsymbol{\xi}}{\partial \mathbf{x}}\right)^T = \left(\frac{\partial \mathbf{X}}{\partial \mathbf{x}} \cdot \frac{\partial \boldsymbol{\xi}}{\partial \mathbf{X}}\right)^T = \left(\frac{\partial \boldsymbol{\xi}}{\partial \mathbf{X}}\right)^T \cdot \left(\frac{\partial \mathbf{X}}{\partial \mathbf{x}}\right)^T$$
$$= \mathbf{F}\left(\tau\right)\mathbf{F}\left(t\right)^{-1}. \tag{3.19}$$

3.2.4   Path Lines

To solve for the path lines of the particles, knowing the velocity field, we integrate the set of equations

$$\frac{d\boldsymbol{\xi}}{d\tau} = \mathbf{u}\left(\boldsymbol{\xi}, \tau\right), \quad \boldsymbol{\xi}\left(\tau\right)|_{\tau=t} = \mathbf{x}. \tag{3.20}$$

When the flow is steady, the velocity field is independent of time, this system is an *autonomous system*.

As an example, consider the case where the velocity is steady and homogeneous:

$$\mathbf{u}(\mathbf{x}) = \mathbf{Lx}. \tag{3.21}$$

The path lines are obtained by solving

$$\dot{\mathbf{x}} = \mathbf{Lx}, \quad \mathbf{x}(0) = \mathbf{X}. \tag{3.22}$$

The solution to this is

$$\mathbf{x} = \boldsymbol{\Phi}(t)\mathbf{X}, \tag{3.23}$$

where $\boldsymbol{\Phi}$ is called the *fundamental matrix*. It obeys

$$\frac{d\boldsymbol{\Phi}}{dt} = \mathbf{L}\boldsymbol{\Phi}, \quad \boldsymbol{\Phi}(0) = \mathbf{I}. \tag{3.24}$$

In fact, we find from (3.13) that $\mathbf{F}$ is indeed the fundamental matrix of system (3.22). The solution to (3.24) is

$$\mathbf{F}(t) = \exp(t\mathbf{L}), \tag{3.25}$$

where the exponential function of a tensor is defined as [12]

$$\exp(\mathbf{A}) = \sum_{n=0}^{\infty} \frac{1}{n!}\mathbf{A}^n = \mathbf{I} + \mathbf{A} + \frac{1}{2}\mathbf{A}^2 + \cdots$$

Thus the path lines are all given by

$$\boldsymbol{\xi}(\tau) = \boldsymbol{\Phi}(\tau)\mathbf{X} = \boldsymbol{\Phi}(\tau)\boldsymbol{\Phi}(t)^{-1}\mathbf{x}(t),$$
$$= e^{(\tau-t)\mathbf{L}}\mathbf{x}. \tag{3.26}$$

The problem of calculating the exponential of a constant tensor falls into three categories, depending on the canonical form of the tensor, see Huilgol and Phan-Thien [39].

3.2.5   Oscillatory Shear Flow

We now consider the oscillatory shear flow, where the velocity depends explicitly on time:

$$u = \omega\gamma_a y \cos\omega t, \quad v = 0, \quad w = 0. \tag{3.27}$$

The equations to solve for the path lines are

$$\dot{x} = \omega \gamma_a y \cos \omega t, \quad \dot{y} = 0, \quad \dot{z} = 0,$$
$$x(0) = X, \quad y(0) = Y, \quad z(0) = Z.$$

Integrate these,

$$x(t) = X + \gamma_a Y \sin \omega t, \quad y(t) = Y, \quad z(t) = Z.$$

The path lines are then given by, note that $\boldsymbol{\xi} = (\xi, \psi, \zeta)$,

$$\xi(\tau) = X + \gamma_a Y \sin \omega \tau, \qquad \psi(\tau) = Y, \, \zeta(\tau) = Z.$$
$$\xi(\tau) = x + \gamma_a y (\sin \omega \tau - \sin \omega t), \, \psi(\tau) = y, \, \zeta(\tau) = z. \tag{3.28}$$

Suppose the path lines have been determined, then the relative deformation gradient and the strain tensors may be calculated from (3.16) and (3.17). We illustrate this with the path lines for the simple shear flow (3.74) (Problem 3.2):

$$\boldsymbol{\xi} = \mathbf{x} + (\tau - t)\,\mathbf{L}\mathbf{x},$$

which leads to the relative deformation gradient,

$$\mathbf{F}_t(\tau) = (\nabla_\mathbf{x}\boldsymbol{\xi})^T = \mathbf{I} + (\tau - t)\,\mathbf{L},$$

from which the Cauchy–Green strain tensor can be calculated as

$$\begin{aligned}
\mathbf{C}_t(\tau) = \mathbf{F}_t(\tau)^T \mathbf{F}_t(\tau) &= (\mathbf{I} + (\tau - t)\,\mathbf{L})^T (\mathbf{I} + (\tau - t)\,\mathbf{L}) \\
&= \mathbf{I} + (\tau - t)\left(\mathbf{L} + \mathbf{L}^T\right) + (\tau - t)^2\,\mathbf{L}^T\mathbf{L}.
\end{aligned} \tag{3.29}$$

3.3 Rivlin–Ericksen Tensors

Suppose the relative Cauchy–Green tensor has been determined. The n-th *Rivlin–Ericksen tensor* is defined as

$$\mathbf{A}_n(t) = \left.\frac{d^n}{d\tau^n}\mathbf{C}_t(\tau)\right|_{\tau=t}, \quad n = 1, 2, \dots \tag{3.30}$$

Since $\mathbf{C}_t(\tau)|_{\tau=t} = \mathbf{I}$, because $\mathbf{F}_t(t) = \mathbf{I}$, we may define

$$\mathbf{A}_0 = \mathbf{I}, \tag{3.31}$$

and extend the definition (3.30) to $n = 0, 1, \ldots$ In effect, the Rivlin–Ericksen tensors are defined as the coefficients of the following Taylor series about t:

$$\mathbf{C}_t(\tau) = \sum_{n=0} \frac{(\tau - t)^n}{n!} \mathbf{A}_n(t). \tag{3.32}$$

Rivlin–Ericksen tensors can be determined directly from the velocity field, without having to find the strain tensor. This is shown below.

First, we note from (3.19),

$$\frac{d}{d\tau}\mathbf{F}_t(\tau) = \frac{d}{d\tau}\left[\mathbf{F}(\tau)\mathbf{F}(t)^{-1}\right] = \mathbf{L}(\tau)\mathbf{F}(\tau)\mathbf{F}(t)^{-1} = \mathbf{L}(\tau)\mathbf{F}_t(\tau).$$

Then,

$$\frac{d}{d\tau}\mathbf{F}_t(\tau)^T = \left[\mathbf{L}(\tau)\mathbf{F}_t(\tau)\right]^T = \mathbf{F}_t(\tau)^T\mathbf{L}(\tau)^T.$$

Therefore, from (3.17)

$$\begin{aligned}
\frac{d}{d\tau}\mathbf{C}_t(\tau) &= \frac{d}{d\tau}\left[\mathbf{F}_t(\tau)^T\mathbf{F}_t(\tau)\right] \\
&= \mathbf{F}_t(\tau)^T\mathbf{L}(\tau)^T\mathbf{F}_t(\tau) + \mathbf{F}_t(\tau)^T\mathbf{L}(\tau)\mathbf{F}_t(\tau). \tag{3.33}
\end{aligned}$$

When $\tau = t$, the relative deformation gradient is the unit tensor, and we have

$$\mathbf{A}_1(t) = \left.\frac{d}{d\tau}\mathbf{C}_t(\tau)\right|_{\tau=t} = \mathbf{L}(t) + \mathbf{L}(t)^T = 2\mathbf{D}(t), \tag{3.34}$$

i.e., the first Rivlin–Ericksen tensor is twice the strain rate tensor. Higher-order Rivlin–Ericksen tensors can be obtained by taking derivatives of (3.33) repeatedly. However, it is more instructive to look at an alternative way of calculating the Rivlin–Ericksen tensors, which also reveals the nature of the tensors.

We start with the length square of a fluid element in the current time:

$$dx(t)^2 = d\mathbf{x} \cdot d\mathbf{x} = \mathbf{F}^T\mathbf{F} : d\mathbf{X}d\mathbf{X} = \mathbf{C}(t) : d\mathbf{X}d\mathbf{X}. \tag{3.35}$$

Similarly

$$d\xi(\tau)^2 = \mathbf{C}(\tau) : d\mathbf{X}d\mathbf{X}. \tag{3.36}$$

Now, since

$$\begin{aligned}
\mathbf{C}_t(\tau) &= \mathbf{F}_t(\tau)^T\mathbf{F}_t(\tau) = \mathbf{F}(t)^{-T}\mathbf{F}(\tau)^T\mathbf{F}(\tau)\mathbf{F}(t)^{-1} \\
&= \mathbf{F}(t)^{-T}\mathbf{C}(\tau)\mathbf{F}(t)^{-1},
\end{aligned}$$

we have

$$\frac{d^n}{d\tau^n}\mathbf{C}_t\left(\tau\right) = \mathbf{F}\left(t\right)^{-T}\frac{d^n}{d\tau^n}\mathbf{C}\left(\tau\right)\mathbf{F}\left(t\right)^{-1}.$$

Thus

$$\mathbf{F}\left(t\right)^T\left[\frac{d^n}{d\tau^n}\mathbf{C}_t\left(\tau\right)\right]\mathbf{F}\left(t\right) = \frac{d^n}{d\tau^n}\mathbf{C}\left(\tau\right).$$

Taking the scalar product of this with $d\mathbf{X}d\mathbf{X}$ and $\tau = t$, and recall (3.36) we obtain

$$d\mathbf{X}^T\mathbf{F}\left(t\right)^T\left[\frac{d^n}{d\tau^n}\mathbf{C}_t\left(\tau\right)\right]\mathbf{F}\left(t\right)d\mathbf{X} = \left[\frac{d^n}{d\tau^n}\mathbf{C}\left(\tau\right)\right] : d\mathbf{X}d\mathbf{X}$$

$$\mathbf{A}_n : d\mathbf{x}d\mathbf{x} = \left.\frac{d^n}{d\tau^n}d\xi\left(\tau\right)^2\right|_{\tau=t},$$

which relates the Rivlin–Ericksen tensors to the high-order stretching rate of a fluid element. For example,

$$\mathbf{A}_1 : d\mathbf{x}d\mathbf{x} = \frac{d}{dt}d\xi\left(t\right)^2. \tag{3.37}$$

A recursive relation can be derived by noting that

$$\mathbf{A}_{n+1} : d\mathbf{x}d\mathbf{x} = \frac{d}{dt}\left(\frac{d^n}{dt^n}d\xi\left(t\right)^2\right)$$

$$= \frac{d}{dt}\left(\mathbf{A}_n : d\mathbf{x}d\mathbf{x}\right)$$

$$= \left(\frac{d}{dt}\mathbf{A}_n : d\mathbf{x}d\mathbf{x} + \mathbf{A}_n : \frac{d}{dt}\left(d\mathbf{x}\right)d\mathbf{x} + \mathbf{A}_n : d\mathbf{x}\frac{d}{dt}\left(d\mathbf{x}\right)\right).$$

But

$$\frac{d}{dt}\left(d\mathbf{x}\right) = \frac{d}{dt}\left(\mathbf{F}d\mathbf{X}\right) = \mathbf{L}\mathbf{F}d\mathbf{X} = \mathbf{L}d\mathbf{x},$$

and therefore

$$\mathbf{A}_{n+1} : d\mathbf{x}d\mathbf{x} = \left(\frac{d}{dt}\mathbf{A}_n + \mathbf{A}_n\mathbf{L} + \mathbf{L}^T\mathbf{A}_n\right) : d\mathbf{x}d\mathbf{x},$$

which leads to the recursive formula due to Rivlin and Ericksen [78]

$$\mathbf{A}_{n+1} = \frac{d}{dt}\mathbf{A}_n + \mathbf{A}_n\mathbf{L} + \mathbf{L}^T\mathbf{A}_n, \quad \mathbf{A}_0 = \mathbf{I}, \quad n = 1, 2, \ldots \tag{3.38}$$

As an example, let's calculate the Rivlin–Ericksen tensors for the simple shear flow (3.72). The first Rivlin–Ericksen tensor is twice the strain rate tensor:

$$[\mathbf{A}_1] = \begin{bmatrix} 0 & \dot{\gamma} & 0 \\ \dot{\gamma} & 0 & 0 \\ 0 & 0 & 0 \end{bmatrix}.$$

The second Rivlin–Ericksen tensor is obtained from the first using (3.38),

$$\mathbf{A}_2 = \mathbf{A}_1 \mathbf{L} + \mathbf{L}^T \mathbf{A}_1$$
$$= \begin{bmatrix} 0 & \dot{\gamma} & 0 \\ \dot{\gamma} & 0 & 0 \\ 0 & 0 & 0 \end{bmatrix} \begin{bmatrix} 0 & \dot{\gamma} & 0 \\ 0 & 0 & 0 \\ 0 & 0 & 0 \end{bmatrix} + \begin{bmatrix} 0 & 0 & 0 \\ \dot{\gamma} & 0 & 0 \\ 0 & 0 & 0 \end{bmatrix} \begin{bmatrix} 0 & \dot{\gamma} & 0 \\ \dot{\gamma} & 0 & 0 \\ 0 & 0 & 0 \end{bmatrix} = \begin{bmatrix} 0 & 0 & 0 \\ 0 & 2\dot{\gamma}^2 & 0 \\ 0 & 0 & 0 \end{bmatrix}.$$

All other higher-order Rivlin–Ericksen tensors are zero for this flow. In fact, in this flow $\mathbf{L}^2 = \mathbf{0}$, and it can be verified from (3.29) that

$$\mathbf{C}_t(\tau) = \mathbf{I} + (\tau - t)\mathbf{A}_1 + \frac{1}{2}(\tau - t)^2 \mathbf{A}_2. \tag{3.39}$$

3.4 Small Strain

When the strain is small, in the sense that the fluid particles remain close to their original positions in the reference configuration at all times, then the strain may be calculated by introducing the displacement function $\mathbf{v}$:

$$\mathbf{v} = \mathbf{x}(\mathbf{X}, t) - \mathbf{X}. \tag{3.40}$$

The deformation gradient is

$$\mathbf{F}(t) = (\nabla_{\mathbf{X}} \mathbf{x})^T = \mathbf{I} + \mathbf{E}(t), \quad \mathbf{E}(t) = (\nabla_{\mathbf{X}} \mathbf{v})^T. \tag{3.41}$$

When the displacement gradient is small, terms of order $\varepsilon^2 = O\left(\|\mathbf{E}\|^2\right)$ and higher can be neglected, one has

$$\mathbf{F}(t)^{-1} = \mathbf{I} - \mathbf{E}(t),$$
$$\mathbf{C}(t) = \mathbf{I} + \mathbf{E}(t) + \mathbf{E}(t)^T,$$
$$\mathbf{F}_t(\tau) = \mathbf{F}(\tau)\mathbf{F}(t)^{-1} = \mathbf{I} + \mathbf{E}(\tau) - \mathbf{E}(t),$$
$$\mathbf{C}_t(\tau) = \mathbf{F}_t(\tau)^T \mathbf{F}_t(\tau) = \mathbf{I} + \mathbf{E}(\tau) + \mathbf{E}(\tau)^T - \mathbf{E}(t) - \mathbf{E}(t)^T. \tag{3.42}$$

In terms of the infinitesimal strain tensor

$$\varepsilon(t) = \frac{1}{2}\left(\mathbf{E}(t) + \mathbf{E}(t)^T\right),\tag{3.43}$$

we have

$$\mathbf{C}(t) = \mathbf{I} + 2\varepsilon(t), \quad \mathbf{C}_t(\tau) = \mathbf{I} + 2\varepsilon(\tau) - 2\varepsilon(t).\tag{3.44}$$

In the right polar decomposition of $\mathbf{F}$, $\mathbf{F} = \mathbf{RU}$, $\mathbf{U}$ is the square root of $\mathbf{C}$,

$$\mathbf{U} = \mathbf{C}^{1/2} = \mathbf{I} + \varepsilon.\tag{3.45}$$

Thus

$$\mathbf{R} = \mathbf{FU}^{-1} = (\mathbf{I} + \mathbf{E})(\mathbf{I} - \varepsilon) = \mathbf{I} + \mathbf{E} - \varepsilon = \mathbf{I} + \omega,$$

where ω is the infinitesimal rotation tensor:

$$\omega = \frac{1}{2}\left(\mathbf{E} - \mathbf{E}^T\right).\tag{3.46}$$

3.5 Equations of Balance

The equations of balance are mathematical statements of the conservation of mass, linear and angular momentum, and energy.

3.5.1 Reynolds Transport Theorem

Theorem 1 Let $\Phi(\mathbf{x}, t)$ be a field (scalar, vector or tensor) defined over a region V occupied by the body $\mathcal{B}$ at time t. The Reynolds transport theorem states that (for a portray of Reynolds, see Fig. 3.5)

$$\frac{d}{dt}\int_V \Phi\, dV = \int_V \left(\frac{d\Phi}{dt} + \Phi\nabla\cdot\mathbf{u}\right) dV = \int_V \left(\frac{\partial\Phi}{\partial t} + \nabla\cdot(\Phi\mathbf{u})\right) dV\tag{3.47}$$

where $\mathbf{u}$ is the velocity field, and d/dt is the material derivative (3.7). This is proved by expressing the volume integral in the reference configuration,

$$\frac{d}{dt}\int_V \Phi\, d\mathbf{x} = \frac{d}{dt}\int_{V_0} \Phi J\, d\mathbf{X},\tag{3.48}$$

Fig. 3.5 Osborne Reynolds (1842–1912) introduced the lubrication theory and formulated the framework for turbulence flow. He was the first Professor in the UK university system to hold the title *Professor of Engineering* at Owens College (now University of Manchester). The Reynolds number and stresses are named after him

where $J = \det\left(\partial \mathbf{x}/\partial \mathbf{X}\right) = \det \mathbf{F}$ is the Jacobian of the transformation (we use $d\mathbf{x}$ interchangeably with dV), and V_0 is the region occupied by the reference configuration.

Lemma We record the following Lemma

$$\frac{dJ}{dt} = J\nabla \cdot \mathbf{u}. \tag{3.49}$$

This is proved by using a result obtained previously, Sect. 1.8.1, or by noting that

$$\begin{aligned}
dJ &= \det\left(\mathbf{F} + d\mathbf{F}\right) - \det\left(\mathbf{F}\right) = \det\left[\mathbf{F}\left(\mathbf{I} + \mathbf{F}^{-1}d\mathbf{F}\right)\right] - \det \mathbf{F} \\
&= \det \mathbf{F}\det\left(\mathbf{I} + \mathbf{F}^{-1}d\mathbf{F}\right) - \det \mathbf{F} \\
&= \det \mathbf{F}\left(1 + \operatorname{tr}\left(\mathbf{F}^{-1}d\mathbf{F}\right)\right) - \det \mathbf{F} \\
&= J\operatorname{tr}\left(\mathbf{F}^{-1}d\mathbf{F}\right).
\end{aligned}$$

Divide both sides by dt, and use $\dot{\mathbf{F}} = \mathbf{LF}$, and note that

$$\operatorname{tr}\left(\mathbf{F}^{-1}\mathbf{LF}\right) = F_{ij}^{-1}L_{jk}F_{ki} = \delta_{kj}L_{jk} = L_{kk} = \nabla \cdot \mathbf{u}$$

leading to the lemma (3.49). This lemma can be used directly in (3.48) to prove (3.47).

Another form for the Reynolds transport theorem which emphasises the flux of Φ into the volume V bounded by the surface S is given below by recognizing that

$$\int_V \left(\frac{d\Phi}{dt} + \Phi \nabla \cdot \mathbf{u} \right) dV = \int_V \left(\frac{\partial \Phi}{\partial t} + \mathbf{u} \cdot \nabla \Phi + \Phi \nabla \cdot \mathbf{u} \right) dV$$

$$= \int_V \left(\frac{\partial \Phi}{\partial t} + \nabla \cdot (\Phi \mathbf{u}) \right) dV$$

$$= \int_V \frac{\partial \Phi}{\partial t} dV + \int_S (\Phi \mathbf{u}) \cdot \mathbf{n} dS$$

Theorem 2

$$\frac{d}{dt} \int_V \Phi d\mathbf{x} = \int_V \frac{\partial \Phi}{\partial t} dV + \int_S (\Phi \mathbf{u}) \cdot \mathbf{n} dS. \tag{3.50}$$

This form allows a physical interpretation of the theorem: the first term on the right represents the rate of creation of the quantity Φ, and the second term, the flux of Φ into the volume V through its bounding surface.

3.5.2 Conservation of Mass

The mass in the volume V is conserved at all time, i.e.,

$$\frac{d}{dt} \int_V \rho dV = 0,$$

where $\rho(\mathbf{x}, t)$ is the density field at time t. From Reynolds transport theorem (3.47),

$$\int_V \left(\frac{\partial \rho}{\partial t} + \nabla \cdot (\rho \mathbf{u}) \right) dV = 0.$$

Since the volume V is arbitrary, a necessary and sufficient condition for the conservation of mass is

$$\frac{\partial \rho}{\partial t} + \nabla \cdot (\rho \mathbf{u}) = \frac{d\rho}{dt} + \rho \nabla \cdot \mathbf{u} = 0. \tag{3.51}$$

For an incompressible material, the density is constant everywhere, and the conservation of mass demands that

$$\nabla \cdot \mathbf{u} = \frac{\partial u_i}{\partial x_i} = \mathrm{tr}\mathbf{L} = \mathrm{tr}\mathbf{D} = \frac{1}{2}\mathrm{tr}\mathbf{A}_1 = 0. \tag{3.52}$$

From a solid point of view, the conservation of mass requires that

$$\rho dV = \rho_R dV_R \qquad \rho_R = \rho J,$$

where $J = \det \mathbf{F}$ is the Jacobian of the deformation. For an incompressible material, we have

$$J = \det \mathbf{F} = 1, \quad \det \mathbf{C} = 1, \tag{3.53}$$

at all time.

Theorem 3 As a corollary to (3.47) and (3.51) we have

$$\frac{d}{dt} \int_V \rho \Phi \, dV = \int_V \rho \frac{d\Phi}{dt} dV. \tag{3.54}$$

This is easily demonstrated by using (3.47) and (3.51) on the left hand side:

$$\frac{d}{dt} \int_V \rho \Phi \, dV = \int_V \left(\Phi \left(\frac{d\rho}{dt} + \rho \nabla \cdot \mathbf{u} \right) + \rho \frac{d\Phi}{dt} \right) dV.$$

3.5.3 Conservation of Momentum

The forces acting on the body are surface forces $\mathbf{t}$ (tractions), and body forces $\mathbf{b}$ (those that act at a distance).

Body Force Density. An example of body force is gravitational. If $\mathbf{b}(\mathbf{x}, t)$ is the body force density defined on V, then the resulting force and moment (about a fixed point O) on V due to the body force field are given respectively by

$$\int_V \rho \mathbf{b} \, dV, \quad \int_V \mathbf{x} \times \mathbf{b} \, dV.$$

Surface Force. Surface traction is a concept due to Cauchy. Consider a particle X occupying the position $\mathbf{x}$ at time t. Construct a surface S_t through this point with unit normal vector $\mathbf{n}(\mathbf{x}, t)$ at point $\mathbf{x}$, which separates the body into two regions: B^+ is the region into which the unit normal $\mathbf{n}$ is directed and B^- on the other side (see Fig. 3.6). $\mathbf{t}(\mathbf{x}, t; \mathbf{n})$ is called the surface force density per unit (current) area if the force and moment (about O) exerted on B^- by B^+ are given respectively by

Fig. 3.6 The traction $\mathbf{t}(\mathbf{x}, t; \mathbf{n})$ is the force per unit area exerted by B^+ on B^-

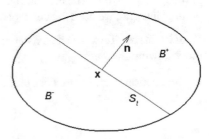

$$\int_{S_t} \mathbf{t}\,(\mathbf{x}, t)\,dS, \qquad \int_{S_t} \mathbf{x} \times \mathbf{t}\,(\mathbf{x}, t)\,dS.$$

Balance of Linear Momentum. Newton's second law (a postulate), as applied to a volume V occupied by the body, requires that

$$\frac{d}{dt} \int_V \rho \mathbf{u}\,dV = \int_S \mathbf{t}\,dS + \int_V \rho \mathbf{b}\,dV. \qquad (3.55)$$

Here S is the bounding surface of V. The first term on the right hand of (3.55) represents the surface force acting on V due the body outside V, and the second term is the net body force on V.

Corollary (3.54) can be used on the left side of the preceding equation, leading to

$$\int_V \rho \frac{d\mathbf{u}}{dt}\,dV = \int_S \mathbf{t}\,dS + \int_V \rho \mathbf{b}\,dV. \qquad (3.56)$$

Balance of Angular Momentum. The balance of angular momentum (a postulate) can likewise be written down as, in the absence of body couple,

$$\frac{d}{dt} \int_V \mathbf{x} \times \rho \mathbf{u}\,dV = \int_S \mathbf{x} \times \mathbf{t}\,dS + \int_V \mathbf{x} \times \rho \mathbf{b}\,dV. \qquad (3.57)$$

Again, corollary (3.54) can be used on the left side of the preceding equation, leading to

$$\int_V \mathbf{x} \times \rho \frac{d\mathbf{u}}{dt}\,dV = \int_S \mathbf{x} \times \mathbf{t}\,dS + \int_V \mathbf{x} \times \rho \mathbf{b}\,dV. \qquad (3.58)$$

Note that the term involving $d\mathbf{x}/dt = \mathbf{u}$ does not contribute to the left hand side of this because of the definition of the cross product. This should be contrasted to the rigid body mechanics case where Newton's second law is a postulate and the balance of angular momentum is a consequence (a theorem). Here two postulates are required (Newton's 2nd law and the balance of angular momentum).

The term

$$\mathbf{a} = \frac{d\mathbf{u}}{dt} = \frac{\partial \mathbf{u}}{\partial t} + \mathbf{u} \cdot \nabla \mathbf{u} \qquad (3.59)$$

is recognised as the acceleration field. To convert the first term on the right into a volume integral, we need the concept of the stress tensor $\mathbf{T}(\mathbf{x}; \mathbf{n})$, also due to Cauchy.

Cauchy Stress Tensor. The existence of the *Cauchy stress tensor* is guaranteed by the following theorem.

• The traction vector (force per unit area) satisfies

$$\mathbf{t}\,(\mathbf{x}, t; -\mathbf{n}) = -\mathbf{t}\,(\mathbf{x}, t; \mathbf{n})\,. \qquad (3.60)$$

Fig. 3.7 Existence of the
stress tensor

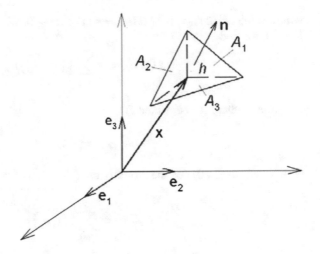

- Further, there exists a second-order tensor field $\mathbf{T}(\mathbf{x}, t)$ with the following properties:

$$\mathbf{t}(\mathbf{x}, t; \mathbf{n}) = \mathbf{T}(\mathbf{x}, t) \cdot \mathbf{n}, \qquad (3.61)$$

with components in the frame $\mathcal{F} = \{\mathbf{e}_1, \mathbf{e}_2, \mathbf{e}_3\}$ given by

$$T_{ij}(\mathbf{x}, t) = \mathbf{t}(\mathbf{x}, t; \mathbf{e}_j) \cdot \mathbf{e}_i. \qquad (3.62)$$

The proof, due to Gurtin [34], lies in the construction of a one-parameter family of tetrahedra, Fig. 3.7, with vertex at point $\mathbf{x}$, height h (the parameter). The face normal to $\mathbf{n}$ has area A; the face normal to $-\mathbf{e}_i$ has area A_i. From the directional cosine of $\mathbf{n}$

$$A_i(h) = A(h) n_i.$$

Furthermore the volume of the tetrahedron is $V_h = \frac{1}{3} h A(h)$.

Applying (3.56) to the tetrahedron (omitting the time argument t for brevity):

$$\int_{V_h} \rho \mathbf{a}(\mathbf{y}) \, dV(\mathbf{y}) = \int_{V_h} \rho \mathbf{b}(\mathbf{y}) \, dV(\mathbf{y}) + \int_{A(h)} \mathbf{t}(\mathbf{y}; \mathbf{n}) \, dS(\mathbf{y})$$

$$+ \sum_{j=1}^{3} \int_{A_i(h)} \mathbf{t}(\mathbf{y}; -\mathbf{e}_j) \, dS(\mathbf{y}).$$

From the continuity of all the field variables, and the mean-value theorem,

$$(\rho \mathbf{a}(\mathbf{x}) + \mu) V_h = (\rho \mathbf{b}(\mathbf{x}) + \alpha) V_h + (\mathbf{t}(\mathbf{x}; \mathbf{n}) + \beta) A(h)$$

$$+ \sum_{j=1}^{3} \left[\mathbf{t}(\mathbf{x}; -\mathbf{e}_j) + \beta_j \right] A(h) n_j, \qquad (3.63)$$

where $\alpha(h), \beta(h), \beta_j(h), \mu(h) = o(1), h \to 0$. Divide (3.63) by $A(h)$ and let $h \to 0$, we find

$$\mathbf{t}(\mathbf{x}; \mathbf{n}) = -\sum_{j=1}^{3} \mathbf{t}(\mathbf{x}; -\mathbf{e}_j) n_j. \tag{3.64}$$

In the case where $\mathbf{n} = \mathbf{e}_i$ (fixed i)

$$\mathbf{t}(\mathbf{x}; -\mathbf{e}_i) = -\mathbf{t}(\mathbf{x}; \mathbf{e}_i),$$

which leads to (3.60) due the arbitrary configuration of the frame of reference. Thus

$$\mathbf{t}(\mathbf{x}; \mathbf{n}) = \sum_{j=1}^{3} \mathbf{t}(\mathbf{x}; \mathbf{e}_j) n_j. \tag{3.65}$$

From (3.65), the components of $\mathbf{t}$ are given by

$$t_i(\mathbf{x}; \mathbf{n}) = \mathbf{t}(\mathbf{x}; \mathbf{n}) \cdot \mathbf{e}_i = \sum_{j=1}^{3} \mathbf{t}(\mathbf{x}; \mathbf{e}_j) \cdot \mathbf{e}_i n_j$$

$$= T_{ij}(\mathbf{x}) n_j, \tag{3.66}$$

where T_{ij} are defined as in (3.62). From the quotient rule (§1.3.10), T_{ij} are indeed the components of a second-order tensor thus proving the existence of the stress tensor (3.61).

Conservation of Linear Momentum. Returning to the balance of linear momentum (3.56) and using the definition of the stress tensor (3.61), we have

$$\int_V \rho \frac{d\mathbf{u}}{dt} dV = \int_S \mathbf{T} \cdot \mathbf{n} dS + \int_V \rho \mathbf{b} dV,$$

$$\int_V \rho \frac{du_i}{dt} dV = \int_S T_{ij} n_j dS + \int_V \rho b_i dV.$$

The surface integral on the right hand of the preceding equation can be converted into a volume integral to obtain

$$\int_V \rho \frac{d\mathbf{u}}{dt} dV = \int_V \nabla \mathbf{T}^T dV + \int_V \rho \mathbf{b} dV,$$

$$\int_V \rho \frac{du_i}{dt} dV = \int_S \frac{\partial T_{ij}}{\partial x_j} dS + \int_V \rho b_i dV.$$

Since the integrand is continuous on an arbitrary V, the conservation of linear momentum becomes

$$\rho \frac{d\mathbf{u}}{dt} = \nabla \mathbf{T}^T + \rho \mathbf{b}, \qquad \rho \frac{du_i}{dt} = \frac{\partial T_{ij}}{\partial x_j} + \rho b_i. \tag{3.67}$$

Conservation of Angular Momentum. Returning to the balance of angular momentum (3.58),

$$\int_V \mathbf{x} \times \rho \frac{d\mathbf{u}}{dt} dV = \int_S \mathbf{x} \times \mathbf{t} dS + \int_V \mathbf{x} \times \rho \mathbf{b} dV$$

and examining the first term on the right,

$$\int_S \varepsilon_{ijk} x_j t_k dS = \int_S \varepsilon_{ijk} x_j T_{kl} n_l dS = \int_V \frac{\partial}{\partial x_l} \left(\varepsilon_{ijk} x_j T_{kl} \right) dV$$

$$= \int_V \left(\varepsilon_{ijk} \delta_{jl} T_{kl} + \varepsilon_{ijk} x_j \frac{\partial T_{kl}}{\partial x_l} \right) dV$$

$$= \int_V \left(t_i^A + \varepsilon_{ijk} x_j \frac{\partial T_{kl}}{\partial x_l} \right) dV,$$

where the "axial vector" is defined as $t_i^A = \varepsilon_{ijk} T_{kj}$. When these results are substituted back into (3.58),

$$\int_V \mathbf{x} \times \rho \frac{d\mathbf{u}}{dt} dV = \int_V \left(\mathbf{t}^A + \mathbf{x} \times \nabla \cdot \mathbf{T}^T \right) dV + \int_V \mathbf{x} \times \rho \mathbf{b} dV$$

$$\int_V \mathbf{t}^A dV = \int_V \mathbf{x} \times \left[\rho \frac{d\mathbf{u}}{dt} - \nabla \cdot \mathbf{T}^T - \rho \mathbf{b} \right] dV$$

From the conservation of linear momentum, and the continuity of the integrands in the arbitrary volume V, it follows that

$$\mathbf{t}^A = \mathbf{0} \qquad \varepsilon_{ijk} T_{kj} = 0 \qquad \mathbf{T} = \mathbf{T}^T \text{ or } \mathbf{T} \text{ is symmetric.} \tag{3.68}$$

A necessary and sufficient condition for the balance of angular momentum, in the absence of body couples, is that the stress be symmetric. In deriving (3.68), both the conservation of mass and linear momentum are needed.

3.5.4 Conservation of Energy

We define the kinetic energy K, and internal energy E as

$$K = \int_V \frac{1}{2} \rho u^2 dV, \quad E = \int_V \rho \varepsilon dV, \tag{3.69}$$

where ε is the specific internal energy per unit mass. The rate of work done on the body due to surface and body forces is given by

$$\int_S \mathbf{t} \cdot \mathbf{u} \, dS + \int_V \rho \mathbf{b} \cdot \mathbf{u} \, dV.$$

If we define $\mathbf{q}$ to be the flux of energy out of S per unit area, and r the amount of energy created per unit mass, then the mathematical statement for the first law of thermodynamics can be expressed as

$$\frac{d}{dt} \int_V \left(\frac{1}{2} \rho u^2 + \rho \varepsilon \right) dV = \int_S (\mathbf{t} \cdot \mathbf{u} - \mathbf{q} \cdot \mathbf{n}) \, dS$$
$$+ \int_V (\rho r + \rho \mathbf{b} \cdot \mathbf{u}) \, dV. \tag{3.70}$$

The left of (3.70) can be expressed as, using Reynolds transport theorem (3.54),

$$\frac{d}{dt} \int_V \left(\frac{1}{2} \rho u^2 + \rho \varepsilon \right) dV = \int_V \rho \left(\mathbf{u} \cdot \mathbf{a} + \dot{\varepsilon} \right) dV.$$

The surface integral on the right of (3.70) is converted into volume integral as

$$\int_S (\mathbf{t} \cdot \mathbf{u} - \mathbf{q} \cdot \mathbf{n}) \, dS = \int_V \left(\frac{\partial}{\partial x_j} \left(u_i T_{ij} \right) - \frac{\partial q_i}{\partial x_i} \right) dV$$
$$= \int_V \left(\mathbf{u} \cdot \nabla \mathbf{T}^T + \mathbf{T} : \mathbf{L} - \nabla \cdot \mathbf{q} \right) dV.$$

Hence (3.70) becomes

$$\int_V \rho \left(\mathbf{u} \cdot \mathbf{a} + \dot{\varepsilon} \right) dV = \int_V \left(\mathbf{u} \cdot \nabla \mathbf{T}^T + \mathbf{T} : \mathbf{L} - \nabla \cdot \mathbf{q} \right) dV$$
$$+ \int_V (\rho r + \rho \mathbf{b} \cdot \mathbf{u}) \, dV.$$

Because of the conservation of linear momentum, and the continuity of the integrands in the arbitrary volume V, the conservation of energy is reduced to

$$\rho \dot{\varepsilon} = \mathbf{T} : \mathbf{D} - \nabla \cdot \mathbf{q} + \rho r. \tag{3.71}$$

In deriving this, all the three balance equations for mass, linear momentum and angular momentum are required. The term $\mathbf{T} : \mathbf{D}$ represents the rate of work done by the stress, or the "stress power". It is seen that

$$\mathbf{T} : \mathbf{D} = -P \operatorname{tr} \mathbf{D} + \mathbf{S} : \mathbf{D}$$

The rate of work done by the pressure for an incompressible fluid is zero, because $\operatorname{tr} \mathbf{D} = 0$.

Problems

Problem 3.1 Using $\mathbf{FF}^{-1} = \mathbf{I}$ for a deformation gradient $\mathbf{F}$, show that

$$\frac{d}{dt}\mathbf{F}^{-1} = -\mathbf{F}^{-1}\mathbf{L}, \quad \mathbf{F}^{-1}(0) = \mathbf{I}.$$

Problem 3.2 For a simple shear flow, where the velocity field takes the form

$$u = \dot{\gamma}y, \quad v = 0, \quad w = 0, \tag{3.72}$$

show that the velocity gradient and its exponent are given by

$$[\mathbf{L}] = \begin{bmatrix} 0 & \dot{\gamma} & 0 \\ 0 & 0 & 0 \\ 0 & 0 & 0 \end{bmatrix}, \quad \exp(\mathbf{L}) = \mathbf{I} + \mathbf{L}. \tag{3.73}$$

Show that the path lines are given by

$$\boldsymbol{\xi}(\tau) = \mathbf{x} + (\tau - t)\,\mathbf{L}\mathbf{x}. \tag{3.74}$$

so that a fluid element $d X$ can only be stretched linearly in time at most.

Problem 3.3 Repeat the same exercise for an elongational flow, where

$$u = ax, \quad v = by, \quad w = cz, \quad a + b + c = 0. \tag{3.75}$$

In this case, show that

$$[\mathbf{L}] = \begin{bmatrix} a & 0 & 0 \\ 0 & b & 0 \\ 0 & 0 & c \end{bmatrix}, \quad \left[e^{\mathbf{L}}\right] = \begin{bmatrix} e^a & 0 & 0 \\ 0 & e^b & 0 \\ 0 & 0 & e^c \end{bmatrix}. \tag{3.76}$$

Show that the path lines are given by

$$[\boldsymbol{\xi}(\tau)] = \begin{bmatrix} \xi \\ \psi \\ \zeta \end{bmatrix} = \begin{bmatrix} e^{a(\tau-t)} & 0 & 0 \\ 0 & e^{b(\tau-t)} & 0 \\ 0 & 0 & e^{c(\tau-t)} \end{bmatrix} \begin{bmatrix} x \\ y \\ z \end{bmatrix}. \tag{3.77}$$

Conclude that exponential flow can stretch the fluid element exponentially fast.

Problem 3.4 Consider a super-imposed oscillatory shear flow:

$$u = \dot{\gamma}_m y, \quad v = 0, \quad w = \omega \gamma_a y \cos \omega t.$$

Show that the path lines are

$$
\begin{aligned}
\xi(\tau) &= x + \dot{\gamma}_m (\tau - t) y, \\
\psi(\tau) &= y, \\
\zeta(\tau) &= z + \gamma_0 y (\sin \omega \tau - \sin \omega t).
\end{aligned}
\tag{3.78}
$$

Problem 3.5 Calculate the Rivlin–Ericksen tensors for the elongational flow (3.75).

Problem 3.6 Calculate the Rivlin–Ericksen tensors for the unsteady flow (3.78).

Problem 3.7 Write down, in component forms the conservation of mass and linear momentum equations, assuming the fluid is incompressible, in Cartesian, cylindrical and spherical coordinate systems.

Chapter 4
Constitutive Equation: General Principles

Basic Principles and Some Classical Constitutive Equations

In isothermal flow where the conservation of energy is not relevant, there are four scalar balance equations (one conservation of mass and three conservation of linear momentum), and there are 10 scalar variables (3 velocity components, one pressure, and 6 independent stress components – thanks to the conservation of angular momentum, the stress tensor is symmetric). Clearly we do not have a mathematically well-posed problem until 6 extra equations are specified. The constitutive equation, or the rheological equation of state, provides the linkage between the stresses and the kinematics and provides the missing information. Modelling a complex fluid, or finding a relevant constitutive equation for the fluid, is the central concern in rheology. In this chapter, we review some of the well-known classical models, and the general principles underlying constitutive modelling.

4.1 Some Well-Known Constitutive Equations

4.1.1 Perfect Gas

The most well known constitutive equation is the perfect gas law, due to Boyle (Fig. 4.1), where the state of the gas is fully specified by its volume V, its pressure P and its temperature T

$$PV = RT, \qquad (4.1)$$

where R is a universal gas constant.

© Springer International Publishing AG 2017
N. Phan-Thien and N. Mai-Duy, *Understanding Viscoelasticity*,
Graduate Texts in Physics, DOI 10.1007/978-3-319-62000-8_4

Fig. 4.1 Robert Boyle
(1627–1691) made several
important contributions to
Physics and Chemistry, the
best known is the Perfect
Gas Law. He employed
Robert Hooke as his assistant
in the investigation of the
behaviour of air. His
experiments led him to
believe in vacuum, and reject
Descartes' concept of ether

Fig. 4.2 Jean d'Alembert
(1717–1783) was a French
mathematician. He made
several important
contributions to mechanics.
He is most famous for the
D'Alembert principle

4.1.2   Inviscid Fluid

The perfect fluid concept of D'Alembert (Fig. 4.2) and Euler (Fig. 3.2) is another
well-known constitutive equation. In our notation, the stress is given by

$$\mathbf{T} = -P\mathbf{I}, \quad T_{ij} = -P\delta_{ij}. \tag{4.2}$$

Here, P is the pressure. The inviscid fluid model fails to account for the pressure
losses in pipe flow, and a better model is needed.

Fig. 4.3 Joseph Fourier
(1768–1830) was a French
historian, administrator and
mathematician. He was
famous for his Fourier's
series. His name is one of the
72 inscribed on the Eiffel
tower

4.1.3   Fourier's Law

Students of thermodynamics would recognise Fourier's law of heat conduction (Fig.
4.3), linking the heat transfer rate **q** to the temperature gradient:

$$\mathbf{q} = -k\nabla\theta, \quad q_i = -k\frac{\partial\theta}{\partial x_i},\tag{4.3}$$

where k is the thermal conductivity and θ is the temperature field.

4.1.4   Hookean Solid

In 1678 Robert Hooke published his now famous law for material behaviour as a
solution to an anagram that he published two years earlier, *ut tensio sic vis*, which
roughly translated as extension is proportional to the force (see Fig. 4.4). This idea has
gone through several revisions by several well-known scientists, including Young,
Poisson, and Navier (Fig. 4.5), who thought that one needs only one elastic constant
(the other constant, the Poisson's ratio was thought to be 0.25). The concept of the
stress tensor was introduced by Cauchy (Fig. 3.4), who also gave the correct version
of the constitutive equation for infinitesimal elasticity in 1827. In our notation, the
stress tensor is given by

$$\mathbf{T} = \mathbf{C}\cdot\varepsilon, \quad T_{ij} = C_{ijkl}\varepsilon_{kl},\tag{4.4}$$

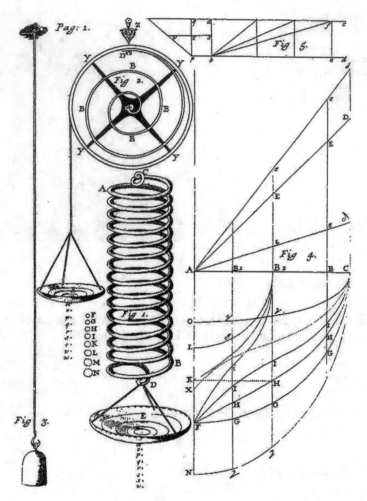

Fig. 4.4 Robert Hooke (1635–1703) was perhaps the foremost experimental scientist in the 17th century. He was a noted architect, an inventor (Hooke universal joint, spring control in watches, reflecting telescope, etc.), a mathematician, a physicist, a chemist and contributed significantly to anatomy, astronomy, botany, chemistry – the term "cell" is due to him. No portrait survived him. He is most well-known for the law of elasticity that bears his name

where

$$\varepsilon = \frac{1}{2}\left(\nabla \mathbf{v} + \nabla \mathbf{v}^T\right), \quad \varepsilon_{ij} = \frac{1}{2}\left(\frac{\partial v_i}{\partial x_j} + \frac{\partial v_j}{\partial x_i}\right) \tag{4.5}$$

is the infinitesimal strain tensor, $\mathbf{v} = \mathbf{x} - \mathbf{X}$ is the displacement vector, $\mathbf{C}$ is a 4th order elasticity tensor, a tensorial material constant, which reflects all the anisotropy of the material. For an isotropic solid, the elastic constants are reduced to just two:

Fig. 4.5 C.L.M.H Navier (1785–1836), a French engineer, developed a particle model for an elastic solid, which has a shear modulus and a Poisson's ratio of 0.25 (see Love [54]). He obtained the Navier–Stokes equations by molecular arguments. His name is one of the 72 inscribed on the Eiffel tower

$$C_{ijkl} = \lambda \delta_{ij} \delta_{kl} + \mu \left(\delta_{ik} \delta_{jl} + \delta_{il} \delta_{jk} \right), \tag{4.6}$$

where λ and μ are called Lamé moduli. Using this in (4.4) produces

$$\mathbf{T} = \lambda \left(\mathrm{tr} \varepsilon \right) \mathbf{I} + 2\mu \varepsilon, \quad T_{ij} = \lambda \varepsilon_{kk} \delta_{ij} + 2\mu \varepsilon_{ij}. \tag{4.7}$$

An alternative form for this can be derived by first taking the traces of both sides of (4.7)

$$\mathrm{tr} \mathbf{T} = (3\lambda + 2\mu) \, \mathrm{tr} \varepsilon,$$

and then substituting this back into (4.7) to obtain

$$\varepsilon = \frac{1}{2\mu} \mathbf{T} - \frac{\lambda}{2\mu \left(3\lambda + 2\mu \right)} \left(\mathrm{tr} \mathbf{T} \right) \mathbf{I}, \quad \varepsilon_{ij} = \frac{1}{2\mu} T_{ij} - \frac{\lambda}{2\mu \left(3\lambda + 2\mu \right)} T_{kk} \delta_{ij}. \tag{4.8}$$

4.1.5 *Newtonian Fluid*

About nine years after Hooke published his paper on elasticity, Newton (Fig. 4.6) introduced the concept of "lack of slipperiness", which is the important quantity that we now call viscosity. Then Navier, in 1827, derived the Navier–Stokes equation,

$$\rho \frac{d\mathbf{u}}{dt} = -\nabla P + \eta \nabla^2 \mathbf{u} + \rho \mathbf{b}. \tag{4.9}$$

In this equation, the terms $\eta \nabla^2 \mathbf{u}$ represent the viscous forces, although Navier did not attach much physical significance to η. Stokes (Fig. 4.7) gave the correct form

Fig. 4.6 Sir Isaac Newton
(1643–1727) was a
dominating personality in
Science. He (and
concurrently Leibnitz)
invented differential and
integral calculus, and the
gravitational theory. He was
appointed the Lucasian
Professor at Cambridge at
the age of 26

Fig. 4.7 George Gabrielle
Stokes (1819–1903) was a
Irish mathematician. He was
appointed the Lucasian
Professor at Cambridge at
the age of 30. He is
remembered for Stokes flow
and his contributions in the
Navier–Stokes equations

to the constitutive equation that we call a Newtonian fluid:

$$\mathbf{T} = -P\mathbf{I} + \Lambda\,(\mathrm{tr}\mathbf{D})\,\mathbf{I} + 2\eta\mathbf{D}, \quad T_{ij} = -P\delta_{ij} + \Lambda D_{kk}\delta_{ij} + 2\eta D_{ij}. \tag{4.10}$$

Here, η is the viscosity, Λ is the bulk viscosity, and $\mathbf{D} = (\nabla\mathbf{u} + \nabla\mathbf{u}^T)/2$ is the strain rate tensor. Stokes assumed that

$$\Lambda = -\frac{2}{3}\eta,$$

so that pure volumetric change does not affect the stress (tr $\mathbf{T}$ is independent of tr $\mathbf{D}$). Furthermore, the terms $\Lambda\,\mathrm{tr}\,\mathbf{D}\delta_{ij}$ can be absorbed in the pressure term. This leads to the familiar constitutive equation for Newtonian fluids:

$$\mathbf{T} = -P\mathbf{I} + 2\eta\mathbf{D}. \tag{4.11}$$

4.1.6 Non-Newtonian Fluid

The term *non-Newtonian fluid* is an all-encompassing term denoting any fluid that does not obey (4.11). To discuss constitutive relations for non-Newtonian fluids, we need a convenient way to classify different flow regimes.

4.2 Weissenberg and Deborah Numbers

Most non-Newtonian fluids have a characteristic time scale λ. In a flow with a characteristic shear rate $\dot{\gamma}$ and a characteristic frequency ω, or characteristic time T, two dimensionless groups can be formed

$$
\begin{aligned}
\text{Deborah number} \quad & \text{De} = \lambda\omega \text{ or } \lambda/T, \\
\text{Weissenberg number} \quad & \text{Wi} = \lambda\dot{\gamma}
\end{aligned} \tag{4.12}
$$

4.2.1 Deborah Number

The Deborah number,[1] the ratio between the fluid relaxation time and the flow characteristic time, represents the transient nature of the flow relative to the fluid time scale. If the observation time scale is large (small De number), the material responds like a fluid, and if it is small (large De number), we have a solid-like response. Under this viewpoint, there is no fundamental difference between solids and liquids; it is only a matter of time scale of observation. In the limit, when De $= 0$ one has a Newtonian liquid, and when De $= \infty$, an elastic solid.

4.2.2 Weissenberg Number

The Weissenberg number is the ratio of elastic to viscous forces. It has been variously defined, but usually as given in (4.12). Thus one can have a flow with a small Wi number and a large De number, and vice versa. We expect a significant non-Newtonian behaviour in a large Wi number flow, and therefore the constitutive

---

[1] The terminology is due to M. Reiner.

Fig. 4.8 Pipkin diagram
delineates different flow
regimes, and relevant
constitutive equations [72]

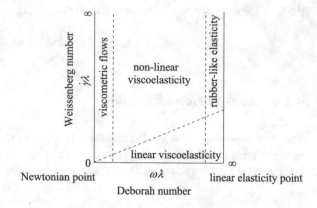

equation must contain the relevant non-Newtonian physics. A different definition of
the Weissenberg number is explored in Problem 4.3.

Pipkin's diagram (Pipkin and Tanner [72]) helps guide the choice of constitutive
equations. In Fig. 4.8, the vertical axis represents the Weissenberg number and the
horizontal axis the Deborah number. Newtonian response is represented by a single
point, at De $= 0 =$ Wi. Elastic response is also represented by a single point, at
De $= \infty$. Nearly steady flows, at low De numbers, can be analysed by assuming
viscometric motion (order fluids), and fast flows, at large De numbers, a rubber-like
response is expected - the relevant constitutive equation here is rubber-like elasticity.
The region at low Wi numbers (at small strain amplitudes) can be handled by a
linear viscoelastic model. The large domain in the middle of the diagram, marked
non-linear viscoelasticity, is the constitutive modeller's haven!

4.3 Some Guidelines in Constitutive Modelling

There are two alternatives for constitutive modelling: the continuum approach and
the microstructure approach. In the continuum approach, the material is assumed to
be a continuum, with no micro-inertial feature. The relevant variables are identified,
and are related in a framework that ensured invariance under a change of frames.
Different restrictions are then imposed to simplify the constitutive equation as far as
practicable. In the microstructure approach, a physical model of the microstructure
representing the material is postulated. Solving the deformation at that level using
well-tested physical principles (Newton's laws, conservation laws, etc.) allows the
average stress and strain to be related producing a constitutive equation. In the continuum approach one is usually left with a general constitutive equation, which may
have some undetermined functions or functionals (loosely speaking, functionals are
functions of functions). The details of these functions or functionals may be furnished
by relevant experiments. In the microstructure approach, the constitutive equations
tend to be more specific and therefore more relevant to the material in question.

In the mid 1950s, there were some intense activities in setting up a rigorous theoretical framework for continuum mechanics. Everything possible is set up in an axiomatic format. This has been good in focusing on what is permissible. However, it has the unfortunate consequence that it leaves the students with the impression that all one needs is a set of relevant variables and some principles – and that would allows us to construct a general constitutive equation for any material.

It is generally believed that relevant constitutive equations should be based on a (simplified) model of the microstructure. When the physics governing the microstructure interactions are complicated, one must not hesitate to introduce elements of continuum modelling, but the continuum approach should not completely replace the microstructure modelling.

4.3.1   Oldroyd Approach

The basic ideas behind constitutive modelling of finite deformation were well understood in the early 1950s, but these ideas have not been extended to all continuous materials undergoing large deformation. It was Oldroyd[2] [64] who clearly enunciated that a constitutive equation must be based on

- the relative motion of the neighbourhood of a particle;
- the history of the metric tensor (i.e., strain tensor) associated with the particle;
- the convected coordinate system embedded in the material and deforming with it;
- and the physical constants defining the symmetry of the material.

It was unfortunate that his work has been grossly overlooked, see Tanner and Walters [86] for an interesting historical account. The later influential work of Noll [61, 62] put these ideas in an axiomatic form that is elegant and appealing to the generation of graduate students at that time. We will go through the principles as detailed by Noll, but always keep in mind the relevance of microstructure models.

4.3.2   Principle of Material Objectivity

Consider a change of frame

$$\mathbf{x}' = \mathbf{c}(t) + \mathbf{Q}(t)\,\mathbf{x}, \tag{4.13}$$

which consists of a spatial translation (by $\mathbf{c}$) and a rotation (through an orthogonal tensor $\mathbf{Q}$). A physical quantity is said to be objective, or frame-invariant, when it is

---

[2]James G. Oldroyd (1921–1982) was a Professor in Applied Mathematics at Universities of Wales and Liverpool. He made several important contributions to the constitutive equation formulation. The Oldroyd fluids (fluid A and fluid B) were named after him.

invariant under the transformation (4.13). Specifically, using the prime to denote the quantity in the new frame,

- a scalar ϕ is invariant when its value remains unchanged under a change of frame (4.13)

$$\phi' = \phi, \tag{4.14}$$

- a vector **u** is invariant when it transforms under a change of frame (4.13) according to

$$\mathbf{u}' = \mathbf{Q}(t)\,\mathbf{u}, \tag{4.15}$$

- a tensor **T** is invariant when it transforms under a change of frame (4.13) according to

$$\mathbf{T}' = \mathbf{Q}(t)\,\mathbf{T}\mathbf{Q}(t)^T. \tag{4.16}$$

Note that only $\mathbf{Q}(t)$ is involved in this transformation.

4.3.3 Objectivity of the Stress

The principle of material objectivity asserts that the stress tensor must be objective under a change of frame (4.13).

This principle pre-supposes that the material has no inertial feature at the microscale, that is, it is a continuous media. With micro-inertia, such as suspensions of microsized particles of a different density than the fluid, there will be a component of stress due to the particle inertia that is not objective. This is a "principle" for inertialess microstructure only. Indeed, if a microstructural model violates objectivity, the reason can always be found in the physics of the microstructure – if the physics are sound, there may be a very good reason for the stress not being objective (Ryskin and Rallison [79]).

4.3.4 Frame Indifference

This principle enunciates that one does not obtain a new constitutive equation every time there is a change in frame of reference: the constitutive operator is the same for all observers in relative motion. The objectivity and frame indifference principles roughly correspond to Oldroyd's third point.

There are kinematic quantities that are not objective. We give a few examples here.

Deformation Gradient Tensor. Consider the deformation gradient tensor,

$$\mathbf{F}(t) = \left[\frac{\partial \mathbf{x}(t)}{\partial \mathbf{X}}\right]^T, \quad F_{ij} = \frac{\partial x_i}{\partial X_j}.$$

Under the change of frame (4.13),

$$F'_{ij} = \frac{\partial x'_i}{\partial X_j} = \frac{\partial x'_i}{\partial x_k}\frac{\partial x_k}{\partial X_j} = Q_{ik}F_{kj}, \quad \mathbf{F}' = \mathbf{Q}\mathbf{F}, \tag{4.17}$$

i.e., $\mathbf{F}$ is not frame-invariant. The relative deformation gradient, $\mathbf{F}_t(\tau) = \mathbf{F}(\tau) \mathbf{F}^{-1}(t)$, is not frame-invariant either,

$$\begin{aligned}\mathbf{F}'_t(\tau) &= \mathbf{F}'(\tau)\mathbf{F}'(t)^{-1} = \mathbf{Q}(\tau)\mathbf{F}(\tau)\left[\mathbf{Q}(t)\mathbf{F}(t)\right]^{-1}\\ &= \mathbf{Q}(\tau)\mathbf{F}(\tau)\mathbf{F}(t)^{-1}\mathbf{Q}(t)^T \\ &= \mathbf{Q}(\tau)\mathbf{F}_t(\tau)\mathbf{Q}(t)^T,\end{aligned} \tag{4.18}$$

since both $\mathbf{Q}(\tau)$ and $\mathbf{Q}(t)$ are involved in this transformation.

Cauchy–Green Tensor. The left Cauchy–Green tensor is objective, but the right Cauchy–Green tensor is not:

$$\begin{aligned}\mathbf{B}' &= \mathbf{F}'\mathbf{F}'^T = \mathbf{Q}\mathbf{F}(\mathbf{Q}\mathbf{F})^T = \mathbf{Q}\mathbf{F}\mathbf{F}^T\mathbf{Q}^T = \mathbf{Q}\mathbf{B}\mathbf{Q}^T,\\ \mathbf{C}' &= \mathbf{F}'^T\mathbf{F}' = (\mathbf{Q}\mathbf{F})^T\mathbf{Q}\mathbf{F} = \mathbf{F}\mathbf{Q}^T\mathbf{Q}\mathbf{F} = \mathbf{C}.\end{aligned} \tag{4.19}$$

The left relative Cauchy–Green tensor is not objective (note the argument of $\mathbf{Q}$):

$$\begin{aligned}\mathbf{B}'_t(\tau) &= \mathbf{F}'_t(\tau)\mathbf{F}'_t(\tau)^T = \mathbf{Q}(\tau)\mathbf{F}_t(\tau)\mathbf{Q}(t)^T\left(\mathbf{Q}(\tau)\mathbf{F}_t(\tau)\mathbf{Q}(t)^T\right)^T\\ &= \mathbf{Q}(\tau)\mathbf{F}_t(\tau)\mathbf{Q}(t)^T\mathbf{Q}(t)\mathbf{F}_t(\tau)^T\mathbf{Q}(\tau)^T \\ &= \mathbf{Q}(\tau)\mathbf{B}_t(\tau)\mathbf{Q}(\tau)^T,\end{aligned} \tag{4.20}$$

but the right relative Cauchy–Green tensor is,

$$\begin{aligned}\mathbf{C}'_t(\tau) &= \mathbf{F}'_t(\tau)^T\mathbf{F}'_t(\tau) = \left(\mathbf{Q}(\tau)\mathbf{F}_t(\tau)\mathbf{Q}(t)^T\right)^T\mathbf{Q}(\tau)\mathbf{F}_t(\tau)\mathbf{Q}(t)^T\\ &= \mathbf{Q}(t)\mathbf{F}_t(\tau)^T\mathbf{Q}(\tau)^T\mathbf{Q}(\tau)\mathbf{F}_t(\tau)\mathbf{Q}(t) \\ &= \mathbf{Q}(t)\mathbf{C}_t(\tau)\mathbf{Q}(t)^T.\end{aligned} \tag{4.21}$$

Velocity Gradient. The velocity gradient is not objective. To see this, we note that the motion is transformed according to,

$$\mathbf{M}'(\mathbf{X}, t) = \mathbf{c}(t) + \mathbf{Q}(t)\mathbf{M}(\mathbf{X}, t), \tag{4.22}$$

since the motion is given by $\mathbf{x}(t) = \mathbf{M}(\mathbf{X}, t)$. Thus the velocity transforms according to

$$\hat{\mathbf{u}}'(\mathbf{X}, t) = \dot{\mathbf{c}}(t) + \mathbf{Q}(t)\hat{\mathbf{u}}(\mathbf{X}, t) + \dot{\mathbf{Q}}(t)\mathbf{x}(\mathbf{X}, t). \tag{4.23}$$

Expressing this in the Eulerian sense

$$\mathbf{u}'(\mathbf{x}, t) = \dot{\mathbf{c}}(t) + \mathbf{Q}(t)\mathbf{u}(\mathbf{x}, t) + \dot{\mathbf{Q}}(t)\mathbf{x}. \tag{4.24}$$

The velocity gradient thus transforms accordingly

$$L'_{ij} = \frac{\partial u'_i}{\partial x'_j} = \frac{\partial u'_i}{\partial x_k}\frac{\partial x_k}{\partial x'_j} = \left(Q_{il}\frac{\partial u_l}{\partial x_k} + \dot{Q}_{ik}\right)Q^T_{kj},$$

that is,

$$\mathbf{L}' = \mathbf{Q}\mathbf{L}\mathbf{Q}^T + \dot{\mathbf{Q}}\mathbf{Q}^T. \tag{4.25}$$

Since $\mathbf{Q}\mathbf{Q}^T = \mathbf{I}$, $\dot{\mathbf{Q}}\mathbf{Q}^T + \mathbf{Q}\dot{\mathbf{Q}}^T = \mathbf{0}$, and $\dot{\mathbf{Q}}\mathbf{Q}^T = -\mathbf{Q}\dot{\mathbf{Q}}^T = -\left(\dot{\mathbf{Q}}\mathbf{Q}^T\right)^T$ is anti-symmetric, and thus the symmetric part of (4.25), or the strain rate tensor, is objective while the anti-symmetric part of (4.25), or the vorticity tensor, is not:

$$\begin{aligned}\mathbf{D}' &= \mathbf{Q}\mathbf{D}\mathbf{Q}^T, \\ \mathbf{W}' &= \mathbf{Q}\mathbf{W}\mathbf{Q}^T + \dot{\mathbf{Q}}\mathbf{Q}^T.\end{aligned} \tag{4.26}$$

Rivlin–Ericksen Tensors. All the Rivlin–Ericksen tensors are objective. We have just seen that the first Rivlin–Ericksen tensor is objective,

$$\mathbf{A}'_1 = \mathbf{Q}\mathbf{A}_1\mathbf{Q}^T. \tag{4.27}$$

Let all the Rivlin–Ericksen tensors up to order $1 \leq n$ be objective. The next Rivlin–Ericksen tensor of order $n + 1$ transforms according to

$$\begin{aligned}\mathbf{A}'_{n+1} &= \frac{d}{dt}\mathbf{A}'_n + \mathbf{A}'_n\mathbf{L}' + \mathbf{L}'^T\mathbf{A}'_n \\ &= \frac{d}{dt}\left(\mathbf{Q}\mathbf{A}_n\mathbf{Q}^T\right) + \mathbf{Q}\mathbf{A}_n\mathbf{Q}^T\left(\mathbf{Q}\mathbf{L}\mathbf{Q}^T + \dot{\mathbf{Q}}\mathbf{Q}^T\right) \\ &\quad + \left(\mathbf{Q}\mathbf{L}\mathbf{Q}^T + \dot{\mathbf{Q}}\mathbf{Q}^T\right)^T\mathbf{Q}\mathbf{A}_n\mathbf{Q}^T \\ &= \mathbf{Q}\left[\frac{d}{dt}\mathbf{A}_n + \mathbf{A}_n\mathbf{L} + \mathbf{L}^T\mathbf{A}_n\right]\mathbf{Q}^T \\ &\quad + \dot{\mathbf{Q}}\mathbf{A}_n\mathbf{Q}^T + \mathbf{Q}\mathbf{A}_n\dot{\mathbf{Q}}^T + \mathbf{Q}\mathbf{A}_n\mathbf{Q}^T\dot{\mathbf{Q}}\mathbf{Q}^T + \mathbf{Q}\dot{\mathbf{Q}}^T\mathbf{Q}\mathbf{A}_n\mathbf{Q}^T.\end{aligned}$$

Since $\mathbf{Q}\mathbf{Q}^T = \mathbf{I}$,

$$\dot{\mathbf{Q}} = -\mathbf{Q}\dot{\mathbf{Q}}^T\mathbf{Q}, \quad \dot{\mathbf{Q}}^T = -\mathbf{Q}^T\dot{\mathbf{Q}}\mathbf{Q}^T,$$

and we find that

$$\mathbf{A}'_{n+1} = \mathbf{Q}\left[\frac{d}{dt}\mathbf{A}_n + \mathbf{A}_n\mathbf{L} + \mathbf{L}^T\mathbf{A}_n\right]\mathbf{Q}^T, \tag{4.28}$$

i.e., the $(n+1)$th-order Rivlin–Ericksen tensor is also objective. By the process of induction, all the Rivlin–Ericksen tensors are objective.

4.3.5   Principle of Local Action

The principle of local action embodies the idea that only particles near a point should be involved in determining the stress at that point. This is consistent with the exclusion of long-range forces, which have already been included in body forces. This is Oldroyd's first point.

4.3.6   Principle of Determinism

This principle states the obvious, that the current stress state in the material is determined by the past history of the motion. Future state of the motion has no say in the current state of the stress; i.e., the material possesses no clairvoyance.

In addition to these principles, there may be restrictions imposed on the constitutive equation because of the symmetry of the material. These symmetry restrictions are discussed separately. This is satisfied by Oldroyd's second and third points.

4.4   Integrity Bases

4.4.1   Isotropic Scalar-Valued Functions

Consider a scalar-valued function of a vector $\mathbf{u}$, and suppose that this function satisfies

$$f(\mathbf{u}) = f(\mathbf{Q}\mathbf{u}), \tag{4.29}$$

for every orthogonal tensor $\mathbf{Q}$. Such a function is called *isotropic*. We are interested in how f depends on $\mathbf{u}$. The answer is simple because only the magnitude of $\mathbf{u}$ is invariant under every orthogonal tensor. Thus f must be a function of $u = |\mathbf{u}|$. In that sense, $u = |\mathbf{u}|$ is called an *integrity basis* for $f(\mathbf{u})$.

Likewise, a scalar-valued function of two vectors $\mathbf{u}$ and $\mathbf{v}$ satisfying

$$f(\mathbf{u}, \mathbf{v}) = f(\mathbf{Qu}, \mathbf{Qv}), \qquad (4.30)$$

for every orthogonal $\mathbf{Q}$, is called an isotropic function of its two arguments. Since

$$\mathbf{u} \cdot \mathbf{u}, \quad \mathbf{v} \cdot \mathbf{v}, \quad \mathbf{u} \cdot \mathbf{v} \qquad (4.31)$$

are the only three invariants under rotation (their lengths and the angle between them are invariant, Weyl [90]),

$$f(\mathbf{u}, \mathbf{v}) = f(\mathbf{u} \cdot \mathbf{u}, \mathbf{v} \cdot \mathbf{v}, \mathbf{u} \cdot \mathbf{v}). \qquad (4.32)$$

The three scalar invariants form the integrity basis for $f(\mathbf{u}, \mathbf{v})$.

Similarly, f is said to be an isotropic function of a tensor $\mathbf{S}$ if for every orthogonal $\mathbf{Q}$,

$$f(\mathbf{S}) = f\left(\mathbf{QSQ}^T\right). \qquad (4.33)$$

Here, f must be a function of the three invariants of $\mathbf{S}$,

$$\operatorname{tr}\mathbf{S}, \quad \operatorname{tr}\mathbf{S}^2, \quad \operatorname{tr}\mathbf{S}^3, \qquad (4.34)$$

or, equivalently, of the three eigenvalues of $\mathbf{S}$. These three invariants form the integrity bases for $f(\mathbf{S})$.

The invariants that can be formed from two tensors $\mathbf{A}$ and $\mathbf{B}$ are

$$\begin{aligned}
&\operatorname{tr}\mathbf{A}, \quad \operatorname{tr}\mathbf{A}^2, \quad \operatorname{tr}\mathbf{A}^3, \quad \operatorname{tr}\mathbf{B}, \quad \operatorname{tr}\mathbf{B}^2, \quad \operatorname{tr}\mathbf{B}^3, \\
&\operatorname{tr}\mathbf{AB}, \quad \operatorname{tr}\mathbf{A}^2\mathbf{B}, \quad \operatorname{tr}\mathbf{AB}^2, \quad \operatorname{tr}\mathbf{A}^2\mathbf{B}^2.
\end{aligned} \qquad (4.35)$$

Thus, an isotropic scalar-valued function of $\mathbf{A}$ and $\mathbf{B}$ must be a function of these 10 invariants. These form the integrity basis for $f(\mathbf{A}, \mathbf{B})$.

We list another integrity basis for a scalar valued, isotropic function of a symmetric second order tensor $\mathbf{S}$ and two vectors $\mathbf{u}$ and $\mathbf{v}$:

$$\begin{aligned}
&\operatorname{tr}\mathbf{S}, \ \operatorname{tr}\mathbf{S}^2, \ \operatorname{tr}\mathbf{S}^3, \ \mathbf{u} \cdot \mathbf{u}, \ \mathbf{u} \cdot \mathbf{v}, \ \mathbf{v} \cdot \mathbf{v}, \ \mathbf{u} \cdot \mathbf{Su}, \ \mathbf{u} \cdot \mathbf{S}^2\mathbf{u}, \\
&\mathbf{v} \cdot \mathbf{Sv}, \ \mathbf{v} \cdot \mathbf{S}^2\mathbf{v}, \ \mathbf{u} \cdot \mathbf{Sv}, \ \mathbf{u} \cdot \mathbf{S}^2\mathbf{v}.
\end{aligned} \qquad (4.36)$$

4.4.2 Isotropic Vector-Valued Functions

Suppose that $\mathbf{w} = \mathbf{g}(\mathbf{v})$ is a vector-valued function of the vector $\mathbf{v}$. It is called isotropic if, for every orthogonal tensor $\mathbf{Q}$,

$$\mathbf{Qg}\,(\mathbf{v}) = \mathbf{g}\,(\mathbf{Qv})\,. \tag{4.37}$$

Now, define a scalar-valued function of two vectors $\mathbf{u}$ and $\mathbf{v}$ through

$$f\,(\mathbf{u}, \mathbf{v}) = \mathbf{u} \cdot \mathbf{g}\,(\mathbf{v})\,.$$

Thus, since $\mathbf{g}$ is isotropic,

$$f\,(\mathbf{Qu}, \mathbf{Qv}) = \mathbf{Qu} \cdot \mathbf{g}\,(\mathbf{Qv}) = \mathbf{Qu} \cdot \mathbf{Qg}\,(\mathbf{v}) = \mathbf{u} \cdot \mathbf{g}\,(\mathbf{v})\,,$$

since $\mathbf{Qu} \cdot \mathbf{Qg} = Q_{ij} u_j Q_{ik} g_k = \delta_{jk} u_j g_k = \mathbf{u} \cdot \mathbf{g}$. Consequently, f is an isotropic function of its two arguments, and thus it is a function of the invariants listed in (4.31). However, by its definition, $f\,(\mathbf{u}, \mathbf{v})$ is linear in its first argument, and therefore

$$f\,(\mathbf{u}, \mathbf{v}) = \mathbf{u} \cdot h\,(\mathbf{v} \cdot \mathbf{v})\,\mathbf{v}\,.$$

It follows that

$$\mathbf{g}\,(\mathbf{v}) = h\,(\mathbf{v} \cdot \mathbf{v})\,\mathbf{v}\,. \tag{4.38}$$

4.4.3 Isotropic Tensor-Valued Functions

A symmetric tensor-valued function of a symmetric tensor $\mathbf{B}$ is isotropic if

$$\mathbf{QG}\,(\mathbf{B})\,\mathbf{Q}^T = \mathbf{G}\,\left(\mathbf{QBQ}^T\right)\,, \tag{4.39}$$

for every orthogonal tensor $\mathbf{Q}$.

Now, define a scalar function of two symmetric tensors through

$$f\,(\mathbf{A}, \mathbf{B}) = \operatorname{tr}\,[\mathbf{AG}\,(\mathbf{B})]\,.$$

From its definition,

$$
\begin{aligned}
f\,\left(\mathbf{QAQ}^T, \mathbf{QBQ}^T\right) &= \operatorname{tr}\,\left[\mathbf{QAQ}^T \mathbf{G}\,\left(\mathbf{QBQ}^T\right)\right] \\
&= \operatorname{tr}\,\left[\mathbf{QAQ}^T \mathbf{QG}\,(\mathbf{B})\,\mathbf{Q}^T\right] \\
&= \operatorname{tr}\,\left[\mathbf{QAG}\,(\mathbf{B})\,\mathbf{Q}^T\right] = Q_{ij} A_{jk} G_{kl} Q_{il} \\
&= \operatorname{tr}\,[\mathbf{AG}\,(\mathbf{B})]\,.
\end{aligned}
$$

That is, f is isotropic in its two arguments. It is therefore a function of the ten invariants listed in (4.35). Since $f\,(\mathbf{A}, \mathbf{B})$ is linear in its first argument,

$$f\,(\mathbf{A}, \mathbf{B}) = \operatorname{tr}\,\left[\mathbf{A}\,\left(g_0\mathbf{I} + g_1\mathbf{B} + g_2\mathbf{B}^2\right)\right]\,,$$

and consequently

$$\mathbf{G}(\mathbf{B}) = g_0\mathbf{I} + g_1\mathbf{B} + g_2\mathbf{B}^2, \tag{4.40}$$

where g_0, g_1, g_2 are scalar-valued functions of the three invariants of $\mathbf{B}$.

The underlying principle is (Pipkin and Rivlin [71]): to find the form for isotropic vector-valued, or a symmetric tensor-valued functions of a vector or a symmetric tensor, first form an artificial scalar product with a second vector or another symmetric tensor, which can be shown to be an isotropic scalar-valued function. Then find the relevant integrity bases for this isotropic scalar-valued function. Finally, because of the linearity in its first arguments, non-linear terms in these arguments can be discarded, arriving at the correct form for the original isotropic function. For functions which are isotropic, or transversely isotropic, or have crystal classes as their symmetric groups, see the review article by Spencer [82]. For functions which are invariant under the full unimodular group, see Fahy and Smith [23].

4.5 Symmetry Restrictions

4.5.1 Unimodular Matrix

Let X and X' be two adjacent particles with positions $\mathbf{X}$ and $\mathbf{X} + d\mathbf{X}$ in the reference configuration. Now consider a change in the local configuration so that X remains at $\mathbf{X}$, while X' goes to $\mathbf{X} + d\mathbf{X}'$. We assume that the gradient of this transformation is $\mathbf{H}$, where

$$d\mathbf{X}' = \mathbf{H}d\mathbf{X}, \tag{4.41}$$

and $\mathbf{H}$ is a proper unimodular matrix, i.e., det $\mathbf{H} = 1$ (the configuration change should not lead to a change in volume).

Now if the motion is such that X goes to $\mathbf{x}$, and X' goes to $\mathbf{x} + d\mathbf{x}$, then we have a new motion M' (with deformation gradient $\mathbf{F}'$):

$$\begin{aligned} M' : X \text{ or } \mathbf{X} &\mapsto \mathbf{x}, \ X' \text{ or } \mathbf{X} + d\mathbf{X}' \mapsto \mathbf{x} + d\mathbf{x}, \\ M : X \text{ or } \mathbf{X} &\mapsto \mathbf{x}, \quad X \text{ or } \mathbf{X} + d\mathbf{X} \mapsto \mathbf{x} + d\mathbf{x}. \end{aligned} \tag{4.42}$$

The two mappings are different, because the two shapes about $\mathbf{X}$ are mapped into the same shape about $\mathbf{x}$ (see Fig. 4.9). Now, since

$$d\mathbf{x} = \mathbf{F}d\mathbf{X} = \mathbf{F}'d\mathbf{X}' = \mathbf{F}'\mathbf{H}d\mathbf{X},$$

we have

$$\mathbf{F}'\mathbf{H} = \mathbf{F} \quad \mathbf{F}' = \mathbf{F}\mathbf{H}^{-1}. \tag{4.43}$$

Fig. 4.9 A change in the local configuration leads to the same shape after deformation

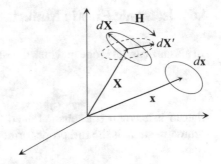

The strain measures for the motion M' can be calculated as

$$\mathbf{B}' = \mathbf{F}'\mathbf{F}'^T = \mathbf{F}\mathbf{H}^{-1}\mathbf{H}^{-T}\mathbf{F}^T, \tag{4.44}$$

$$\mathbf{C}' = \mathbf{F}'^T\mathbf{F}' = \mathbf{H}^{-T}\mathbf{F}^T\mathbf{F}\mathbf{H}^{-1} = \mathbf{H}^{-T}\mathbf{C}\mathbf{H}^{-1}, \tag{4.45}$$

$$\mathbf{F}'_t(\tau) = \mathbf{F}'(\tau)\mathbf{F}'(t)^{-1} = \mathbf{F}(\tau)\mathbf{H}^{-1}\mathbf{H}\mathbf{F}(t)^{-1} = \mathbf{F}_t(\tau), \tag{4.46}$$

$$\mathbf{C}'_t(\tau) = \mathbf{F}'_t(\tau)^T\mathbf{F}'_t(\tau) = \mathbf{C}_t(\tau). \tag{4.47}$$

The relative strain measure $\mathbf{C}_t(\tau)$ is not sensitive to the unimodular changes about X.

4.5.2 Symmetry Group

Suppose that we are interested in a certain constitutive property $\wp$, say the stress tensor, that depends on the kinematics. Moreover, suppose that some unimodular changes of the local shape about X leave this quantity unchanged. Let

$$\mathcal{G}_\wp = \{\mathbf{I}, \mathbf{H}_1, \mathbf{H}_2, \ldots\} \tag{4.48}$$

be the set of all the unimodular transformations that preserves $\wp$, then $\mathcal{G}_\wp$ is a group, called the $\wp$-symmetry group, i.e., the group of unimodular changes in the neighbourhood of X that leaves $\wp$ invariant.

4.5.3 Isotropic Materials

Different materials have different symmetry groups: there are isotropic groups, transversely isotropic groups, etc. We are mainly concerned with isotropic materials, where the symmetry group is the proper orthogonal group, $\mathbf{H}^{-1} = \mathbf{H}^T$.

4.6 Isotropic Elastic Materials

An elastic material is one in which the stress is a function of the deformation gradient:

$$T = f(F),$$ (4.49)

where f is a symmetric tensor-valued function of F. For an isotropic material, the symmetry group is the full proper orthogonal group, that is, for every orthogonal H, we have

$$f(F) = f(FH).$$ (4.50)

Isotropic Constraint. Now F has the unique polar decomposition

$$F = VR,$$

where R is orthogonal. Thus

$$F' = FH = VRH,$$

and because H is orthogonal, RH is orthogonal. In other words, F' can be chosen in the set $\{V, VQ_1, VQ_2, \ldots\}$, where Q_i are orthogonal. Thus f is a function of V alone. Since $B = FF^T = V^2$, f is a function of the strain B:

$$T = f(F) = f(V) = f(B).$$ (4.51)

Objectivity Constraint. Objectivity imposes the following constraint on T:

$$T' = QTQ^T,$$ (4.52)

for every orthogonal Q. Since $T' = f(B') = f(QBQ^T)$ (frame indifference of the stress, objectivity of B) the requirement (4.52) becomes

$$Qf(B)Q^T = f(QBQ^T).$$ (4.53)

Thus f is an isotropic function of B (see (4.39)). The general form for f has been found in (4.40), thus

$$f(B) = \alpha_0 I + \alpha_1 B + \alpha_2 B^2,$$ (4.54)

where the scalar coefficients are functions of the three invariants of B. One may use the Cayley–Hamilton theorem and express B^2 in terms of B and B^{-1}.

Mooney and neo-Hookean Materials. Therefore, the general constitutive equation for an isotropic elastic solid is given by

$$T = \beta_0 I + \beta_1 B - \beta_2 B^{-1},$$ (4.55)

where β_i are functions of the three invariants of $\mathbf{B}$. The term $\beta_0 \mathbf{I}$ can be absorbed in the hydrostatic pressure. The case when β_1, β_2 are constant is called the *Mooney material* [58] or sometimes the Mooney–Rivlin material, in deference to Rivlin. In addition, if $\beta_2 = 0$ it is called the *neo-Hookean material* [87].

If the material behaves like a Mooney model, in a uniaxial deformation (Problem 4.6), the plot of $T_{ZZ}/(\lambda - \lambda^2)$ against λ^{-1} should be a straight line, with slope β_2 and intercept β_1. The data from Rivlin and Saunder (see [87]) showed that the Mooney model is inadequate: in compression, their data indicated that $\beta_2 \approx 0$, whereas in tension, their data showed that β_2/β_1 varies from 0.3 to 1. However, the Mooney model should be reasonable for most qualitative purposes.

4.7 The Simple Material

Noll [62] defined a simple material (solid or liquid) as one in which the current stress is a functional (function of function) of the history of the deformation gradient $\mathbf{F}(\tau)$, $-\infty < \tau \leq t$:

$$\mathbf{T}(t) = \mathcal{G}(\mathbf{F}(\tau)), \quad -\infty < \tau \leq t. \tag{4.56}$$

Objectivity and frame indifference require,

$$\mathbf{Q}(t)\,\mathcal{G}(\mathbf{F}(\tau))\,\mathbf{Q}(t)^T = \mathcal{G}(\mathbf{Q}(\tau)\mathbf{F}(\tau)), \tag{4.57}$$

for all rotational histories $\mathbf{Q}(\tau)$, $-\infty < \tau \leq t$.

Recall now the polar decomposition for $\mathbf{F}$, $\mathbf{F}(\tau) = \mathbf{R}(\tau)\mathbf{U}(\tau)$. Since $\mathbf{Q}$ is arbitrary, we choose $\mathbf{Q}(\tau) = \mathbf{R}(\tau)^T$. Thus (4.57) requires

$$\mathcal{G}(\mathbf{F}(\tau)) = \mathbf{R}(t)\,\mathcal{G}(\mathbf{U}(\tau))\,\mathbf{R}(t)^T$$
$$= \mathbf{F}(t)\mathbf{U}(t)^{-1}\,\mathcal{G}(\mathbf{U}(\tau))\,\mathbf{U}(t)^{-1}\mathbf{F}(t)^T.$$

Using the definition $\mathbf{C} = \mathbf{U}^2$, we define a new functional through

$$\mathcal{F}(\mathbf{C}(\tau)) = \mathbf{U}(t)^{-1}\,\mathcal{G}(\mathbf{U}(\tau))\,\mathbf{U}(t)^{-1}. \tag{4.58}$$

Thus for a simple material,

$$\mathbf{T}(t) = \mathbf{F}(t)\,\mathcal{F}(\mathbf{C}(\tau))\,\mathbf{F}(t)^T, \quad -\infty < \tau \leq t. \tag{4.59}$$

In addition, from Problem 4.7 (4.104),

$$\mathbf{T}(t) = \mathcal{G}(\mathbf{F}(\tau)) = \mathbf{F}(t)\,\mathcal{F}\left(\mathbf{F}(t)^T\,\mathbf{C}_t(\tau)\,\mathbf{F}(t)\right)\mathbf{F}(t)^T, \quad -\infty < \tau \leq t. \tag{4.60}$$

This says that the current stress is a functional of the history of the right relative Cauchy–Green tensor, and the current value of the deformation gradient. That is, we may define a new functional

$$\mathcal{G}\left(\mathbf{F}\left(\tau\right)\right) = \mathcal{H}\left(\mathbf{C}_t\left(\tau\right), \mathbf{F}\left(t\right)\right). \tag{4.61}$$

Replacing $\mathbf{F}(\tau)$ by $\mathbf{F}(\tau)\mathbf{H}$, where $\mathbf{H}$ is unimodular, leaves $\mathbf{C}_t\left(\tau\right)$ unchanged, refer to (4.47), and thus,

$$\mathcal{H}\left(\mathbf{C}_t\left(\tau\right), \mathbf{F}\left(t\right)\right) = \mathcal{H}\left(\mathbf{C}_t\left(\tau\right), \mathbf{F}\left(t\right)\mathbf{H}\right) \tag{4.62}$$

In addition, this functional must obey the objectivity restriction (4.57):

$$\mathbf{Q}\left(t\right)\mathcal{H}\left(\mathbf{C}_t\left(\tau\right), \mathbf{F}\left(t\right)\right)\mathbf{Q}\left(t\right)^T = \mathcal{H}\left(\mathbf{Q}\left(t\right)\mathbf{C}_t\left(\tau\right)\mathbf{Q}\left(t\right)^T, \mathbf{Q}\left(t\right)\mathbf{F}\left(t\right)\right) \tag{4.63}$$

noting the objectivity of the relative right Cauchy–Green strain tensor (4.21).

4.7.1 Simple Fluid

If the material is an isotropic fluid (note that fluid is isotropic in Noll's definition), the stress is invariant under the full orthogonal group. Noll showed in this case, the current stress is given by

$$\mathbf{T}\left(t\right) = \mathcal{F}\left(\mathbf{C}_t\left(\tau\right), \rho\left(t\right)\right), \quad -\infty < \tau \leq t. \tag{4.64}$$

This functional must satisfy objectivity,

$$\mathbf{Q}\left(t\right)\mathcal{F}\left(\mathbf{C}_t\left(\tau\right), \rho\left(t\right)\right)\mathbf{Q}\left(t\right)^T = \mathcal{F}\left(\mathbf{Q}\left(t\right)\mathbf{C}_t\left(\tau\right)\mathbf{Q}\left(t\right)^T, \rho\left(t\right)\right), \tag{4.65}$$

for all orthogonal tensors $\mathbf{Q}$.

4.7.2 Incompressible Simple Fluid

Incompressibility has been introduced as a simplification of real material behaviour. This demands that

$$\rho\left(t\right) = \rho_R, \quad \det \mathbf{F}\left(t\right) = 1, \quad \det \mathbf{C}\left(t\right) = 1, \quad \nabla \cdot \mathbf{u} = 0. \tag{4.66}$$

Its adoption implies that the constitutive relation can only determine the stress up to an isotropic part (the hydrostatic pressure); this hydrostatic pressure must be determined by the equations of balance. We write

$$\mathbf{T} = -P\mathbf{I} + \mathbf{S},$$
$$\mathbf{S}(t) = \mathcal{F}(\mathbf{C}_t(\tau)), \quad -\infty < \tau \le t. \tag{4.67}$$

$\mathbf{S}$ is called the extra stress. Of course the functional $\mathcal{F}$ must satisfy objectivity:

$$\mathbf{Q}(t)\,\mathcal{F}(\mathbf{C}_t(\tau))\,\mathbf{Q}(t)^T = \mathcal{F}\left(\mathbf{Q}(t)\,\mathbf{C}_t(\tau)\,\mathbf{Q}(t)^T\right). \tag{4.68}$$

4.7.3 Fading Memory

The idea of fading memory embodies the notion that distant events in the past (large τ) should have less bearing on the current stress than events in recent past. This idea can be implemented in the functional in various ways, through integral or differential operators. We close this chapter with two classes of constitutive relations obtained by assuming that the fluid memory is instantaneous, and that the deformation is small in some sense, i.e., the relative strain tensor hardly departs from the unit tensor.

4.8 Order Fluids

When the fluid memory is catastrophic (i.e., instantaneous), we can assume that stress is an isotropic function of the Rivlin–Ericksen tensors $\mathbf{A}_n$. Recall that the relative strain tensor can be expressed as a Taylor series with coefficients $\mathbf{A}_n$,

$$\mathbf{S} = \mathbf{f}(\mathbf{A}_1, \mathbf{A}_2, \ldots, \mathbf{A}_N). \tag{4.69}$$

The physical dimensions of $\mathbf{A}_n$ are T^{-n}, where T is the time. A sequence of approximations to $\mathbf{f}$, correct to order T^n, $n = 1, 2, \ldots$ can be developed.
To first order, the Newtonian fluid:

$$\mathbf{S} = \mathbf{S}^{(1)} = \eta_0 \mathbf{A}_1. \tag{4.70}$$

To second order, the second-order fluid model:

$$\mathbf{S} = \mathbf{S}^{(2)} = \mathbf{S}^{(1)} + (\nu_1 + \nu_2)\,\mathbf{A}_1^2 - \frac{\nu_1}{2}\mathbf{A}_2. \tag{4.71}$$

To third order, the third-order fluid model:

$$\mathbf{S} = \mathbf{S}^{(3)} = \mathbf{S}^{(2)} + \alpha_0\left(\mathrm{tr}\mathbf{A}_1^2\right)\mathbf{A}_1 + \alpha_1\left(\mathbf{A}_1\mathbf{A}_2 + \mathbf{A}_2\mathbf{A}_1\right) + \alpha_3\mathbf{A}_3. \tag{4.72}$$

Higher-order fluids can be developed in the same manner. These order fluids possess no memory, and using them to describe memory phenomena may lead to disaster.

4.8.1   Unsteady Motion

To see that the order fluids are unsuitable for discussing unsteady motion, we consider the flow of a second-order fluid (4.71) in a channel of width h,

$$u = u\,(y,t)\,, \quad v = 0, \quad w = 0. \tag{4.73}$$

The first Rivlin–Ericksen tensor and its square are given by

$$[\mathbf{A}_1] = \begin{bmatrix} 0 & \partial u/\partial y & 0 \\ \partial u/\partial y & 0 & 0 \\ 0 & 0 & 0 \end{bmatrix}, \quad [\mathbf{A}_1^2] = \begin{bmatrix} (\partial u/\partial y)^2 & 0 & 0 \\ 0 & (\partial u/\partial y)^2 & 0 \\ 0 & 0 & 0 \end{bmatrix}.$$

The second Rivlin–Ericksen tensor is

$$[\mathbf{A}_2] = \left[\dot{\mathbf{A}}_1\right] + [\mathbf{A}_1\mathbf{L}] + \left[\mathbf{L}^T\mathbf{A}_1\right]$$

$$= \begin{bmatrix} 0 & \partial^2 u/\partial t\partial y & 0 \\ \partial^2 u/\partial t\partial y & 0 & 0 \\ 0 & 0 & 0 \end{bmatrix} + \begin{bmatrix} 0 & 0 & 0 \\ 0 & 2\,(\partial u/\partial y)^2 & 0 \\ 0 & 0 & 0 \end{bmatrix}.$$

The stress components are

$$S_{xy} = \eta_0 \frac{\partial u}{\partial y} - \frac{\nu_1}{2}\frac{\partial^2 u}{\partial t\partial y}, \quad S_{xx} = (\nu_1 + \nu_2)\left(\frac{\partial u}{\partial y}\right)^2,$$

$$S_{yy} = \nu_2 \left(\frac{\partial u}{\partial y}\right)^2, \quad S_{zz} = 0.$$

Now, the equations of balance read

$$\rho\frac{\partial u}{\partial t} = -\frac{\partial P}{\partial x} + \frac{\partial S_{xx}}{\partial x} + \frac{\partial S_{xy}}{\partial y} + \frac{\partial S_{xz}}{\partial z},$$

$$0 = -\frac{\partial P}{\partial y} + \frac{\partial S_{yx}}{\partial x} + \frac{\partial S_{yy}}{\partial y} + \frac{\partial S_{yz}}{\partial z},$$

$$0 = -\frac{\partial P}{\partial z} + \frac{\partial S_{zx}}{\partial x} + \frac{\partial S_{zy}}{\partial y} + \frac{\partial S_{zz}}{\partial z}.$$

With no pressure gradient, the equations of motion reduce to

$$\rho\frac{\partial u}{\partial t} = \eta_0 \frac{\partial^2 u}{\partial y^2} - \frac{\nu_1}{2}\frac{\partial^3 u}{\partial t\partial y^2}, \quad u\,(0,t) = 0, \quad u\,(h,t) = 0. \tag{4.74}$$

This is a linear third-order partial differential equation; we look for a solution by separation of variables:

$$u(y, t) = \sum_{n=1} \phi_n(t) \psi_n(y),$$

where

$$\rho \dot{\phi}_n \psi_n = \eta_0 \phi_n \psi_n'' - \frac{1}{2} \nu_1 \dot{\phi}_n \psi_n''.$$

We can try the Fourier series for ψ_n:

$$\psi_n = a_n \sin \frac{n\pi y}{h},$$

which leads to

$$\left[\rho - \frac{1}{2} \nu_1 \frac{n^2 \pi^2}{h^2} \right] \dot{\phi}_n = -\eta_0 \frac{n^2 \pi^2}{h^2} \phi_n.$$

This has the solution

$$\phi_n(t) = \phi_n^0 \exp \left(\frac{\eta_0 t}{\nu_1/2 - \rho h^2/n^2 \pi^2} \right). \tag{4.75}$$

Clearly, for given material properties and geometry, we can always find n so that the exponent is positive (a positive exponent implies instability: the solution is unbounded in time); that is the solution (4.75) is unstable to any disturbance of the form stated. Only when $\nu_1 = 0$ (Newtonian fluid) is the solution (4.75) stable. Thus all unsteady flows are too fast for the second-order fluid to handle (and indeed for all order fluids).

4.8.2 Velocity Field in a Second-Order Fluid

The Newtonian pressure field p_N obeys

$$\nabla p_N = \eta_0 \nabla \cdot \mathbf{A}_1 - \rho \mathbf{a}, \tag{4.76}$$

where $\mathbf{a}$ is the acceleration field. If the same Newtonian velocity field were to occur in the second-order fluid, then an additional pressure term p_S will arise and this has to satisfy

$$\nabla p_S = \nabla \cdot \left[(\nu_1 + \nu_2) \mathbf{A}_1^2 - \frac{1}{2} \nu_1 \mathbf{A}_2 \right]. \tag{4.77}$$

There are a few special classes of flows for which the right side of (4.77) can be expressed as a gradient of a scalar. We will consider two special cases.

Potential Flows. When the velocity field is a potential flow and, consequently $\mathbf{u} = \nabla\phi$, incompressibility demands that $\phi_{,ii} = 0$, where the comma denotes a partial derivative. Since

$$(\mathbf{A}_1)_{ij} = 2\phi_{,ij}, \tag{4.78}$$

it follows that

$$\nabla \cdot \mathbf{A}_1 = \mathbf{0}. \tag{4.79}$$

Hence, the Newtonian pressure field is given by

$$p_N = -\rho[\phi_{,t} + \frac{1}{2}\mathbf{u} \cdot \mathbf{u}]. \tag{4.80}$$

The results from Problem 4.10, (4.111), show that in potential flows, the Newtonian and second-order fluid velocity fields are identical, with the second-order pressure term given by

$$p_S = \frac{1}{8}\,(\nu_1 + 4\nu_2)\,\mathrm{tr}\,\mathbf{A}_1^2. \tag{4.81}$$

Plane Creeping Flows. The second case where the Newtonian velocity field is also the second-order fluid velocity field is the steady plane flow, where the velocity takes the form $\mathbf{u} = u(x, y)\mathbf{i} + v(x, y)\mathbf{j}$. The Cayley–Hamilton theorem and incompressibility together show that

$$\mathbf{A}_1^2 + (\det \mathbf{A}_1)\mathbf{1} = \mathbf{0}. \tag{4.82}$$

Thus,

$$\det \mathbf{A}_1 = -\frac{1}{2}\mathrm{tr}\,\mathbf{A}_1^2. \tag{4.83}$$

Hence, we have the result

$$\nabla \cdot \mathbf{A}_1^2 = \nabla\left(\frac{1}{2}\mathrm{tr}\,\mathbf{A}_1^2\right). \tag{4.84}$$

To show that the divergence of $\mathbf{A}_2$ can be expressed as a gradient of a scalar, we need the results of Giesekus [31] and Pipkin [70]

$$\nabla \cdot \mathbf{A}_2 = \nabla\left(\frac{1}{\eta_0}\frac{dp_N}{dt} + \frac{3}{4}\mathrm{tr}\,\mathbf{A}_1^2\right). \tag{4.85}$$

Thus, a plane creeping flow in a Newtonian fluid is also a solution to the plane creeping flow problem in a second order fluid.

4.9 Green–Rivlin Expansion

Green and Rivlin [32] proposed an expansion based on the integral of the strain history

$$\mathbf{G}\,(s) = \mathbf{C}_t\,(t - s) - \mathbf{I}, \quad 0 \le s < \infty, \tag{4.86}$$

which is regarded to be small in some sense (i.e., the relative strain $\mathbf{C}_t\,(t - s)$ is near to identity, the undeformed state). The expansion is quite unwieldy, consisting of multiple integral terms. We record the first term here

$$\mathbf{S}\,(t) = \int_0^\infty \mu\,(s)\,\mathbf{G}\,(s)\,ds, \tag{4.87}$$

which is called the finite linear viscoelasticity integral model. Here, $\mu(s)$ is the memory kernel, a decreasing function of s (distant past is less important – fluid has fading memory). Usually an exponential memory function is chosen

$$\mu\,(s) = -\frac{G}{\lambda}e^{-s/\lambda}, \tag{4.88}$$

where G is a modulus and λ is a relaxation time. A multiple relaxation (exponential) mode is sometimes used as well.

Problems

Problem 4.1 In a simple shear deformation of a linear elastic material (4.7), the displacement field takes the form

$$v_1 = \gamma y, \quad v_2 = 0, \quad v_3 = 0, \tag{4.89}$$

where γ is the amount of shear. Find the elastic stress, in particular, the shear stress. This justifies calling μ the shear modulus.

Problem 4.2 In a uni-axial extension of a linear elastic material (4.7), the displacement field is given by

$$v_1 = \varepsilon x, \quad v_2 = -\nu \varepsilon y, \quad v_3 = -\nu \varepsilon z, \tag{4.90}$$

where ε is the elongational strain, and ν is the amount of lateral contraction due to the axial elongation, called Poisson's ratio. If the lateral stresses are zero, show that

$$\nu = \frac{\lambda}{2\,(\lambda + \mu)}, \quad \mu = \frac{E}{2\,(1 + \nu)}, \tag{4.91}$$

where E is the Young's modulus, i.e., $T_{xx} = E\varepsilon$.

Problem 4.3 Show that a material element $d\mathbf{X} = dX\mathbf{P}$, where $\mathbf{P}$ is a unit vector, is stretched according to

$$dx^2 = dX^2 \mathbf{C} : \mathbf{PP},$$

where $\mathbf{C}$ is the right Cauchy-Green tensor. Thus when $\mathbf{P}$ is randomly distributed in space, the average amount of stretch is

$$\langle dx^2 \rangle = \frac{1}{3}dX^2 \mathrm{tr}\mathbf{C},$$

and therefore $\frac{1}{3}\mathrm{tr}\mathbf{C}$ can be used as a definition of the Weissenberg number.

Problem 4.4 Let $\mathbf{f}$ be a vector-valued, isotropic polynomial of a symmetric tensor $\mathbf{S}$ and a vector $\mathbf{v}$. Use the integrity basis in (4.36) to prove that

$$\mathbf{f}(\mathbf{S}, \mathbf{v}) = [f_0\mathbf{1} + f_1\mathbf{S} + f_2\mathbf{S}^2]\mathbf{v},$$

where the scalar valued coefficients are polynomials in the six invariants involving only $\mathbf{S}$ and $\mathbf{v}$ in the list (4.36).

Problem 4.5 Consider a simple shear deformation of a rubber-like material (4.55), where

$$x = X + \gamma Y, \quad y = Y, \quad z = Z. \tag{4.92}$$

Show that the Finger strain tensor $\mathbf{B}$ and its inverse are given by

$$[\mathbf{B}] = \begin{bmatrix} 1 + \gamma^2 & \gamma & 0 \\ \gamma & 1 & 0 \\ 0 & 0 & 1 \end{bmatrix}, \quad [\mathbf{B}^{-1}] = \begin{bmatrix} 1 & -\gamma & 0 \\ -\gamma & 1 + \gamma^2 & 0 \\ 0 & 0 & 1 \end{bmatrix}. \tag{4.93}$$

Consequently, show that the stress tensor is given by

$$[\mathbf{T}] = -P\,[\mathbf{I}] + \begin{bmatrix} \beta_1\,(1 + \gamma^2) - \beta_2 & (\beta_1 + \beta_2)\,\gamma & 0 \\ (\beta_1 + \beta_2)\,\gamma & \beta_1 - \beta_2\,(1 + \gamma^2) & 0 \\ 0 & 0 & \beta_1 - \beta_2 \end{bmatrix}, \tag{4.94}$$

where P is the hydrostatic pressure. Thus, show that the shear stress and the normal stress differences are

$$S = (\beta_1 + \beta_2)\,\gamma, \quad N_1 = (\beta_1 + \beta_2)\,\gamma^2, \quad N_2 = -\beta_2\gamma^2. \tag{4.95}$$

Deduce that the linear shear modulus of elasticity is

$$G = \lim_{\gamma \to 0} (\beta_1 + \beta_2). \tag{4.96}$$

The ratio

$$\frac{N_1}{S} = \gamma \tag{4.97}$$

is independent of the material properties. Such a relation is called *universal*.

Problem 4.6 In a uniaxial elongational deformation of a rubber-like material (4.55), where (in cylindrical coordinates)

$$R = \lambda^{1/2}r, \quad \Theta = \theta, \quad Z = \lambda^{-1}z, \tag{4.98}$$

show that the inverse deformation gradient is

$$\left[\mathbf{F}^{-1}\right] = \begin{bmatrix} \frac{\partial R}{\partial r} & \frac{1}{r}\frac{\partial R}{\partial \theta} & \frac{\partial R}{\partial z} \\ 0 & \frac{R}{r} & 0 \\ \frac{\partial Z}{\partial r} & \frac{1}{r}\frac{\partial Z}{\partial \theta} & \frac{\partial Z}{\partial z} \end{bmatrix} = \begin{bmatrix} \lambda^{1/2} & 0 & 0 \\ 0 & \lambda^{1/2} & 0 \\ 0 & 0 & \lambda^{-1} \end{bmatrix}. \tag{4.99}$$

Consequently, the Finger strain tensor $\mathbf{B}$ and its inverse $\mathbf{B}^{-1}$ are

$$\left[\mathbf{B}^{-1}\right] = \left[\mathbf{F}^{-T}\mathbf{F}^{-1}\right] = \begin{bmatrix} \lambda & 0 & 0 \\ 0 & \lambda & 0 \\ 0 & 0 & \lambda^{-2} \end{bmatrix}, \quad [\mathbf{B}] = \begin{bmatrix} \lambda^{-1} & 0 & 0 \\ 0 & \lambda^{-1} & 0 \\ 0 & 0 & \lambda^2 \end{bmatrix}. \tag{4.100}$$

Thus the total stress tensor for a rubber-like material (4.55) is

$$[\mathbf{T}] = -P\,[\mathbf{I}] + \begin{bmatrix} \beta_1\lambda^{-1} - \beta_2\lambda & 0 & 0 \\ 0 & \beta_1\lambda^{-1} - \beta_2\lambda & 0 \\ 0 & 0 & \beta_1\lambda^2 - \beta_2\lambda^{-2} \end{bmatrix}. \tag{4.101}$$

Under the condition that the lateral tractions are zero, i.e., $T_{rr} = 0$, the pressure can be found, and thus show that the tensile stress is

$$T_{zz} = -P + \beta_1\lambda^2 - \beta_2\lambda^{-2} = \beta_1\lambda^2 - \beta_2\lambda^{-2} - \beta_1\lambda^{-1} + \beta_2\lambda$$
$$= \left(\lambda^2 - \lambda^{-1}\right)\left(\beta_1 + \beta_2\lambda^{-1}\right). \tag{4.102}$$

This tensile stress is the force per unit area in the deformed configuration. As $r = \lambda^{1/2}R$, the corresponding force per unit area in the undeformed configuration is

$$T_{ZZ} = T_{zz}\lambda^{-1} = \left(\lambda - \lambda^{-2}\right)\left(\beta_1 + \beta_2\lambda^{-1}\right). \tag{4.103}$$

Problem 4.7 Show that

$$\mathbf{C}\left(\tau\right) = \mathbf{F}\left(t\right)^{T}\mathbf{C}_{t}\left(\tau\right)\mathbf{F}\left(t\right). \tag{4.104}$$

Problem 4.8 Consider a simple shear flow

$$u = \dot{\gamma}y, \quad v = 0, \quad w = 0. \tag{4.105}$$

Show that the stress tensor in the second-order model is given by

$$[\mathbf{S}] = \eta_0 \begin{bmatrix} 0 & \dot{\gamma} & 0 \\ \dot{\gamma} & 0 & 0 \\ 0 & 0 & 0 \end{bmatrix} + (\nu_1 + \nu_2) \begin{bmatrix} \dot{\gamma}^2 & 0 & 0 \\ 0 & \dot{\gamma}^2 & 0 \\ 0 & 0 & 0 \end{bmatrix} - \frac{\nu_1}{2} \begin{bmatrix} 0 & 0 & 0 \\ 0 & 2\dot{\gamma}^2 & 0 \\ 0 & 0 & 0 \end{bmatrix}. \tag{4.106}$$

Thus, the three viscometric functions are

$$S = \eta_0\dot{\gamma}, \quad N_1 = S_{11} - S_{22} = \nu_1\dot{\gamma}^2, \quad N_2 = S_{22} - S_{33} = \nu_2\dot{\gamma}^2. \tag{4.107}$$

Problem 4.9 Consider a second-order fluid in an elongational flow

$$u = \dot{\varepsilon}x, \quad v = -\frac{\dot{\varepsilon}}{2}y, \quad w = -\frac{\dot{\varepsilon}}{2}z. \tag{4.108}$$

Show that the stress is given by

$$[\mathbf{S}] = \eta_0 \begin{bmatrix} 2\dot{\varepsilon} & 0 & 0 \\ 0 & -\dot{\varepsilon} & 0 \\ 0 & 0 & -\dot{\varepsilon} \end{bmatrix} + (\nu_1 + \nu_2) \begin{bmatrix} 4\dot{\varepsilon}^2 & 0 & 0 \\ 0 & \dot{\varepsilon}^2 & 0 \\ 0 & 0 & \dot{\varepsilon}^2 \end{bmatrix} - \frac{\nu_1}{2} \begin{bmatrix} 4\dot{\varepsilon}^2 & 0 & 0 \\ 0 & \dot{\varepsilon}^2 & 0 \\ 0 & 0 & \dot{\varepsilon}^2 \end{bmatrix}. \tag{4.109}$$

Consequently the elongational viscosity is given by

$$\eta_E = \frac{S_{xx} - S_{yy}}{\dot{\varepsilon}} = 3\eta_0 + 3\left(\frac{\nu_1}{2} + \nu_2\right)\dot{\varepsilon}. \tag{4.110}$$

Problem 4.10 Show that, for potential flows,

$$\nabla \cdot \mathbf{A}_1^2 = \frac{1}{2}\nabla(tr\,\mathbf{A}_1^2), \quad \nabla \cdot \mathbf{A}_2 = \frac{3}{4}\nabla(tr\,\mathbf{A}_1^2). \tag{4.111}$$

Problem 4.11 For steady two-dimensional incompressible flows, a stream function $\psi = \psi(x, y)$ can be defined such that the velocity components u and v can be expressed as

$$u = \frac{\partial\psi}{\partial y}, \quad v = -\frac{\partial\psi}{\partial x}. \tag{4.112}$$

Obtain the stresses in a second-order fluid in terms of ψ. Substitute the stresses into the equations of motion and eliminating the pressure term through the use of the equality of the mixed partial derivatives, i.e., $p_{,xy} = p_{,yx}$, to obtain

$$\eta \Delta^2 \psi - \frac{\nu_1}{2} \mathbf{v} \cdot \nabla(\Delta^2 \psi) = 0, \tag{4.113}$$

where Δ^2 is the two-dimensional biharmonic operator. Deduce that a Newtonian velocity field is also a velocity field for the second-order fluid. The result is due to Tanner [83].

Problem 4.12 In a simple shear flow,

$$u = \dot{\gamma}(t) y, \quad v = 0, \quad w = 0,$$

show that the path lines $\boldsymbol{\xi}(\tau) = (\xi, \psi, \zeta)$ are given by

$$\xi(\tau) = x + y\gamma(t, \tau), \quad \psi(\tau) = y, \quad \zeta(\tau) = z, \tag{4.114}$$

where

$$\gamma(t, \tau) = \int_t^\tau \dot{\gamma}(s)\, ds. \tag{4.115}$$

Find the relative strain tensor, the stress tensor for a finite viscoelastic integral fluid, and show that the shear stress and the normal stress differences are given by

$$S_{12}(t) = \int_0^\infty \mu(s)\gamma(t, t - s)\, ds, \tag{4.116}$$

$$N_1(t) = \int_0^\infty \mu(s)[\gamma(t, t - s)]^2\, ds = -N_2. \tag{4.117}$$

Investigate the case where the shear rate is constant and sinusoidal in time for the memory function in (4.88).

Chapter 5
Inelastic Models and Linear Viscoelasticity

Some Practical Engineering Models

We have seen some of the classical constitutive equations introduced in the last three centuries, and explored the general formulation of constitutive equations in the last chapter. There, we mention that the general constitutive principles should be taken as guidelines only; they should emerge from the physics of the fluids' microstructures. In engineering, the emphasis is to produce the analyses for the design process. Therefore we prefer simple models with the "right" physics to be included. "Right" physics here means to correctly account for the flow process to be modelled: if the flow process does not call for certain behaviour, then it may be left out in the constitutive modelling. In addition, engineers do not have any qualm in supplementing a constitutive equation with empirical data, as long as the correct physical framework has already been incorporated in the constitutive equation. We are aiming to strive for simplicity in a correct constitutive framework, with enough empirical inputs as needed, to ensure a quantitative prediction to the physical flow phenomenon that we are modelling.

5.1 Inelastic Fluids

When the flow phenomena are dominated by viscosity effects, then it makes sense to model the viscosity function accurately. *Inelastic*, or *generalised Newtonian, fluids* are those for which the extra stress tensor is proportional to the strain rate tensor, but the "constant" of proportionality (the viscosity) is allowed to depend on the strain rate:

$$\mathbf{S} = 2\eta\left(\dot{\gamma}\right)\mathbf{D}, \tag{5.1}$$

where $\dot{\gamma} = \sqrt{2\mathrm{tr}\,\mathbf{D}^2}$ is called the generalised strain rate (in a simple shear flow, this quantity reduces to the magnitude of the shear rate). Inelastic model possesses neither

© Springer International Publishing AG 2017
N. Phan-Thien and N. Mai-Duy, *Understanding Viscoelasticity*,
Graduate Texts in Physics, DOI 10.1007/978-3-319-62000-8_5

Fig. 5.1 Different non-Newtonian behaviours

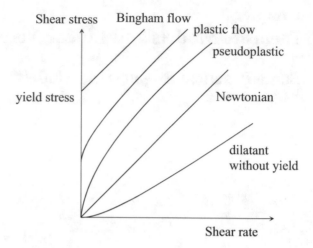

memory nor elasticity, and therefore it is unsuitable for transient flows, or flows that call for elasticity effects. It is only useful in steady viscometric flows where an accurate representation of the viscosity is paramount. Depending on the functional form of $\eta(\dot{\gamma})$, one can get different non-Newtonian behaviours, see Fig. 5.1.

Bingham fluids are those that can support a yield stress. When the shear stress exceeds this yield value, the fluid flows like a Newtonian fluid, with a constant viscosity. Plastic fluids are yield-stress types of fluids, with flow viscosity decreasing with shear rate (shear thinning). The term "pseudoplastic" means that the viscosity decreases with shear rates (shear thinning). The opposite of pseudoplastic is dilatant (shear thickening).

5.1.1   Carreau Model

Different forms for the viscosity have been proposed for pseudoplastic fluids; the most popular one is the Carreau model:

$$\eta(\dot{\gamma}) = \eta_\infty + \frac{\eta_0 - \eta_\infty}{\left(1 + \Gamma^2\dot{\gamma}^2\right)^{(1-n)/2}}. \tag{5.2}$$

There are four parameters: η_∞ is the infinite-shear-rate viscosity, η_0 is the zero-shear-rate viscosity, n is called the power-law index ($n-1$ is the slope of $(\eta - \eta_\infty)/(\eta_0 - \eta_\infty)$ versus $\dot{\gamma}$ in log-log plot), and Γ is a constant bearing the dimension of time (this constant has no relation to the relaxation time of the fluid). A graph of (5.2) is illustrated in Fig. 5.2.

The model is meant for shear-thinning fluids and therefore $0 < n \le 1$. $n = 1$ represents the Newtonian behaviour.

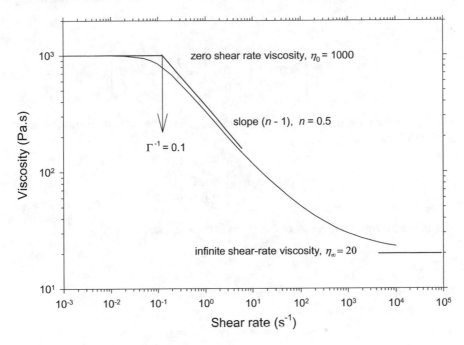

Fig. 5.2 A typical plot of the Carreau viscosity

5.1.2   Power-Law Model

Included in the Carreau model is the power-law model:

$$\eta\left(\dot{\gamma}\right) = k \left|\dot{\gamma}\right|^{n-1} , \tag{5.3}$$

where k is called the consistency and n the power-law index. When $\eta_\infty = 0$ and at high shear rate, the Carreau model (5.2) reduces to the power-law model with power-law index n and consistency $k = \eta_0 \Gamma^{n-1}$. The power-law model breaks down in regions where the shear rate is zero - in these regions the stress is unbounded.

Simple Shear Flow. In a simple shearing flow, where the fluid is confined between two plates and the top plate is moving with a velocity U, the velocity field takes the form

$$u = \dot{\gamma}y, \quad v = 0, \quad w = 0, \quad \dot{\gamma} = U/h.$$

The velocity gradient and the strain rate tensors are

$$[\mathbf{L}] = \left[\nabla \mathbf{u}^T\right] = \begin{bmatrix} 0 & \dot{\gamma} & 0 \\ 0 & 0 & 0 \\ 0 & 0 & 0 \end{bmatrix}, \quad [\mathbf{D}] = \frac{1}{2}\left(\mathbf{L} + \mathbf{L}^T\right) = \frac{1}{2}\begin{bmatrix} 0 & \dot{\gamma} & 0 \\ \dot{\gamma} & 0 & 0 \\ 0 & 0 & 0 \end{bmatrix},$$

and

$$[\mathbf{D}^2] = \frac{1}{4} \begin{bmatrix} 0 & \dot{\gamma} & 0 \\ \dot{\gamma} & 0 & 0 \\ 0 & 0 & 0 \end{bmatrix} \begin{bmatrix} 0 & \dot{\gamma} & 0 \\ \dot{\gamma} & 0 & 0 \\ 0 & 0 & 0 \end{bmatrix} = \frac{1}{4} \begin{bmatrix} \dot{\gamma}^2 & 0 & 0 \\ 0 & \dot{\gamma}^2 & 0 \\ 0 & 0 & 0 \end{bmatrix}.$$

The generalised shear rate is $\dot{\gamma} = \sqrt{2\mathrm{tr}\,\mathbf{D}^2}$, and the stress tensor for the inelastic model (5.1) is

$$[\mathbf{S}] = \eta\,(\dot{\gamma}) \begin{bmatrix} 0 & \dot{\gamma} & 0 \\ \dot{\gamma} & 0 & 0 \\ 0 & 0 & 0 \end{bmatrix}. \tag{5.4}$$

The only non-zero component of the stress is the shear stress,

$$S_{xy} = \eta\,(\dot{\gamma})\,\dot{\gamma}. \tag{5.5}$$

Elongational Flow. In an elongational flow, where

$$u = \dot{\varepsilon}x, \quad v = -\frac{\dot{\varepsilon}}{2}y, \quad w = -\frac{\dot{\varepsilon}}{2}z,$$

we have

$$[\mathbf{L}] = \begin{bmatrix} \dot{\varepsilon} & 0 & 0 \\ 0 & -\dot{\varepsilon}/2 & 0 \\ 0 & 0 & -\dot{\varepsilon}/2 \end{bmatrix} = [\mathbf{D}]\,, \quad [\mathbf{D}^2] = \begin{bmatrix} \dot{\varepsilon}^2 & 0 & 0 \\ 0 & \dot{\varepsilon}^2/4 & 0 \\ 0 & 0 & \dot{\varepsilon}^2/4 \end{bmatrix}, \quad \dot{\gamma} = \sqrt{3}\dot{\varepsilon}.$$

Thus the stress tensor for the inelastic model (5.1) is

$$[\mathbf{S}] = 2\eta\left(\sqrt{3}\dot{\varepsilon}\right) \begin{bmatrix} \dot{\varepsilon} & 0 & 0 \\ 0 & -\dot{\varepsilon}/2 & 0 \\ 0 & 0 & -\dot{\varepsilon}/2 \end{bmatrix} \tag{5.6}$$

and the elongational viscosity is given by

$$\eta_E = \frac{N_1}{\dot{\varepsilon}} = 3\eta\left(\sqrt{3}\dot{\varepsilon}\right). \tag{5.7}$$

It has been known that the elongational viscosity and the shear viscosity have dissimilar shape. Therefore, in-elastic models may not be suitable in processes where there is a mixture of shear and elongational flow components.

5.2 Linear Viscoelasticity

The concept of linear viscoelasticity was originated with Maxwell (Fig. 5.3), who proposed the following equation in 1867–68,

$$\frac{d\sigma}{dt} = E\frac{d\varepsilon}{dt} - \frac{\sigma}{\lambda}, \tag{5.8}$$

where σ is the (one-dimensional) stress, ε is the (one-dimensional) strain, E is the modulus of elasticity and λ is a time constant. When the relaxation time is zero, keeping the product $\eta = \lambda E$ constant, the Newtonian model is recovered. And when the relaxation is infinitely large, a further integration yields the Hookean model.

Years later, Meyer (1874) introduced the equation

$$\sigma = G\gamma + \eta\frac{d\gamma}{dt}, \tag{5.9}$$

combining the solid response (G is the elastic modulus, γ is the shear strain) and liquid response (η is the viscosity) in one equation. This equation is now known as the Kelvin–Voigt body – it should be called Kelvin–Meyer–Voigt (a full account can be found in Tanner and Walter [86]).

Boltzmann (Fig. 5.4) criticised the lack of generality in Maxwell's and Meyer's work and proposed that the stress at the current time depends not only on the current strain, but on the past strains as well. It was assumed that a strain at a distant past contributes less to the stress than a more recent strain. This is recognised as the familiar concept of fading memory. Furthermore, linear superposition was assumed: supposing that the strain between times t' and $t' + dt'$, say $d\gamma(t')$, contributes

Fig. 5.3 J.C. Maxwell (1831–1879) published his first scientific paper when he was fourteen. He set up Cavendish Laboratory at Cambridge in 1874, and died of cancer at an early age of 48. He united electricity and magnetism into the concept of electromagnetism. He also introduced the concept of stress relaxation in the kinetic theory of gases

Fig. 5.4 Ludwig Boltzmann
(1844–1906) was most
famous for his atomic
viewpoint and his invention
of statistical mechanics. He
is said never to have failed
any student taking his course

$G\left(t - t'\right) d\gamma\left(t'\right)$ to the stress, then the total stress at time t is

$$\sigma\left(t\right) = \int_{-\infty}^{t} G\left(t - t'\right) d\gamma\left(t'\right) = \int_{-\infty}^{t} G\left(t - t'\right) \dot{\gamma}\left(t'\right) dt'. \tag{5.10}$$

Here, $G(t)$ is a decreasing function of time, the *relaxation modulus*, and $\dot{\gamma}$ is the shear rate. A three-dimensional version of this is

$$\mathbf{S}\left(t\right) = 2 \int_{-\infty}^{t} G\left(t - t'\right) \mathbf{D}\left(t'\right) dt', \tag{5.11}$$

$$S_{ij}\left(t\right) = 2 \int_{-\infty}^{t} G\left(t - t'\right) D_{ij}\left(t'\right) dt'.$$

The Newtonian liquid is recovered with the delta memory function

$$G\left(t - t'\right) = \eta_0 \delta\left(t - t'\right)$$

$$\mathbf{S}\left(t\right) = 2 \int_{-\infty}^{t} \eta_0 \delta\left(t - t'\right) \mathbf{D}\left(t'\right) dt' = 2\eta_0 \mathbf{D}\left(t\right). \tag{5.12}$$

The most-often used relaxation modulus function is the *Maxwell discrete relaxation spectrum*:

$$G\left(t\right) = \sum_{j=1}^{N} G_j e^{-t/\lambda_j}, \tag{5.13}$$

which consists of a discrete spectrum of relaxation times $\{G_j, \lambda_j\}$. The linear viscoelastic constitutive model (5.11) is not objective; it is only valid at vanishingly small strains.

5.2.1 Simple Shear Flow

In a flow with constant strain rate $\mathbf{D}$, the linear viscoelastic stress (5.11) is

$$\mathbf{S}(t) = 2\eta(t)\mathbf{D}, \quad \eta(t) = \int_{-\infty}^{t} G(t - t')\, dt'. \tag{5.14}$$

With the Maxwell relaxation modulus (5.13),

$$\mathbf{S}(t) = \sum_{j=1}^{N} 2\eta_j \left(1 - e^{-t/\lambda_j}\right) \mathbf{D}, \quad \eta_j = G_j \lambda_j. \tag{5.15}$$

Now consider an oscillatory shear flow between two parallel plates at a distance h apart, where the bottom plate is stationary and the top plate is sinusoidally displaced by a small amount $\delta \sin \omega t$, $\delta \ll h$. The top plate velocity is

$$U(t) = \delta\omega \cos \omega t.$$

The shear rate and the shear strain are, respectively,

$$\dot{\gamma}(t) = \frac{\delta}{h}\omega \cos \omega t = \dot{\gamma}_0 \cos \omega t,$$

$$\gamma(t) = \frac{\delta}{h} \sin \omega t = \gamma_0 \sin \omega t, \quad \gamma_0 = \frac{\delta}{h} \ll 1, \quad \dot{\gamma}_0 = \omega \gamma_0. \tag{5.16}$$

The only non-zero component of the stress is the shear stress,

$$S_{12} = \int_{-\infty}^{t} G(t - t')\, \dot{\gamma}_0 \cos \omega t'\, dt',$$

$$= \int_{0}^{\infty} \dot{\gamma}_0 G(s) \cos \omega (t - s)\, ds,$$

$$= \int_{0}^{\infty} \dot{\gamma}_0 G(s) \left[\cos \omega t \cos \omega s + \sin \omega t \sin \omega s\right] ds,$$

$$= G'(\omega)\, \gamma_0 \sin \omega t + \eta'(\omega)\, \dot{\gamma}_0 \cos \omega t,$$

where the coefficients in the strain, $G'(\omega)$, the *storage modulus*, and in the strain rate, $\eta'(\omega)$, the *dynamic viscosity*, are material functions of the frequency; they are defined by

$$G'(\omega) = \int_0^\infty \omega G(s) \sin \omega s\, ds, \quad \eta'(\omega) = \int_0^\infty G(s) \cos \omega s\, ds. \tag{5.17}$$

The other two related quantities, the *loss modulus* G'', and the *storage viscosity* η'' are defined as

$$G''(\omega) = \omega \eta'(\omega), \quad \eta''(\omega) = \frac{G'(\omega)}{\omega}. \tag{5.18}$$

Sometimes it is more convenient to work with complex numbers, and the complex modulus G^* and the complex viscosity η^* are thus defined as

$$G^*(\omega) = G'(\omega) + iG''(\omega), \quad \eta^*(\omega) = \eta'(\omega) - i\eta''(\omega). \tag{5.19}$$

One can denote the shear rate as $\dot{\gamma}^* = \dot{\gamma}_0 e^{i\omega t}$, then

$$\dot{\gamma}(t) = \dot{\gamma}_0 \mathrm{Re}\left(e^{i\omega t}\right), \quad S_{12}(t) = \mathrm{Re}\left(\eta^* \dot{\gamma}_0^*\right). \tag{5.20}$$

It should be remembered that the linear viscoelasticity model is valid only at small strains, that is, both G' and η' are independent of the strain amplitude γ_0; this ought to be tested before a frequency sweep is done. The strain amplitude for which linearity holds could be as large as 20% for polymer melts and solutions, and as small as 0.1% for biological materials, such as bread dough.

Figure 5.5 is a plot of the storage modulus and dynamic viscosity of a water-dough as functions of the frequency, where the strain amplitude was kept at 0.1%.

By performing the inverse Fourier transform of (5.17), the relaxation modulus can be obtained from dynamic data as

$$G(s) = \frac{2}{\pi} \int_0^\infty \eta'(\omega) \cos \omega s\, d\omega = \frac{2}{\pi} \int_0^\infty \frac{G'(\omega)}{\omega} \sin \omega s\, d\omega. \tag{5.21}$$

Inverting the dynamic properties to obtain the relaxation modulus using (5.21) may be an ill-conditioned problem (see Problem 5.3).

5.2.2   Step Strain

Consider the stress relaxation experiment after a step strain, Fig. 5.6. Suppose a shear strain of magnitude γ is imposed at time t_0 within a short period Δ, as sketched in Fig. 5.6. During this period the strain rate can be assumed constant, given by $\dot{\gamma} = \gamma/\Delta$. The stress is (for $t > t_0 + \Delta$)

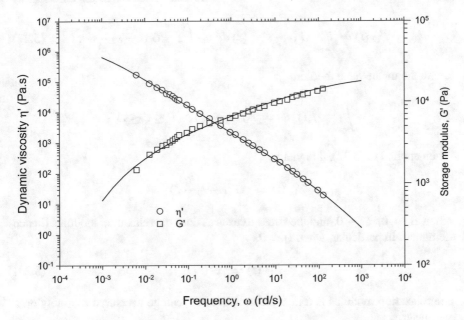

Fig. 5.5 Dynamic properties of flour-water dough. The strain amplitude was set at 0.1%

Fig. 5.6 Stress relaxation after a step strain

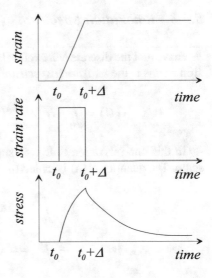

$$S_{12}(t) = \int_{-\infty}^{t} G(t-s)\,\dot{\gamma}(s)\,ds = \int_{t_0}^{t_0+\Delta} G(t-s)\,\frac{\gamma}{\Delta}ds. \tag{5.22}$$

Recall the mean-value theorem,

$$\int_{t_0}^{t_0+\Delta} f(s)\,ds = \Delta f(t_0+\zeta),\quad 0 \le \zeta \le \Delta.$$

When applied to (5.22), this yields

$$S_{12}(t) = \gamma G(t-t_0-\zeta).$$

When $\Delta \to 0$, $\zeta \to 0$, and the stress relaxes as does the relaxation modulus (hence the name). In particular, when $t_0 = 0$,

$$S_{12}(t) = \gamma G(t). \tag{5.23}$$

The relaxation modulus is a material function and can be measured routinely on a rheometer.

5.2.3   Relaxation Spectrum

We have met the discrete Maxwell relaxation spectrum, (5.13). The continuous version of this is the *relaxation spectrum H*, defined by the relation

$$G(t) = \int_{0}^{\infty} H(\lambda)\,e^{-t/\lambda}\frac{d\lambda}{\lambda} = \int_{-\infty}^{\infty} H(\lambda)\,e^{-t/\lambda}d\ln\lambda. \tag{5.24}$$

In this, the relaxation time λ is supposedly evenly distributed on a logarithmic scale. The quantity $H(\lambda)$ is called the relaxation spectrum. Note that

$$G'(\omega) = \int_{0}^{\infty} \omega\sin\omega s \int_{0}^{\infty} H(\lambda)\,e^{-s/\lambda}\frac{d\lambda}{\lambda}ds$$

$$= \int_{0}^{\infty} \omega H(\lambda)\,\frac{d\lambda}{\lambda}\int_{0}^{\infty} e^{-s/\lambda}\sin\omega s\,ds. \tag{5.25}$$

The last integral can be evaluated, giving

$$G'(\omega) = \int_{-\infty}^{\infty} \frac{\omega^2\lambda^2}{1+\omega^2\lambda^2}H(\lambda)\,d\ln\lambda. \tag{5.26}$$

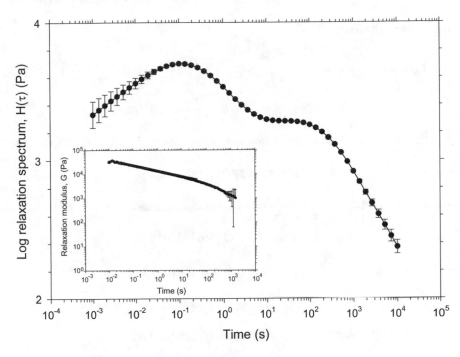

Fig. 5.7 Relaxation modulus and spectrum

Similarly,

$$\eta'(\omega) = \int_0^\infty \frac{H(\lambda)}{1 + \omega^2 \lambda^2} d\lambda. \tag{5.27}$$

These results should be compared to those obtained with the discrete Maxwell relaxation modulus, Problem 5.4. Inversion of the dynamic data to find H according to (5.26) or (5.27) is an ill-conditioned problem (Problem 5.3). In Fig. 5.7 the relaxation modulus, and its spectrum are shown for a dough-water flour, using a regularisation method of Weese [89].

5.3 Correspondence Principle

5.3.1 Quasi-Static Approximation

To solve boundary-value problems for a linear viscoelastic fluid, the quasi-static approximation is used. Here, one ignores the inertia terms $\rho\dot{\mathbf{u}}$. This is possible if the characteristic frequency is not too high (De $\ll$ 1). Let suppose the flow starts from time zero, before which the stress is zero. The linear viscoelastic stress and the equations of motion are

$$\mathbf{T}(t) = -P\mathbf{I} + \mathbf{S} = -P\mathbf{I} + \int_0^t G\left(t - t'\right)\left(\nabla\mathbf{u}\left(t'\right) + \nabla\mathbf{u}\left(t'\right)^T\right)dt', \quad (5.28)$$

$$\nabla \cdot \mathbf{T} = -\nabla P + \nabla \cdot \mathbf{S} = \mathbf{0}, \quad \nabla \cdot \mathbf{u} = 0, \quad \mathbf{x} \in V, \quad (5.29)$$

subjected to a relevant boundary condition on the bounding surface S, for example,

$$\mathbf{u}(\mathbf{x}, t) = \mathbf{u}_0(t), \quad \mathbf{x} \in S. \quad (5.30)$$

We can take the Laplace transform of all the above equations to arrive at

$$\bar{\mathbf{T}}(s) = -\bar{P}\mathbf{I} + \bar{G}\left(\nabla\bar{\mathbf{u}} + \nabla\bar{\mathbf{u}}^T\right), \quad (5.31)$$

$$-\nabla\bar{P} + \bar{G}\nabla^2\bar{\mathbf{u}} = \mathbf{0}, \quad \nabla \cdot \bar{\mathbf{u}} = 0, \quad (5.32)$$

$$\bar{\mathbf{u}}(\mathbf{x}, s) = \bar{\mathbf{u}}_0(s), \quad \mathbf{x} \in S, \quad (5.33)$$

where the overbar denotes a Laplace transform variable, i.e.,

$$\bar{\phi}(s) = \int_0^\infty e^{-st}\phi(t)\,dt. \quad (5.34)$$

Equations (5.31)–(5.33) are identical to those of the corresponding Newtonian (Stokes) flow problem, except that the viscosity is now replaced by $\bar{G}$. Thus the solution in the Laplace transform domain matches the Stokes solution. This is the essence of the *Correspondence Principle*. The solution in the physical domain is then obtained by inverting the Laplace transform. We give an example for the start-up of a circular Couette flow.

5.3.2 Circular Couette Flow

For a circular Couette flow, the velocity field in cylindrical coordinates (Fig. 5.8) is

$$\mathbf{u} = \{0, r\Omega(r, t), 0\}, \quad (5.35)$$

where the boundary conditions on the inner cylinder (angular velocity Ω_i, radius R_i) and the outer cylinder (stationary, radius R_o) are

$$\Omega(R_i, t) = \Omega_i, \quad \Omega(R_o, t) = 0. \quad (5.36)$$

Fig. 5.8 Circular Couette flow

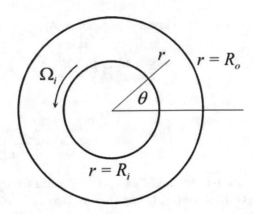

The strain rate tensor is

$$[\mathbf{D}] = \frac{1}{2} \begin{bmatrix} 0 & r\dfrac{\partial \Omega}{\partial r} & 0 \\ r\dfrac{\partial \Omega}{\partial r} & 0 & 0 \\ 0 & 0 & 0 \end{bmatrix}.$$

The only non-zero component of the linear viscoelastic stress is the shear component

$$S_{r\theta} = \int_0^t G\left(t - t'\right) r \frac{\partial \Omega}{\partial r}\left(r, t'\right) dt'.$$

The balance of linear momentum requires

$$\frac{\partial}{\partial r}\left(r^2 S_{r\theta}\right) = 0, \qquad S_{r\theta} = \frac{M\left(t\right)}{2\pi r^2},$$

where $M(t)$ is a "constant" of integration. The torque on the inner cylinder is

$$\Gamma = \int_{R_i}^{R_o} 2\pi r^2 S_{r\theta} dr = M\left(R_o - R_i\right). \tag{5.37}$$

Taking the Laplace transform,

$$\bar{S}_{r\theta} = \bar{G} r \frac{\partial \bar{\Omega}}{\partial r} \qquad \frac{\partial \bar{\Omega}}{\partial r} = \frac{\bar{M}}{2\pi \bar{G} r^3}.$$

Integrating

$$\bar{\Omega} = C - \frac{\bar{M}}{4\pi r^2 \bar{G}},$$

where C is an integration constant. Applying the boundary conditions (5.36),

$$\bar{\Omega} = \frac{\bar{M}}{4\pi R_o^2 \bar{G}}\left(1 - \frac{R_o^2}{r^2}\right), \quad \bar{\Omega}_i = \frac{\bar{M}}{4\pi R_o^2 \bar{G}}\left(1 - \frac{R_o^2}{R_i^2}\right). \tag{5.38}$$

Note that

$$\frac{\bar{\Omega}}{\bar{\Omega}_i} = \frac{1 - R_o^2/r^2}{1 - R_o^2/R_i^2}, \tag{5.39}$$

which is the Stokes solution (in Laplace transform domain). In the Newtonian case, $G(t) = \eta_0 \delta(t)$, giving $\bar{G} = \eta_0$. Hence,

$$M_N = \frac{4\pi R_o^2 \eta_0 \Omega_i}{1 - R_o^2/R_i^2}. \tag{5.40}$$

For the linear viscoelastic case,

$$\bar{M} = \frac{4\pi R_o^2}{1 - R_o^2/R_i^2}\bar{G}\bar{\Omega}_i,$$

$$M(t) = \frac{4\pi R_o^2}{1 - R_o^2/R_i^2}\int_0^t G\left(t - t'\right)\Omega_i\left(t'\right) dt'. \tag{5.41}$$

5.4 Mechanical Analogs

In the older rheology literature, one finds mechanical analogs for linear viscoelastic behaviours, springs for solid Hookean behaviour and dashpots for viscous Newtonian behaviour. We illustrate this with a few popular models in Fig. 5.9.

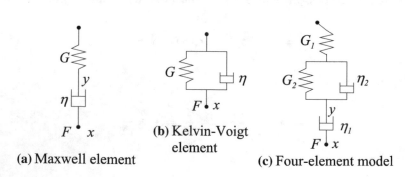

(a) Maxwell element   **(b)** Kelvin-Voigt element   **(c)** Four-element model

Fig. 5.9 Mechanical analogs of linear viscoelastic behaviours

Maxwell Element. Figure 5.9a shows the Maxwell model, where the spring element represents a Hookean behaviour, and the dashpot element, a Newtonian viscous behaviour. These elements are arranged in series with the understanding that the displacements (strains) are additive, and the forces (stresses) are equal across the elements. Then

$$\dot{x} = \dot{y} + (\dot{x} - \dot{y})$$
$$= \frac{\dot{F}}{G} + \frac{F}{\eta},$$

or

$$F + \frac{\eta}{G}\dot{F} = \eta\dot{x}.$$

Thus, if F is identified with the stress S_{ij} and x, the strain γ_{ij}, then one obtains the Maxwell model

$$S_{ij} + \frac{\eta}{G}\dot{S}_{ij} = \eta\dot{\gamma}_{ij}, \tag{5.42}$$

where $\lambda = \eta/G$ may be identified as the relaxation time. The Maxwell model is a fluid (it cannot support a shear stress without deforming).

Kelvin-Voigt-Meyer Element. Figure 5.9b shows the Kelvin-Voigt-Meyer model, where the spring and the dashpot elements are arranged in parallel. Across these elements, the displacements (strains) are equal, and the forces (stresses) are additive. Thus

$$Gx + \eta\dot{x} = F.$$

Identify x with the strain γ_{ij}, and F with the stress S_{ij}, one obtains the Kelvin–Voigt–Meyer model:

$$G\gamma_{ij} + \eta\dot{\gamma}_{ij} = S_{ij}. \tag{5.43}$$

The Kelvin–Voigt–Meyer material is a solid (it can support a shear stress indefinitely without deforming). Similar to the Maxwell model, $\lambda = \eta/G$ is the relaxation time of the model.

Four-Element Model. In a similar manner, Fig. 5.9c shows a so-called four-element model. The displacements are additive, the forces are the same across series elements; and across parallel elements, the displacements are the same, the forces are additive. Working in Laplace transform domain, with zero initial conditions, and then convert back to time domain, it may be shown that (Problem 5.6)

$$F + a_1\dot{F} + a_2\ddot{F} = b_1\dot{x} + b_2\ddot{x},$$

where

$$a_1 = \frac{\eta_1}{G_2}\left(1 + \frac{G_2}{G_1} + \frac{\eta_2}{\eta_1}\right), \quad a_2 = \frac{\eta_1\eta_2}{G_1G_2}, \quad b_1 = \eta_1, \quad b_2 = \frac{\eta_1\eta_2}{G_2},$$

leading to the following stress-strain relation for the four-element model

$$S_{ij} + a_1\dot{S}_{ij} + a_2\ddot{S}_{ij} = b_1\dot{\gamma}_{ij} + b_2\ddot{\gamma}_{ij}. \tag{5.44}$$

Multimode Models. To each of the one-relaxation time models, we could define a corresponding multimode model. For example, the multimode Kelvin–Voigt–Meyer model is written as

$$\gamma_{ij} = \sum_{n=1}^{N}\gamma_{ij}^{(n)}, \quad \gamma_{ij}^{(n)} + \lambda_n\dot{\gamma}_{ij}^{(n)} = \frac{S_{ij}}{G_n}, \tag{5.45}$$

where λ_n, $n = 1, \ldots, N$, are the relaxation times.

Problems

Problem 5.1 Show that, with the relaxation modulus function (5.13), the relation (5.11) is equivalent to

$$\mathbf{S} = \sum_{j=1}^{N}\mathbf{S}^{(j)}, \tag{5.46}$$

$$\mathbf{S}^{(j)} + \lambda_j\dot{\mathbf{S}}^{(j)} = 2\eta_j\mathbf{D}, \quad \eta_j = G_j\lambda_j.$$

This relation is called the linear Maxwell equation. Equation (5.46) is equivalent to (5.9).

Problem 5.2 Show that the shear stress for (5.11) in an oscillatory flow, where the shear rate is $\dot{\gamma} = \dot{\gamma}_0\cos(\omega t)$, can be expressed as

$$S_{12} = |G^*|\sin(\omega t + \phi), \quad \tan\phi = \frac{G''}{G'}, \tag{5.47}$$

where $\tan\phi$ is called the loss tangent.

Problem 5.3 Verify that for the spectrum

$$H(\lambda) = \cos^2(n\lambda),$$

$$\eta'(\omega) = \int_0^\infty \frac{H(\omega)}{1 + \lambda^2 \omega^2} d\omega = \frac{\pi}{4\omega} \left(1 - e^{-2n/\omega}\right).$$

At large n, the data η' is smooth, but the spectrum is highly oscillatory. Conclude that the inverse problem of finding $H(\lambda)$, given the data η' in the chosen form is ill-conditioned – that is, a small variation in the data (in the exponentially small term) may lead to a large variation in the solution.

Problem 5.4 For the Maxwell discrete relaxation spectrum (5.13), show that

$$G(t) = \sum_{j=1}^N G_j e^{-t/\lambda_j}, \ G'(\omega) = \sum_{j=1}^N \frac{G_j \omega^2 \lambda_j^2}{1 + \omega^2 \lambda_j^2}, \ \eta'(\omega) = \sum_{j=1}^N \frac{G_j \lambda_j}{1 + \omega^2 \lambda_j^2}. \quad (5.48)$$

In particular, with one relaxation mode $\lambda = \lambda_1$,

$$\tan \phi = \frac{1}{\omega \lambda} \quad (5.49)$$

deduce that as $\omega = 0 \to \infty$, the response goes from fluid ($\phi = \pi/2$) to solid behaviour ($\phi = 0$).

Problem 5.5 Suppose we have a Maxwell material with one relaxation time,

$$G(t) = \frac{\eta_0}{\lambda} e^{-t/\lambda}.$$

and Ω_i = constant. Show that the solution to the circular Couette flow problem considered in Sect. 5.3.2 is

$$\frac{M(t)}{M_N} = 1 - e^{-t/\lambda}. \quad (5.50)$$

Problem 5.6 Working in Laplace transform domain, show that the mechanical analog of Fig. 5.9c leads to (5.44).

Chapter 6
Steady Viscometric Flows

Shear Flows

There is a class of flows of the simple fluid, equivalent to the simple shearing flow, for which the kinematics and the stress can be completely determined. Ericksen [21] called them *laminar shear flows*, but the current term used to describe these flows is *viscometric flows* [16]. We review this class of flows here.

6.1 Kinematics

First, consider a simple shear flow with the kinematics

$$u = \dot{\gamma}y, \quad v = 0, \quad w = 0, \tag{6.1}$$

where the shear rate $\dot{\gamma}$ is a constant. This flow has the velocity gradient tensor

$$[\mathbf{L}] = \begin{bmatrix} 0 & \dot{\gamma} & 0 \\ 0 & 0 & 0 \\ 0 & 0 & 0 \end{bmatrix},$$

which obeys $\mathbf{L}^2 = \mathbf{0}$. For this flow, the only non-trivial Rivlin–Ericksen tensors are $\mathbf{A}_1$ and $\mathbf{A}_2$; the rest of the Rivlin–Ericksen tensors are nil,

$$[\mathbf{A}_1] = \begin{bmatrix} 0 & \dot{\gamma} & 0 \\ \dot{\gamma} & 0 & 0 \\ 0 & 0 & 0 \end{bmatrix}, \quad [\mathbf{A}_2] = \begin{bmatrix} 0 & 0 & 0 \\ 0 & 2\dot{\gamma}^2 & 0 \\ 0 & 0 & 0 \end{bmatrix}.$$

Consequently, the relative right Cauchy–Green tensor is quadratic in the time lapse, recall (3.32),

© Springer International Publishing AG 2017
N. Phan-Thien and N. Mai-Duy, *Understanding Viscoelasticity*,
Graduate Texts in Physics, DOI 10.1007/978-3-319-62000-8_6

$$\mathbf{C}_t\left(t-s\right) = \mathbf{I} - s\mathbf{A}_1 + \frac{s^2}{2}\mathbf{A}_2, \quad 0 \le s. \tag{6.2}$$

Ericksen [21] referred to flows obeying (6.2) as *laminar shear flows*, Coleman [16] called them *viscometric flows*. Yin and Pipkin [92] embarked on a search for all such flows and now our knowledge of them is essentially complete. One can write for (6.1),

$$\mathbf{u} = \dot{\gamma}\left(\mathbf{b} \cdot \mathbf{x}\right)\mathbf{a}, \tag{6.3}$$

where $\mathbf{a} = \mathbf{e}_1$ and $\mathbf{b} = \mathbf{e}_2$. We now define viscometric flows as those where the velocity field obeys (6.3), for three mutually orthogonal unit vectors $\mathbf{a}$, $\mathbf{b}$, and $\mathbf{c}$, refer to Fig. 6.1.

The three directions $\mathbf{a}$, $\mathbf{b}$, and $\mathbf{c}$ are called the shear axes; $\mathbf{a}$ is the direction of shear, $\mathbf{b}$ is the direction of shear rate, and $\mathbf{c}$ is the vorticity axis. The motion can be visualised as the relative sliding motion of a stack of playing cards, each card represents a slip surface $\mathbf{b} \cdot \mathbf{x} = \text{constant}$.

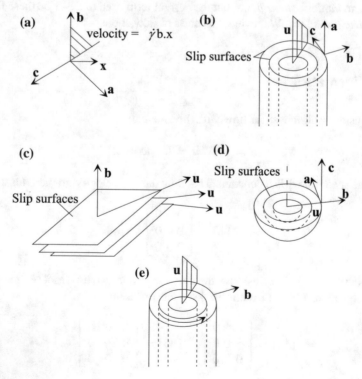

Fig. 6.1 Viscometric flows

In Fig. 6.1, all such flows are sketched: (a) simple shearing flow, (b) steady parallel flow, (c) rectilinear flow, (d) circular flow, and (e) helical flow. Shear flow has already been considered, we now briefly look at the rest.

6.1.1 Steady Parallel Flow

In the steady parallel flow (Fig. 6.1b), the velocity field takes the form

$$\mathbf{u} = w\,(x,\,y)\,\mathbf{k}, \tag{6.4}$$

where $\mathbf{a} = \mathbf{k}$ is a unit vector in the z-direction, the slip surfaces are cylinders with constant $w\,(x,\,y)$. The velocity gradient is

$$(\nabla\mathbf{u})^T = \frac{\partial w}{\partial x}\mathbf{ki} + \frac{\partial w}{\partial y}\mathbf{kj} = \dot{\gamma}\mathbf{ab}, \tag{6.5}$$

with

$$\dot{\gamma}\mathbf{b} = \frac{\partial w}{\partial x}\mathbf{i} + \frac{\partial w}{\partial y}\mathbf{j}, \quad \dot{\gamma}^2 = \left(\frac{\partial w}{\partial x}\right)^2 + \left(\frac{\partial w}{\partial y}\right)^2. \tag{6.6}$$

The material derivative of $\dot{\gamma}$ is zero, i.e., the shear rate is constant along each streamline.

6.1.2 Rectilinear Flow

Another class of viscometric flows is the rectilinear flow in which the velocity field takes the form (Fig. 6.1c)

$$\mathbf{u} = u\,(z)\,\mathbf{i} + v\,(z)\,\mathbf{j}. \tag{6.7}$$

These flows have parallel plane surfaces $z = $ constant, like a pack of playing cards. The velocity gradient is

$$(\nabla\mathbf{u})^T = \frac{\partial u}{\partial z}\mathbf{ik} + \frac{\partial v}{\partial z}\mathbf{jk} = \dot{\gamma}\mathbf{ab}, \tag{6.8}$$

where the direction of the shear rate is $\mathbf{b} = \mathbf{k}$, and

$$\dot{\gamma}\mathbf{a} = \frac{\partial u}{\partial z}\mathbf{i} + \frac{\partial v}{\partial z}\mathbf{j}, \quad \dot{\gamma}^2 = \left(\frac{\partial u}{\partial z}\right)^2 + \left(\frac{\partial v}{\partial z}\right)^2. \tag{6.9}$$

Again the shear rate remains constant along each streamline.

6.1.3 Axial Fanned Flow

In the axial fanned flow, the velocity field takes the form

$$\mathbf{u} = c\theta\mathbf{k}, \quad 0 \leq \theta \leq 2\pi, \quad \dot{\gamma} = c/r, \tag{6.10}$$

where c is a constant and $\theta = \tan^{-1}(y/x)$.

6.1.4 Helical Flow

In the axial translation, rotation and screw motions of coaxial circular slip surfaces (Fig. 6.1e), the velocity field takes the form

$$\mathbf{u} = r\omega(r)\,\mathbf{e}_\theta + u(r)\,\mathbf{e}_z. \tag{6.11}$$

The velocity gradient is

$$(\nabla\mathbf{u})^T = r\frac{\partial\omega}{\partial r}\mathbf{e}_\theta\mathbf{e}_r + \frac{\partial u}{\partial r}\mathbf{e}_z\mathbf{e}_r = \dot{\gamma}\mathbf{a}\mathbf{b}, \tag{6.12}$$

where $\mathbf{b} = \mathbf{e}_r$, and

$$\dot{\gamma}\mathbf{a} = r\frac{\partial\omega}{\partial r}\mathbf{e}_\theta + \frac{\partial u}{\partial r}\mathbf{e}_z, \quad \dot{\gamma}^2 = r^2\left(\frac{\partial\omega}{\partial r}\right)^2 + \left(\frac{\partial u}{\partial r}\right)^2. \tag{6.13}$$

The shear rate remains constant along each streamline. These flows include

1. Circular pipe flow, or Poiseuille flow, when the flow occurs in a circular pipe, or annular flow, when it occurs between two concentric cylinders. Here $\omega = 0$.
2. Circular Couette flow, when the flow occurs between concentric cylinders, one or both rotating, and $u = 0$.
3. Helical flow, when both rotational and translational components are present. Since the angular velocity is ω, a particle covers 2π radians in $2\pi/\omega$ seconds, while rising $2\pi u/\omega$. This rise is constant on each cylinder surface.

6.1.5 Helicoidal Flow

If all the helical paths in the previous flow have the same rise per turn, then the slip surfaces need not be helical but can be general helicoids. The velocity field takes the form

$$\mathbf{u} = (r\mathbf{e}_\theta + c\mathbf{e}_z)\,\omega(r, z - c\theta), \tag{6.14}$$

where c is a constant and ω is a function of r and $z - c\theta$. All the helices have the same rise per turn, i.e., $2\pi/c$. The velocity gradient is

$$
\begin{aligned}
(\nabla \mathbf{u})^T &= \left[\nabla \left(r\omega \mathbf{e}_\theta + c\omega \mathbf{e}_z \right) \right]^T = \left(\nabla \omega \left(r\mathbf{e}_\theta + c\mathbf{e}_z \right) + \omega \left(\mathbf{e}_r \mathbf{e}_\theta - \mathbf{e}_\theta \mathbf{e}_r \right) \right)^T \\
&= \left(r\mathbf{e}_\theta + c\mathbf{e}_z \right) \nabla \omega + \omega \left(\mathbf{e}_\theta \mathbf{e}_r - \mathbf{e}_r \mathbf{e}_\theta \right) \\
&= \dot{\gamma} \mathbf{a} \mathbf{b}.
\end{aligned}
\tag{6.15}
$$

6.2 Stresses in Steady Viscometric Flows

It is remarkable that the stresses in steady viscometric flows can be determined completely for isotropic simple fluids. Take, for example, the simple shear flow where $\mathbf{u} = \dot{\gamma} y \mathbf{i}$. The general form of the stress tensor is

$$
[\mathbf{S}] = \begin{bmatrix} S_{xx} & S_{xy} & 0 \\ S_{xy} & S_{yy} & 0 \\ 0 & 0 & S_{zz} \end{bmatrix}.
$$

The shear stress S_{xy} must be an odd function of the shear rate based on physical grounds alone.[1] Thus we can write

$$
S_{xy} = \dot{\gamma} \eta \left(\dot{\gamma} \right),
\tag{6.16}
$$

where the viscosity, defined as $\eta = S_{xy}/\dot{\gamma}$, must be an even function of the shear rate:

$$
\eta \left(-\dot{\gamma} \right) = \eta \left(\dot{\gamma} \right).
\tag{6.17}
$$

Reversing the direction of shear will not change the normal stress components, and therefore these will be even functions of the shear rate. The arbitrary pressure can be eliminated by taking the differences between these normal stresses. We thus define the first and the second normal stress differences by

$$
S_{xx} - S_{yy} = N_1 \left(\dot{\gamma} \right), \quad S_{yy} - S_{zz} = N_2 \left(\dot{\gamma} \right),
\tag{6.18}
$$

respectively. These normal stress differences are even function of the shear rate, and they vanish when the shear rate is zero. This is made explicit by writing

---

[1]The fluid has no way of knowing that the experimenter has suddenly changed his mind and re-defined x_1 as $-x_1$. It will continue merrily reporting the same shear rate and stress. To the experimenter, however, he will notice that the shear rate and the shear stress have the same magnitudes as before, but they have changed signs. He therefore concludes that the shear stress is an odd function of the shear rate. The same story applies to normal stresses; they are even functions of the shear rate.

$$N_1\left(\dot{\gamma}\right) = \dot{\gamma}^2 \nu_1\left(\dot{\gamma}\right), \quad N_2\left(\dot{\gamma}\right) = \dot{\gamma}^2 \nu_2\left(\dot{\gamma}\right), \tag{6.19}$$

where ν_1 and ν_2 are called the first and second normal stress coefficients. They are even functions of the shear rate. Collectively, $\eta\left(\dot{\gamma}\right)$, $N_1\left(\dot{\gamma}\right)$ and $N_2\left(\dot{\gamma}\right)$ are called *viscometric functions*. They are material properties for the fluid.

In a general steady viscometric flow, the above reasoning continues to hold, and we write the stress tensor using the base vectors $\mathbf{a}, \mathbf{b}, \mathbf{c}$ as

$$\mathbf{T} = -P\mathbf{I} + \dot{\gamma}\eta\left(\mathbf{ab} + \mathbf{ba}\right) + \left(N_1 + N_2\right)\left(\mathbf{aa} + \mathbf{bb}\right) - N_1\mathbf{bb}.$$

Here, P is the hydrostatic pressure, $\mathbf{I} = \mathbf{aa} + \mathbf{bb} + \mathbf{cc}$ is the unit tensor, and

$$\mathbf{A}_1 = \dot{\gamma}\left(\mathbf{ab} + \mathbf{ba}\right), \quad \mathbf{A}_1^2 = \dot{\gamma}^2\left(\mathbf{aa} + \mathbf{bb}\right), \tag{6.20}$$

$$\mathbf{A}_2 = \mathbf{A}_1\mathbf{L} + \mathbf{L}^T\mathbf{A}_1 = \dot{\gamma}^2\left(\mathbf{ab} + \mathbf{ba}\right)\mathbf{ab} + \dot{\gamma}^2\mathbf{ba}\left(\mathbf{ab} + \mathbf{ba}\right) = 2\dot{\gamma}^2\mathbf{bb},$$

Writing the total stress tensor as $\mathbf{T} = -P\mathbf{I} + \mathbf{S}$, the extra stress tensor $\mathbf{S}$ can be written as

$$\mathbf{S} = \eta\mathbf{A}_1 + \left(\nu_1 + \nu_2\right)\mathbf{A}_1^2 - \frac{\nu_1}{2}\mathbf{A}_2. \tag{6.21}$$

This resembles the second-order fluid model (4.71), but with important differences. The second-order fluid model is a slow-flow approximation to the simple fluid, and all the coefficients in the model are constant, whereas (6.21) is a restriction on the simple fluid in steady viscometric flows, and therefore is valid *only* in steady viscometric flows for *all* simple fluids. All the coefficients in (6.21) are functions of the strain rate. More importantly, (6.21) is not a model of a fictitious fluid, but is a proven theorem for steady viscometric flows [17].

6.2.1 Controllable and Partially Controllable Flows

If the velocity field can be fully determined (with or without inertia), no matter what form the viscometric functions may take, then the flow is said to be *controllable*. There are flows in which the kinematics are fully determined by the viscosity function alone – the normal stress differences do not influence the velocity field. Such flows are called *partially controllable*.

Problems

Problem 6.1 Show that the velocity gradient for (6.3) is

$$\mathbf{L} = \dot{\gamma}\mathbf{ab}. \tag{6.22}$$

Fig. 6.2 Shear flow between inclined planes

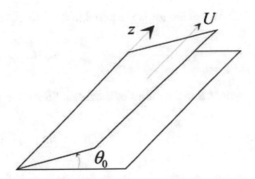

Consequently, show that all the flows represented by (6.3) are isochoric. Show that the shear rate is $|\dot{\gamma}|$.

Problem 6.2 Show that the shear rate for the helicoidal flow (6.14) is

$$\dot{\gamma}^2 = \left(r^2 + c^2\right) \nabla\omega \cdot \nabla\omega. \qquad (6.23)$$

Problem 6.3 Consider the shear flow between two tilted plates: the first plate is at rest and the second plate, tilted at an angle θ_0 to the first plate, is moving with a velocity U in the $\mathbf{k}$−direction, as shown in Fig. 6.2. Show that

$$\mathbf{u} = U\frac{\theta}{\theta_0}\mathbf{e}_z. \qquad (6.24)$$

Show that the stress is given by

$$\mathbf{T} = -P\mathbf{I} + \eta\dot{\gamma}\left(\mathbf{e}_z\mathbf{e}_\theta + \mathbf{e}_\theta\mathbf{e}_z\right) + \left(N_1 + N_2\right)\mathbf{e}_z\mathbf{e}_z + N_2\mathbf{e}_\theta\mathbf{e}_\theta, \qquad (6.25)$$

where the shear rate is $\dot{\gamma} = U/r\theta_0$, and

$$P = P\left(r_0\right) + I_2\left(\dot{\gamma}\right) - I_2\left(\dot{\gamma}_0\right), \quad I_2\left(\dot{\gamma}\right) = \int_0^{\dot{\gamma}} \dot{\gamma}\nu_2 d\dot{\gamma}. \qquad (6.26)$$

Suggest a way to measure N_2 based on this.

Problem 6.4 The flow between two parallel, coaxial disks is called *torsional flow*. In this flow, the bottom disk is fixed, and the top disk rotates at an angular velocity of Ω. The distance between the disks is h. Neglecting the fluid inertia, show that

$$\mathbf{u} = \Omega r\frac{z}{h}\mathbf{e}_\theta, \quad \dot{\gamma} = \Omega\frac{r}{h}. \qquad (6.27)$$

Show that the torque required to turn the top disk is

$$M = 2\pi \int_0^R \dot{\gamma}\eta\,(\dot{\gamma})\,r^2 dr, \qquad (6.28)$$

where R is the radius of the disks. Show that the pressure is

$$P\,(r) = \int_{\dot{\gamma}}^{\dot{\gamma}_R} \dot{\gamma}\,(\nu_1 + \nu_2)\,d\dot{\gamma}. \qquad (6.29)$$

From the axial stress, show that the normal force on the top disk is

$$F = \pi R^2 \dot{\gamma}_R^{-2} \int_0^{\dot{\gamma}_R} \dot{\gamma}\,(N_1 - N_2)\,d\dot{\gamma}, \qquad (6.30)$$

where $\dot{\gamma}_R = \Omega R/h$ is the shear rate at the rim $r = R$.

By normalizing the torque and the force as

$$m = \frac{M}{2\pi R^3}, \quad f = \frac{F}{\pi R^2}, \qquad (6.31)$$

show that

$$\eta\,(\dot{\gamma}_R) = \frac{m}{\dot{\gamma}_R}\left[3 + \frac{d\ln m}{d\ln \dot{\gamma}_R}\right], \qquad (6.32)$$

and

$$N_1\,(\dot{\gamma}_R) - N_2\,(\dot{\gamma}_R) = f\left(2 + \frac{d\ln f}{d\ln \dot{\gamma}_R}\right). \qquad (6.33)$$

Relations (6.32), (6.33) are the basis for the operation of the parallel-disk viscometer.

Problem 6.5 In a pipe flow, of radius R and pressure drop/unit length $\Delta P/L$, show that the flow rate is

$$Q = 8\pi \left(\frac{L}{\Delta P}\right)^3 \int_0^{\tau_w} \frac{\tau^3}{\eta}\,d\tau, \qquad (6.34)$$

where τ is the shear stress, and τ_w is the shear stress at the wall. In terms of the reduced discharge rate,

$$q = \frac{Q}{\pi R^3}, \qquad (6.35)$$

show that

$$\frac{dq}{d\tau_w} = \frac{1}{\eta\,(\tau_w)} - \frac{3q}{\tau_w}, \qquad (6.36)$$

or

$$\dot{\gamma}_w = q\,(\tau_w)\left[3 + \frac{d \ln q}{d \ln \tau_w}\right].$$

(6.37)

The relation (6.37) is due to Rabinowitch [74] and is the basis for capillary viscometry.

Chapter 7
Polymer Solutions

From Atoms to Flows

In the microstructure approach to the quest for a relevant constitutive equation for the complex-structure fluid, a relevant model for the microstructure is postulated, and the consequences of the micromechanics are then explored at the macrostructural level, with appropriate averages being taken to smear out the details of the microstructure. The advantage of this is that the resulting constitutive equation is expected to be relevant to the material concerned; and if a particular phenomenon is not well modelled, the microstructural model can be revisited and the relevant physics put in place. This iterative model-building process is always to be preferred over the continuum approach. In this chapter, we will concentrate on the constitutive modelling of dilute polymer solutions.

7.1 Characteristics of a Polymer Chain

Viscoelastic fluids are predominantly suspensions of long-chain polymer molecules in a solvent. We will not be concerned with aspects of polymer chemistry here. It is sufficient for us to know that a typical polymer, with a molecular weight of the order 10^7 grams per mole, has about 10^5 repeating (monomer) units, and its monomer molecular weight is of the order 10^2 grams per mole.

7.1.1 Random-Walk Model

The simplest representation of a polymer molecule is a *freely rotating chain* with N segments, as illustrated in Fig. 7.1. Each segment has a constant bond length b, but is randomly oriented in space. The segments are not physical entities; they can cross over each other in space and can be freely rotating irrespective of the neighbouring

© Springer International Publishing AG 2017
N. Phan-Thien and N. Mai-Duy, *Understanding Viscoelasticity*,
Graduate Texts in Physics, DOI 10.1007/978-3-319-62000-8_7

Fig. 7.1 A random-walk
model of a polymer chain

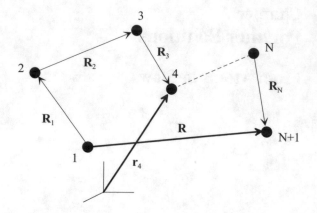

segments. Its end-to-end vector is given by

$$\mathbf{R} = \sum_{j=1}^{N} \mathbf{R}_j. \tag{7.1}$$

On the average, we expect that a randomly oriented vector will have zero mean, and
its square is constant as a consequence of the constant bond length:

$$\langle \mathbf{R}_j \rangle = \mathbf{0}, \quad \langle \mathbf{R}_j \cdot \mathbf{R}_j \rangle = b^2 \quad \text{(no sum)}. \tag{7.2}$$

Here and elsewhere, the angular brackets denote the average with respect to the
probability density function of the variable concerned. Thus, if $P(\mathbf{R}, t)\, d\mathbf{R}$ is the
probability of finding a segment of configuration between $\mathbf{R}$ and $\mathbf{R} + d\mathbf{R}$ at time t
then the n-th moment of $\mathbf{R}$ is defined as

$$\left\langle \underbrace{\mathbf{R}\mathbf{R}\ldots\mathbf{R}}_{n \text{ times}} \right\rangle = \int \underbrace{\mathbf{R}\mathbf{R}\ldots\mathbf{R}}_{n \text{ times}}\, P(\mathbf{R}, t)\, d\mathbf{R}. \tag{7.3}$$

Thus, a chain has zero end-to-end vector on average,

$$\langle \mathbf{R} \rangle = \sum_{j=1}^{N} \langle \mathbf{R}_j \rangle = \mathbf{0}, \tag{7.4}$$

and its mean square is

$$\langle \mathbf{R} \cdot \mathbf{R} \rangle = \langle R^2 \rangle = \sum_{j=1}^{N}\sum_{k=1}^{N} \langle \mathbf{R}_j \cdot \mathbf{R}_k \rangle = \sum_{j=1}^{N} \langle \mathbf{R}_j \cdot \mathbf{R}_j \rangle + \sum_{j=1}^{N}\sum_{k \neq j}^{N} \langle \mathbf{R}_j \cdot \mathbf{R}_k \rangle.$$

Since $\mathbf{R}_j$ is independent of $\mathbf{R}_k$, $k \neq j$, $\langle \mathbf{R}_j \cdot \mathbf{R}_k \rangle = 0$, the last double sum on the right side is zero. From (7.2), we have

$$\langle R^2 \rangle = Nb^2. \tag{7.5}$$

Non-freely-rotating chains, for example, chains where the bond angle between successive segments remains fixed, has been considered [26] – they all lead to

$$\langle R^2 \rangle = kNb^2, \tag{7.6}$$

where k is a constant depending on the geometry.

Strong Flow. Thus, in the random walk model, a chain of extended contour length Nb is expected to have a linear dimension of $O\left(\sqrt{Nb}\right)$. It is difficult to unravel a polymer molecule, and flows that can do this are called strong.

Diffusion Equation. For a given chain of N segments, the end-to-end vector is a stochastic quantity[1] and must be characterised by its probability density function $P(\mathbf{R}; N)$. Now consider a chain of N segments, with an end-to-end vector of $\mathbf{R} - \mathbf{b}$. The probability of $N + 1$ segments having an end-to-end vector $\mathbf{R}$ is precisely the probability of the last segment $(N + 1)$ having a bond vector $\mathbf{b}$, conditional on the first N segments having an end-to-end vector of $\mathbf{R} - \mathbf{b}$:

$$P(\mathbf{R}; N + 1) = \int P(\mathbf{R} - \mathbf{b}; N) \, P_b(\mathbf{b}) \, d\mathbf{b}. \tag{7.7}$$

This is the property of Markovian processes [52] – namely what happens at any given instant depends only on the instantaneous state of the system, not on its previous history; $P_b(\mathbf{b})$ is called the transition probability, which depends on both current and next states of the process. Here we assume that $\mathbf{b}$ is completely independent of the current state. The equation can be expanded in a Taylor series for $|\mathbf{b}| \ll |\mathbf{R}|$,

$$P(\mathbf{R}; N) + \frac{\partial}{\partial N} P(\mathbf{R}; N) + \cdots$$
$$= \int P_b(\mathbf{b}) \left[P(\mathbf{R}; N) - \mathbf{b} \cdot \nabla P(\mathbf{R}; N) + \frac{1}{2!} \mathbf{b}\mathbf{b} : \nabla\nabla P(\mathbf{R}; N) + \cdots \right] d\mathbf{b}.$$

The average with respect to $\mathbf{b}$ is taken, noting that its distribution is purely random,

$$\int P_b(\mathbf{b}) \, d\mathbf{b} = 1, \quad \int \mathbf{b} P_b(\mathbf{b}) \, d\mathbf{b} = 0, \quad \int \mathbf{b}\mathbf{b} P_b(\mathbf{b}) \, d\mathbf{b} = \frac{b^2}{3}\mathbf{I}.$$

---

[1] A stochastic process is a family of random variables $X(t)$, where t is the time, X is a random variable, and $X(t)$ is the value observed at time t. The totality of $\{X(t), \ t \in \mathbb{R}\}$ is said to be a random function or a stochastic process.

Thus, we obtain the following diffusion equation for the process $\mathbf{R}$:

$$\frac{\partial}{\partial N} P(\mathbf{R}; N) = \frac{b^2}{6} \nabla^2 P(\mathbf{R}; N). \tag{7.8}$$

The solution of this, subjected to the "initial condition"

$$\lim_{N \to 0} P(\mathbf{R}; N) = \delta(\mathbf{R})$$

is the Gaussian distribution

$$P(\mathbf{R}; N) = \left(\frac{3}{2\pi N b^2}\right)^{3/2} \exp\left(-\frac{3R^2}{2N b^2}\right). \tag{7.9}$$

This distribution is unrealistic in the sense that there is a finite probability for $R > Nb$; a more exact treatment produces the *Langevin distribution* [26], which vanishes at $R \geq Nb$ as required.

7.2 Forces on a Chain

In simple models for dilute polymer solutions, such as the Rouse bead-spring model shown in Fig. 7.2a, a polymer chain is discretised into several effective segments, called *Kuhn segments*,[2] each of which has a point mass (bead) undergoing some motion in a solvent (which is treated as a continuum). Each Kuhn segment may contain several monomer units. Each bead accelerates in response to the forces exerted on it by the solvent, the flow process, and the surrounding beads, and consequently the chain will adopt a configuration. The task here is to relate the microstructure information to a constitutive description of the fluid. When there are only two beads, the model is called the *elastic dumbbell model* [46], Fig. 7.2b, which has been most popular in elucidating the main features of the rheology of dilute polymer solutions.

The forces acting on the beads include:

Hydrodynamic forces: These arise from the average hydrodynamic resistance of the motion of the polymer through a viscous solvent. Since the relevant Reynolds number based on the size of the polymer chain is negligibly small, the average motion of the chain is governed by Stokes equations, and Stokes resistance can be used to model this. The chain is usually treated as a number of discrete points of resistance, each having a frictional coefficient. In the simplest model, a frictional force of

$$\mathbf{F}_i^{(d)} = \zeta(\mathbf{u}_i - \dot{\mathbf{r}}_i), \tag{7.10}$$

---

[2] W. Kuhn (1899–1963) was a Professor at the Technische Hochschule in Karlruhe, and later on, in Basel, Switzerland. He is most famous for the f-summation theorem in quantum mechanics.

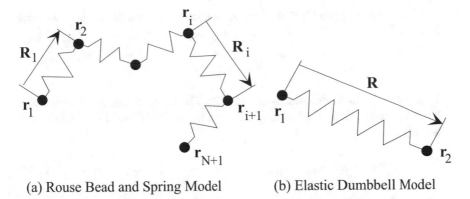

(a) Rouse Bead and Spring Model   (b) Elastic Dumbbell Model

Fig. 7.2 (a) Rouse model and (b) the elastic dumbbell model of a polymer chain

is assumed to be acting on the i-th bead, which has a velocity $\dot{\mathbf{r}}_i - \mathbf{u}_i$ relative to the solvent, and ζ is a constant frictional coefficient, which is usually taken as $6\pi\eta_s a$ (Stokes drag on the bead), where η_s is the viscosity of the solvent, and a represents the size of the bead. To obtain a more realistic model of the nature of the dependence of the frictional forces on the configuration and the deformation of the polymer chain, ζ can be allowed to depend on the length of the segment i, or indeed it may be considered to be a second-order tensor, which reflects the physical idea that the resistance to the motion perpendicular to the chain is much higher than that along the chain. The simple frictional model (7.10) neglects hydrodynamic interaction with other beads, and is usually called the *free-draining* assumption. Hydrodynamic interaction arises because of the solvent velocity that appears in (7.10) contains disturbance terms due to the presence of other beads.

Tension in the chain: A chain in equilibrium will tend to curl up into a spherical configuration, with the most probable state of zero end-to-end vector. However, if the chain ends are forcibly extended, then there is a tension or a spring force, arising in the chain, solely due to the fewer configurations available to the chain. To find the expression for the chain tension, we recall that the probability density function is proportional to the number of configurations available to the chain (i.e., the entropy), and thus the Helmholtz free energy of the chain is [26]

$$F_r\left(\mathbf{r}_j\right) = A\left(T\right) - kT \ln P\left(\mathbf{r}_j\right), \tag{7.11}$$

where $A\left(T\right)$ is a function of the temperature alone. The entropic spring force acting on bead i is

$$\mathbf{F}_i^{(s)} = -\frac{\partial F_r}{\partial \mathbf{r}_i} = kT \frac{\partial \ln P}{\partial \mathbf{r}_i}. \tag{7.12}$$

For the Gaussian chain (7.9), the tension required to extend the chain by a vector $\mathbf{R}$ is

$$\mathbf{F} = \frac{3kT}{Nb^2}\mathbf{R}. \tag{7.13}$$

This applies to individual beads of Fig. 7.2a. Thus if each segment consists of n Kuhn segments, each of bond length b, then the force on bead i due to the chain tension is

$$\mathbf{F}_i^{(s)} = \frac{3kT}{nb^2}\left(\mathbf{r}_{i+1} - \mathbf{r}_i + \mathbf{r}_{i-1} - \mathbf{r}_i\right) = \frac{3kT}{nb^2}\left(\mathbf{R}_i - \mathbf{R}_{i-1}\right). \tag{7.14}$$

This implies that the beads are connected by linear springs of stiffness

$$H = \frac{3kT}{nb^2}. \tag{7.15}$$

A distribution, which better accounts for the finite segment length, is the Langevin distribution,[3] and this results in the so-called inverse Langevin spring law for the chain tension:

$$L\left(\frac{bF}{kT}\right) = \coth\left(\frac{bF}{kT}\right) - \frac{bF}{kT} = \frac{r}{nb}, \tag{7.16}$$

where F is the magnitude of the force, r the magnitude of the extension, and the Langevin function is defined as $L(x) = \coth x - x$. A useful approximation of the Langevin spring law is the Warner spring [6]

$$H_i = \frac{3kT}{nb^2}\frac{1}{1 - (R_i/L_i)^2}, \tag{7.17}$$

where $L_i = nb$ is the maximum extended length of segment i. This stiffness approaches infinity as $R_i \to L_i$.

Brownian forces: Brownian forces are the cumulative effect of the bombardment of the chain by the solvent molecules.[4] These forces have a small correlation time scale, typically the vibration period of a solvent molecule, of the order 10^{-13}s for water molecules. If we are interested in time scales considerably larger than

---

[3]P. Langevin (1872–1940) introduced the stochastic DE (7.20) in 1908, and showed that the particle obeys the same diffusion equation as described by Einstein (1905).

[4]The random zig-zag motion of small particles (less than about 10 μm) is named after R. Brown (1773–1858), an English botanist, who mistook this as a sign of life. He travelled with Matthew Flinders to Australia in 1801 on the ship *Investigator* as a naturalist. The correct explanation of the phenomenon was given by Perrin (Fig. 7.3). Brownian particles are those undergoing a random walk, or Brownian motion.

Fig. 7.3 The French physicist Jean Baptiste Perrin (1870–1942) gave the correct explanation to the random motion of small particles as observed by Brown and confirmed the theoretical calculations by Einstein. For this work he was awarded the Nobel Prize for Physics in 1926. He was also the founder of the Centre National de la Recherche Scientifique

this correlation time scale, then the Brownian forces acting on bead i, $\mathbf{F}_i^{(b)}$, can be considered as white noise having a zero mean and a delta autocorrelation function[5]:

$$\left\langle \mathbf{F}_i^{(b)}(t) \right\rangle = \mathbf{0}, \quad \left\langle \mathbf{F}_i^{(b)}(t+s)\, \mathbf{F}_j^{(b)}(t) \right\rangle = 2\delta_{ij}\delta(s)\,\mathbf{f}. \qquad (7.18)$$

This autocorrelation states that the strength of the Brownian forces is the measure of the integral correlation function over a time scale which is considerably greater than the correlation time scale of the Brownian forces:

$$2\mathbf{f} = \int_{-\infty}^{\infty} \left\langle \mathbf{F}_i^{(b)}(t+s)\, \mathbf{F}_i^{(b)}(t) \right\rangle ds. \qquad (7.19)$$

The strength of the Brownian forces is not an arbitrary quantity determined by a constitutive modelling process; it is in fact related to the mobility of the Brownian particle - this is the essence of the fluctuation-dissipation theorem which we discuss next.

---

[5]This approximation is called white noise, i.e., Gaussian noise of all possible frequencies uniformly distributed. Sometimes it is called "rain-on-the-roof" approximation: two (or more) rain drops do not fall on the same spot on the roof.

7.3 Fluctuation-Dissipation Theorem

7.3.1 Langevin Equation

There are several fluctuation-dissipation theorems [47], relating the strength of the fluctuating quantity to the macroscopic "mobility" of the phenomenon concerned. The following development is patterned after Hinch [36].

All micro-mechanical models for a polymer chain in a dilute solution can be written as

$$\mathbf{m}\ddot{\mathbf{x}} + \zeta\dot{\mathbf{x}} + \mathbf{K}\mathbf{x} = \mathbf{F}^{(b)}(t), \tag{7.20}$$

called the *Langevin equation* (for a portray of Langevin, see Fig. 7.4), where the system state is represented by the finite-dimensional vector $\mathbf{x}$, such that its kinetic energy is $\frac{1}{2}\mathbf{m} : \dot{\mathbf{x}}\dot{\mathbf{x}}$, and its generalised linear momentum is $\mathbf{m}\dot{\mathbf{x}}$, $\mathbf{m}$ being a generalised inertia tensor. The inertia tensor $\mathbf{m}$ is defined through the kinetic energy, and therefore there is no loss of generality in considering only symmetric $\mathbf{m}$. The system is acted on by a frictional force, which is linear in its state velocities, a restoring force (possibly nonlinear in $\mathbf{x}$), and a Brownian force $\mathbf{F}^{(b)}(t)$. We assume that the frictional tensor coefficient ζ is symmetric. This system can be conveniently started from rest at time $t = 0$.

Since the Brownian force has only well-defined statistical properties, the Langevin equation (7.20) must be understood as a *stochastic differential equation* [52]. It can only be "solved" by specifying the probability distribution $W(\mathbf{u}, \mathbf{x}, t)$ of the process $\{\mathbf{u} = \dot{\mathbf{x}}, \mathbf{x}\}$ defined so that $W(\mathbf{u}, \mathbf{x}, t)\,d\mathbf{u}d\mathbf{x}$ is the probability of finding the

Fig. 7.4 Paul Langevin (1872–1946) was a Professor of Physics at College de France. He studied under Lord Kelvin and Pierre Curie

process at the state between $\{\mathbf{u}, \mathbf{x}\}$ and $\{\mathbf{u} + d\mathbf{u}, \mathbf{x} + d\mathbf{x}\}$ at time t. Prescribing the initial conditions $\mathbf{u}(0) = \mathbf{u}_0$, $\mathbf{x}(0) = \mathbf{x}_0$ for (7.20) is equivalent to specifying a delta probability at time $t = 0$:

$$W(\mathbf{u}, \mathbf{x}, 0) = \delta(\mathbf{u} - \mathbf{u}_0)\,\delta(\mathbf{x} - \mathbf{x}_0).$$

The distribution $W(\mathbf{u}, \mathbf{x}, t)$ is the *phase space description* of the stochastic process $\{\mathbf{u}, \mathbf{x}\}$. The dependence of W on $\mathbf{x}$ or $\mathbf{u}$ can be eliminated by integrating out the unwanted independent variable. Then, we have either a *velocity space*, or a *configuration space* description, respectively.

7.3.2 Equi-Partition of Energy

The existence of the temperature T of the surrounding fluid demands that the distribution in the velocity space must satisfy the *equi-partition energy principle*:

$$\lim_{t \to \infty} \langle \dot{\mathbf{x}}(t)\,\dot{\mathbf{x}}(t) \rangle = kT\mathbf{m}^{-1}, \tag{7.21}$$

as demanded by the kinetic theory of gases, i.e., each mode of vibration is associated with a kinetic energy of $\frac{1}{2}kT$. We now explore the consequence of this on the Langevin system (7.20).

There are three time scales in this system:

1. τ_r the relaxation time scale of the chain in its lowest mode; this time scale is of the order $|\boldsymbol{\zeta}\mathbf{K}^{-1}|$;
2. τ_i the much shorter inertial relaxation time scale of the chain; this time scale is of the order $|\mathbf{m}\boldsymbol{\zeta}^{-1}|$;
3. τ_c the still shorter correlation time scale of the Brownian force – this is of the same order as the relaxation time scale of a solvent molecule.

7.3.3 Fluctuation-Dissipation Theorem

In general we have $\tau_c \ll \tau_i \ll \tau_r$, but the estimate of τ_i can vary considerably. To derive a fluctuation-dissipation theorem for the Langevin equation (7.20), it is sufficient to consider only events on the time scale τ_i. In this time scale, $\mathbf{m}(\mathbf{x})$ and $\boldsymbol{\zeta}(\mathbf{x})$ can be replaced by their local values, i.e., regarded as constant, and the state vector can be re-defined to eliminate $\mathbf{K}\mathbf{x}$ so that (7.20) becomes

$$\ddot{\mathbf{x}} + \mathbf{m}^{-1}\boldsymbol{\zeta}\dot{\mathbf{x}} = \mathbf{m}^{-1}\mathbf{F}^{(b)}(t), \tag{7.22}$$

subjected to the initial rest state

$$\dot{\mathbf{x}}(0) = \mathbf{0} = \mathbf{x}(0). \tag{7.23}$$

Note that $\mathbf{A} \cdot e^{\mathbf{A}t} = e^{\mathbf{A}t} \cdot \mathbf{A}$,

$$\frac{d}{dt} \left(e^{\mathbf{A}t} \mathbf{v} \right) = e^{\mathbf{A}t} \left(\dot{\mathbf{v}} + \mathbf{A}\mathbf{v} \right).$$

Identify $\mathbf{A} = \mathbf{m}^{-1}\boldsymbol{\zeta}$ and $\mathbf{v} = \dot{\mathbf{x}}$, the solution to (7.22) is therefore given by

$$\dot{\mathbf{x}}(t) = \int_0^t \exp\left\{ \mathbf{m}^{-1}\boldsymbol{\zeta} \left(t' - t \right) \right\} \mathbf{m}^{-1} \mathbf{F}^{(b)} \left(t' \right) dt'. \tag{7.24}$$

This leads to the expectation

$$\langle \dot{\mathbf{x}}(t) \, \dot{\mathbf{x}}(t) \rangle = \int_0^t \int_0^t \exp\left\{ \mathbf{m}^{-1}\boldsymbol{\zeta} \left(t' - t \right) \right\} \mathbf{m}^{-1} \left\langle \mathbf{F}^{(b)} \left(t' \right) \mathbf{F}^{(b)} \left(t'' \right) \right\rangle$$
$$\cdot \, \mathbf{m}^{-1} \exp\left\{ \boldsymbol{\zeta}\mathbf{m}^{-1} \left(t'' - t \right) \right\} dt'dt''.$$

If $\tau_c \ll \tau_i$, the white noise assumption for the Brownian force can be used, so that

$$\left\langle \mathbf{F}^{(b)} \left(t' \right) \mathbf{F}^{(b)} \left(t'' \right) \right\rangle = 2\delta \left(t' - t'' \right) \mathbf{f},$$

giving

$$\langle \dot{\mathbf{x}}(t) \, \dot{\mathbf{x}}(t) \rangle = 2 \int_0^t \exp\left\{ \mathbf{m}^{-1}\boldsymbol{\zeta} \left(t' - t \right) \right\} \mathbf{m}^{-1} \mathbf{f} \mathbf{m}^{-1} \exp\left\{ \boldsymbol{\zeta}\mathbf{m}^{-1} \left(t' - t \right) \right\} dt'.$$

This can be integrated by parts to yield

$$\langle \dot{\mathbf{x}}(t) \, \dot{\mathbf{x}}(t) \rangle = -2 \exp\left\{ -\mathbf{m}^{-1}\boldsymbol{\zeta}\tau \right\} \boldsymbol{\zeta}^{-1} \mathbf{f} \mathbf{m}^{-1} \exp\left\{ -\boldsymbol{\zeta}\mathbf{m}^{-1}t \right\}$$
$$+ 2\boldsymbol{\zeta}^{-1} \mathbf{f} \mathbf{m}^{-1} - \boldsymbol{\zeta}^{-1}\mathbf{m} \langle \dot{\mathbf{x}}(t) \, \dot{\mathbf{x}}(t) \rangle \boldsymbol{\zeta}\mathbf{m}^{-1}.$$

In the limit of $t \to \infty$ (i.e., $t \gg \tau_i$ but $t \ll \tau_r$ so that the equation of state remains linear), the equi-partition of energy (7.21) holds, and we have

$$kT\mathbf{m}^{-1} = 2\boldsymbol{\zeta}^{-1} \mathbf{f} \mathbf{m}^{-1} - kT\mathbf{m}^{-1},$$

or

$$\mathbf{f} = kT\boldsymbol{\zeta}. \tag{7.25}$$

This is the fluctuation-dissipation theorem, relating the strength of the Brownian force to the mobility of the Brownian system; any dependence on the configuration of $\mathbf{f}$ is inherited from that of ζ.

7.3.4 Diffusivity Stokes–Einstein Relation

The diffusivity of a Brownian particle is defined by

$$\mathbf{D} = \lim_{t \to \infty} \frac{1}{2} \frac{d}{dt} \langle \mathbf{x}(t) \mathbf{x}(t) \rangle. \tag{7.26}$$

This is equivalent to

$$\mathbf{D} = \lim_{t \to \infty} \frac{1}{2} \int_0^t \langle \dot{\mathbf{x}}(t) \dot{\mathbf{x}}(t - \tau) + \dot{\mathbf{x}}(t - \tau) \dot{\mathbf{x}}(t) \rangle \, d\tau. \tag{7.27}$$

In Problem 7.1, it can be shown that

$$\mathbf{D} = kT\zeta^{-1}. \tag{7.28}$$

7.3.5 Fokker–Planck Equation

As mentioned earlier, the Langevin equation can only be considered solved when the probability function of the process is specified. In the limit $\mathbf{m} \to \mathbf{0}$ it can be shown that the configuration probability density function $\phi(\mathbf{x}, t)$ satisfies

$$\frac{\partial \phi}{\partial t} = \lim_{\Delta t \to 0} \frac{\partial}{\partial \mathbf{x}} \cdot \left[\frac{\langle \Delta \mathbf{x} \Delta \mathbf{x} \rangle}{2 \Delta t} \cdot \frac{\partial \phi}{\partial \mathbf{x}} - \frac{\langle \Delta \mathbf{x} \rangle}{\Delta t} \phi \right]. \tag{7.29}$$

This is the Fokker–Planck,[6] or Smoluchowski[7] diffusion equation. A clear exposition of this can be found in Chandrasekhar [14], Fig. 7.5.

It can be shown that (Problem 7.2), in the limit of $\mathbf{m} \to \mathbf{0}$, the Fokker–Planck equation is

$$\frac{\partial \phi}{\partial t} = \frac{\partial}{\partial \mathbf{x}} \cdot \left[kT\zeta^{-1} \frac{\partial \phi}{\partial \mathbf{x}} + \zeta^{-1} \cdot \mathbf{K}\mathbf{x}\phi \right]. \tag{7.30}$$

---

[6] A.D. Fokker derived the diffusion equation for a Brownian particle in velocity space in 1914. The general case was considered by M. Planck (1858–1947) in 1917.

[7] The general solution to the random walk problem in one dimension was obtained by M. von Smoluchowski in 1906.

Fig. 7.5 Subrahmanyan
Chandrasekhar (1910–95)
was an outstanding Indian
astrophysicist. He worked on
various aspects of stellar
dynamics and was awarded
the Nobel Prize in 1983

7.3.6 Smoothed-Out Brownian Force

Since the probability must satisfy

$$\int \phi(\mathbf{x}, t)\, d\mathbf{x} = 1, \tag{7.31}$$

an application of the Reynolds transport theorem yields

$$\frac{\partial \phi}{\partial t} + \frac{\partial}{\partial \mathbf{x}} \cdot (\dot{\mathbf{x}} \phi) = 0, \tag{7.32}$$

assuming that we deal with an equivalent "deterministic" system $\mathbf{x}$. By comparing
(7.32)–(7.30), the velocity of this equivalent system must satisfy

$$\dot{\mathbf{x}} = -\zeta^{-1} \cdot \mathbf{K}\mathbf{x} - kT\zeta^{-1} \cdot \frac{\partial \ln \phi}{\partial \mathbf{x}}.$$

That is,

$$\zeta \dot{\mathbf{x}} + \mathbf{K}\mathbf{x} = -kT \frac{\partial \ln \phi}{\partial \mathbf{x}}. \tag{7.33}$$

Comparing this to (7.20), it is as though the Brownian force has been replaced by

$$\mathbf{F}^{(b)}(t) = -kT \frac{\partial \ln \phi}{\partial \mathbf{x}}. \tag{7.34}$$

Of course, this equation is not mathematically meaningful: the left side is a stochastic, and the right side is a deterministic quantity. It is so defined for the sole purpose of getting the correct diffusion equation (7.30). Written in the form (7.34), this force is called the *smoothed-out Brownian force*; it is a device employed in most texts dealing with kinetic theories of polymers, e.g., Bird et al. [6].

7.4 Stress Tensor

There are several ways to derive the expression for the stress tensor contributed by the polymer chains in a dilute solution. One is the probabilistic approach (Bird et al. [6]), where the number of polymer chains straddling a surface and the net force acting on that surface by the chains are calculated. The force per unit area can be related to the stress tensor. Another approach is to calculate the free energy of the chain from its entropy, and the rate of work done can be related to the dissipation due to the presence of the chains from which the expression for the stress tensor can be derived [48]. We present a simple mechanistic approach here to derive the expression for the stress tensor.

Consider the bead-spring model for a polymer chain, as shown in Fig. 7.6. The tension in the i-th Kuhn segment is denoted by $\mathbf{f}_i$. If the chain is Gaussian, then

$$\mathbf{f}_i = H_i \mathbf{R}_i, \text{(no sum)} \tag{7.35}$$

where $\mathbf{R}_i$ is the end-to-end vector, and H_i is the stiffness of segment i, given in (7.15). This bead-spring model is also called the Rouse model, and the chain is known as the Rouse chain. Using the approach of Landau and Lifshitz [47] and Batchelor [4], the fluid is taken as an effective continuum made up of a homogeneous suspension of Rouse chains. Its effective stress is simply the volume-averaged stress:

$$\langle \mathbf{T} \rangle = \frac{1}{V} \int_V \mathbf{T} dV = \frac{1}{V} \int_{V_s} \mathbf{T} dV + \frac{1}{V} \sum_p \int_{V_p} \mathbf{T} dV, \tag{7.36}$$

Fig. 7.6 Connector force in a Rouse chain

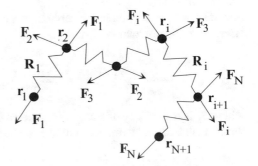

where $\mathbf{T}$ is the total stress, V is a representative volume containing several chains, and is made up of a solvent volume V_s and a polymer volume $\sum V_p$. In the solvent volume, the stress is simply the solvent stress, and we have

$$\frac{1}{V} \int_{V_s} \mathbf{T}^{(s)} dV = \frac{1}{V} \int_V \mathbf{T}^{(s)} dV - \frac{1}{V} \sum_p \int_{V_p} \mathbf{T}^{(s)} dV. \qquad (7.37)$$

With a Newtonian solvent, the first term on the right of (7.37) is simply a Newtonian stress,

$$\frac{1}{V} \int_V \mathbf{T}^{(s)} dV = -p_1 \mathbf{I} + 2\eta_s \mathbf{D}, \qquad (7.38)$$

where p_1 is the hydrostatic pressure, η_s the solvent viscosity, and $\mathbf{D}$ the strain rate tensor. The second term on the right of (7.37) is

$$\frac{1}{V} \sum_p \int_{V_p} \mathbf{T}^{(s)} dV = \frac{1}{V} \sum_p \int_{V_p} \left[-p\mathbf{I} + \eta_s \left(\nabla \mathbf{u} + \nabla \mathbf{u}^T \right) \right] dV$$

$$= -p_2 \mathbf{I} + \frac{\eta_s}{V} \sum_p \int_{S_p} (\mathbf{un} + \mathbf{nu}) \, dS. \qquad (7.39)$$

Since the chain is modelled as a series of discrete beads, where the interaction with the solvent and other segments takes place, the surface of the chain p consists of the surfaces of the beads. The connectors are entirely fictitious, they are allowed to cross one another; thus the model is sometimes called the phantom chain model.

On the bead surface, the velocity is regarded as uniform, and can be taken out of the integral. Thus

$$\int_{S_p} \mathbf{un} dS = \mathbf{u} \int_{S_p} \mathbf{n} dS = \mathbf{0},$$

by an application of the divergence theorem. The contribution from (7.39) is therefore only an isotropic stress, which can be lumped into the hydrostatic pressure.

Next, if we consider the chain as a continuum as well, then from the force equilibrium we must have $\nabla \cdot \mathbf{T} = \mathbf{0}$ in the chain, and thus $T_{ik} = \partial(T_{ij} x_k)/\partial x_j$. The volume integral can be converted into a surface integral, and the contribution to the effective stress from the polymer chains is

$$\frac{1}{V} \sum_p \int_{V_p} \mathbf{T} dV = \frac{1}{V} \sum_p \int_{S_p} \mathbf{x} \mathbf{T} \cdot \mathbf{n} dS = \nu \int_{S_p} \mathbf{x} \mathbf{T} \cdot \mathbf{n} dS, \qquad (7.40)$$

where S_p is the surface of a representative chain in V, $\mathbf{T} \cdot \mathbf{n}$ is the traction arising in the chain due to the interaction with the flow, and ν is the number density of the chain (number of chains per unit volume); the passage to the second equality is permissible because of the homogeneity assumption which allows us to just consider one generic chain. We can now replace the integral in (7.40) by a sum of integrals over the beads:

$$\int_{S_p} \mathbf{xT} \cdot \mathbf{n} dS = \sum_i \int_{\text{bead } i} \mathbf{xT} \cdot \mathbf{n} dS.$$

On bead i, $\mathbf{x}$ can be replaced by $\mathbf{r}_i$ and taken outside the integral, and the remaining integral of the traction on the surface of bead i is therefore the drag force, which bead i exerts on the solvent, and is proportional to the velocity of the bead relative to the solvent:

$$\mathbf{F}_i^{(d)} = -\zeta \left(\dot{\mathbf{r}}_i - \mathbf{u}_i \right).$$

In the absence of inertia, this force is equal to the connector forces plus the Brownian forces acting on the beads:

$$\int_{S_p} \mathbf{xT} \cdot \mathbf{n} dS = -\mathbf{r}_1 \mathbf{f}_1 + \mathbf{r}_2 \left(\mathbf{f}_1 - \mathbf{f}_2 \right) + \cdots + \mathbf{r}_{N+1} \mathbf{f}_N - \sum_{i=1}^{N+1} \mathbf{r}_i \mathbf{F}_i^{(b)} \qquad (7.41)$$

$$= \sum_{i=1}^{N} \mathbf{R}_i \mathbf{f}_i - \sum_{i=1}^{N+1} \mathbf{r}_i \mathbf{F}_i^{(b)}.$$

Next, the ensemble average with respect to the distribution function of $\mathbf{R}_i$ is taken. The contribution from the Brownian forces is only an isotropic stress, as can be shown either by using the expression for the smoothed-out Brownian forces, or by integrating the Langevin equations directly. This is demonstrated using the expression for the smoothed-out Brownian force,

$$\sum_{i=1}^{N+1} \mathbf{r}_i \mathbf{F}_i^{(b)} = -kT \sum_{i=1}^{N+1} \int \mathbf{r}_i \frac{\partial \ln \phi}{\partial \mathbf{r}_i} \phi d\mathbf{r}_1 \ldots d\mathbf{r}_{N+1}$$

$$= -kT \sum_{i=1}^{N+1} \int \mathbf{r}_i \frac{\partial \phi}{\partial \mathbf{r}_i} d\mathbf{r}_1 \ldots d\mathbf{r}_{N+1}.$$

The integral can be evaluated by parts,

$$\int \mathbf{r}_i \frac{\partial \phi}{\partial \mathbf{r}_i} d\mathbf{r}_1 \ldots d\mathbf{r}_{N+1} = \int \frac{\partial}{\partial \mathbf{r}_i} \left(\mathbf{r}_i \phi \right) d\mathbf{r}_1 \ldots d\mathbf{r}_{N+1} - \mathbf{I} \int \phi d\mathbf{r}_1 \ldots d\mathbf{r}_{N+1}.$$
$$(7.42)$$

The first integral on the right of (7.42) is a volume integral over an unbounded domain. It can be converted into surface integral at infinity. On this surface, $\phi \to 0$ and therefore the resulting surface integral vanishes. The second integral on the right of (7.42) is unity, because ϕ is the probability density function. Thus,

$$\sum_{i=1}^{N+1} \mathbf{r}_i \mathbf{F}_i^{(b)} = kT(N+1)\mathbf{I}. \tag{7.43}$$

7.4.1 Kramers Form

With all isotropic stresses absorbed in the pressure, the polymer-contributed stress (7.41) is

$$\mathbf{S}^{(p)} = \nu \sum_{i=1}^{N} \langle \mathbf{R}_i \mathbf{f}_i \rangle = \nu \sum_{i=1}^{N} \langle H_i \mathbf{R}_i \mathbf{R}_i \rangle. \tag{7.44}$$

This is called the Kramers form for the polymer-contributed stress. The total stress tensor in a dilute polymer solution is

$$\langle \mathbf{T} \rangle = -p\mathbf{I} + \mathbf{S}^{(s)} + \mathbf{S}^{(p)} = -p\mathbf{I} + 2\eta_s \mathbf{D} + \nu \sum_{i=1}^{N} \langle H_i \mathbf{R}_i \mathbf{R}_i \rangle. \tag{7.45}$$

7.5 Elastic Dumbbell Model

The simplest model designed to capture the slowest, and in many ways, the most important relaxation mode of a polymer chain, is the elastic dumbbell model first proposed by Kuhn [46] (Fig. 7.2b). Here we care only about the end-to-end vector of the polymer chain, and all interactions between the solvent and the chain are localised at two beads, located at the chain ends, $\mathbf{r}_1$ and $\mathbf{r}_2$. Each bead is associated with a frictional factor ζ and a negligible mass m. We will assume a Gaussian chain, with the constant spring stiffness

$$H = \frac{3kT}{Nb^2},$$

where N is the number of effective Kuhn segments in the dumbbell, each of which has an extended length b. Furthermore, the frictional coefficient $\zeta = 6\pi\eta_s a$ is assumed to be constant, where η_s is the solvent viscosity and a represents the radius of the beads. The model is also called the *linear elastic dumbbell* model to emphasise the linear force law being used. Although the general equations have been developed in the previous section, it is instructive to write down all the equations again, for this particular case.

7.5.1 *Langevin Equations*

The equations of motion are,

$$m\ddot{\mathbf{r}}_1 = \zeta \left(\mathbf{u}_1 - \dot{\mathbf{r}}_1 \right) + H \left(\mathbf{r}_2 - \mathbf{r}_1 \right) + \mathbf{F}_1^{(b)} \left(t \right),$$
$$m\ddot{\mathbf{r}}_2 = \zeta \left(\mathbf{u}_2 - \dot{\mathbf{r}}_2 \right) + H \left(\mathbf{r}_1 - \mathbf{r}_2 \right) + \mathbf{F}_2^{(b)} \left(t \right), \tag{7.46}$$

where $\mathbf{u}_i = \mathbf{u}\left(\mathbf{r}_i \right)$ is the fluid velocity evaluated at the location of the bead i, and $\mathbf{F}_i^{(b)}\left(t \right)$ is the Brownian force acting on bead i. The fluctuation-dissipation theorem (7.25) can be used to relate the strength of the Brownian forces to the mobility of the beads:

$$\left\langle \mathbf{F}_j^{(b)} \left(t \right) \right\rangle = \mathbf{0}, \quad \left\langle \mathbf{F}_i^{(b)} \left(t + s \right) \mathbf{F}_j^{(b)} \left(t \right) \right\rangle = 2kT\zeta\delta\left(s \right)\delta_{ij}\mathbf{I}. \tag{7.47}$$

Let us now define the centre of gravity and the end-to-end vector of the dumbbell respectively by

$$\mathbf{R}^{(c)} = \frac{1}{2} \left(\mathbf{r}_2 + \mathbf{r}_1 \right), \quad \mathbf{R} = \mathbf{r}_2 - \mathbf{r}., \tag{7.48}$$

The solvent velocity can be expanded about the centre of gravity,

$$\mathbf{u}_1 = \mathbf{u}^{(c)} - \tfrac{1}{2}\mathbf{R} \cdot \nabla\mathbf{u}^{(c)} + \tfrac{1}{8}\mathbf{R}\mathbf{R} : \nabla\nabla\mathbf{u}^{(c)} + O\left(R^3 \right),$$
$$\mathbf{u}_2 = \mathbf{u}^{(c)} + \tfrac{1}{2}\mathbf{R} \cdot \nabla\mathbf{u}^{(c)} + \tfrac{1}{8}\mathbf{R}\mathbf{R} : \nabla\nabla\mathbf{u}^{(c)} + O\left(R^3 \right), \tag{7.49}$$

where the superscript c denotes an evaluation at the centre of gravity. From (7.46),

$$m\ddot{\mathbf{R}}^{(c)} = \zeta \left(\mathbf{u}^{(c)} - \dot{\mathbf{R}}^{(c)} \right) + \tfrac{1}{8}\zeta\mathbf{R}\mathbf{R} : \nabla\nabla\mathbf{u}^{(c)} + \mathbf{F}^{(b,c)} \left(t \right),$$
$$m\ddot{\mathbf{R}} \quad = \zeta \left(\mathbf{L}\mathbf{R} - \dot{\mathbf{R}} \right) - 2H\mathbf{R} + \mathbf{F}^{(b)} \left(t \right), \tag{7.50}$$

where $\mathbf{L} = \left(\nabla\mathbf{u}^{(c)} \right)^T$ is the velocity gradient evaluated at the centre of gravity of the dumbbell, and

$$\mathbf{F}^{(b,c)} = \frac{1}{2} \left(\mathbf{F}_1^{(b)} + \mathbf{F}_2^{(b)} \right), \quad \mathbf{F}^{(b)} = \mathbf{F}_2^{(b)} - \mathbf{F}_1^{(b)} \tag{7.51}$$

are the Brownian forces acting on the centre of gravity and the end-to-end vector. From (7.47),

$$\left\langle \mathbf{F}^{(b,c)} \left(t \right) \right\rangle = \mathbf{0}, \quad \left\langle \mathbf{F}^{(b,c)} \left(t + s \right) \mathbf{F}^{(b,c)} \left(t \right) \right\rangle = kT\zeta\delta\left(s \right)\mathbf{I},$$
$$\left\langle \mathbf{F}^{(b)} \left(t \right) \right\rangle = \mathbf{0}, \quad \left\langle \mathbf{F}^{(b)} \left(t + s \right) \mathbf{F}^{(b)} \left(t \right) \right\rangle = 4kT\zeta\delta\left(s \right)\mathbf{I}. \tag{7.52}$$

With negligible mass, the Langevin equations (7.50) become

$$\dot{\mathbf{R}}^{(c)} = \mathbf{u}^{(c)} + \tfrac{1}{8}\mathbf{RR} : \nabla\nabla\mathbf{u}^{(c)} + \zeta^{-1}\mathbf{F}^{(b,c)}(t),$$
$$\dot{\mathbf{R}} = \mathbf{LR} - 2\zeta^{-1}H\mathbf{R} + \zeta^{-1}\mathbf{F}^{(b)}(t). \tag{7.53}$$

7.5.2 Average Motion

If the flow is homogeneous, i.e., $\mathbf{L}$ is constant, $\nabla\nabla\mathbf{u}^{(c)} = \mathbf{0}$, and the dumbbell's centre of gravity drifts just like a particle of fluid,

$$\langle \dot{\mathbf{R}}^{(c)} \rangle = \langle \mathbf{u}^{(c)} \rangle = \mathbf{L}\langle \mathbf{R}^{(c)} \rangle = \mathbf{u}\left(\langle \mathbf{R}^{(c)} \rangle\right). \tag{7.54}$$

A migration from the streamline of the centre of gravity will be induced by a non-homogeneous flow field. This migration is

$$\langle \dot{\mathbf{R}}^{(c)} - \mathbf{u}^{(c)} \rangle = \frac{1}{8}\langle \mathbf{RR} \rangle : \nabla\nabla\mathbf{u}^{(c)}. \tag{7.55}$$

The average end-to-end vector evolves in time according to

$$\langle \dot{\mathbf{R}} \rangle = \mathbf{L}\langle \mathbf{R} \rangle - 2H\zeta^{-1}\langle \mathbf{R} \rangle, \tag{7.56}$$

which consists of a flow-induced stretching (first term on the right) plus a restoring mechanism (second term on the right) due to the connector spring force. The parameter

$$\lambda = \frac{\zeta}{4H} = \frac{\zeta N b^2}{12 kT} \tag{7.57}$$

is called the *Rouse relaxation time*.

7.5.3 Strong and Weak Flows

Equation (7.56) has been used as a basis for delineating between strong and weak flows: strong flows are those in which the flow-induced deformation overcomes the restoring force allowing the microstructure (as represented by $\langle \mathbf{R} \rangle$) to grow exponentially in time. Otherwise the flow is weak.

Since we are more interested in the end-to-end vector, the process $\mathbf{R}^{(c)}$ can now be discarded. The Fokker–Planck equation for the density distribution function for $\mathbf{R}$ reads

$$\frac{\partial \phi}{\partial t} = \frac{\partial}{\partial \mathbf{R}} \cdot \left[\frac{2kT}{\zeta} \frac{\partial \phi}{\partial \mathbf{R}} - \left(\mathbf{LR} - \frac{2H}{\zeta} \mathbf{R} \right) \phi \right]. \tag{7.58}$$

In many cases, there is no need to find the full probability distribution – all we want is $\langle \mathbf{RR} \rangle$, or the equation governing the evolution of $\langle \mathbf{RR} \rangle$, since this quantity is related to the stress. This can be accomplished without solving for ϕ.

First, from (7.53)

$$\frac{d}{dt} \mathbf{RR} = \dot{\mathbf{R}}\mathbf{R} + \mathbf{R}\dot{\mathbf{R}} \tag{7.59}$$

$$= \mathbf{L} \cdot \mathbf{RR} + \mathbf{RR} \cdot \mathbf{L}^T - 4H\zeta^{-1}\mathbf{RR} + \zeta^{-1} \left(\mathbf{RF}^{(b)} + \mathbf{F}^{(b)}\mathbf{R} \right).$$

Secondly, $\mathbf{R}$ and $\mathbf{F}^{(b)}$ have widely different time scales. Thus

$$\langle \mathbf{RF}^{(b)} \rangle = \langle \mathbf{R}(\Delta t) \mathbf{F}^{(b)}(\Delta t) \rangle$$

$$= \left\langle \left(\left[\mathbf{LR} - 2H\zeta^{-1}\mathbf{R} \right] \Delta t + \zeta^{-1} \int_0^{\Delta t} \mathbf{F}^{(b)}(t) \, dt \right) \mathbf{F}^{(b)}(\Delta t) \right\rangle$$

$$= \zeta^{-1} \int_0^{\Delta t} \langle \mathbf{F}^{(b)}(t) \mathbf{F}^{(b)}(\Delta t) \rangle \, dt = 4kT\mathbf{I} \int_0^{\Delta t} \delta(t - \Delta t) \, dt$$

$$= 2kT\mathbf{I}.$$

Thus, from (7.59),

$$\frac{d}{dt} \langle \mathbf{RR} \rangle = \mathbf{L} \langle \mathbf{RR} \rangle + \langle \mathbf{RR} \rangle \mathbf{L}^T - \frac{4H}{\zeta} \langle \mathbf{RR} \rangle + \frac{4kT}{\zeta}\mathbf{I}.$$

Re-arranging, and recalling (7.57),

$$\langle \mathbf{RR} \rangle + \lambda \left\{ \frac{d}{dt} \langle \mathbf{RR} \rangle - \mathbf{L} \langle \mathbf{RR} \rangle - \langle \mathbf{RR} \rangle \mathbf{L}^T \right\} = \frac{1}{3} Nb^2 \mathbf{I}. \tag{7.60}$$

Next using (7.45),

$$\langle \mathbf{T} \rangle = -p\mathbf{I} + \mathbf{S}^{(s)} + \mathbf{S}^{(p)} = -p\mathbf{I} + 2\eta_s \mathbf{D} + \nu H \langle \mathbf{RR} \rangle. \tag{7.61}$$

7.5.4 Upper-Convected Maxwell Model

Since the polymer-contributed stress is $\nu H \langle \mathbf{RR} \rangle$, we can multiply (7.60) with νH to generate the equation for $\mathbf{S}^{(p)}$:

$$\mathbf{S}^{(p)} + \lambda \left\{ \frac{d}{dt} \mathbf{S}^{(p)} - \mathbf{LS}^{(p)} - \mathbf{S}^{(p)}\mathbf{L}^T \right\} = G\mathbf{I}, \tag{7.62}$$

where

$$G = \frac{1}{3}Nb^2\nu H = \nu kT. \tag{7.63}$$

The derivative operator implicitly define in the braces (acting on $\mathbf{S}^{(p)}$) on the left side of (7.62) is one of the many derivatives introduced by Oldroyd [64] to guarantee the stress tensor objectivity. It is called the *upper-convected derivative*, $\delta/\delta t$, and is defined by

$$\frac{\delta \mathbf{A}}{\delta t} = \frac{d\mathbf{A}}{dt} - \mathbf{L}\mathbf{A} - \mathbf{A}\mathbf{L}^T. \tag{7.64}$$

We customarily re-define the polymer-contributed stress as

$$\mathbf{S}^{(p)} = G\mathbf{I} + \boldsymbol{\tau}^{(p)}. \tag{7.65}$$

Then, since

$$\frac{\delta}{\delta t}\mathbf{I} = -\left(\mathbf{L} + \mathbf{L}^T\right) = -2\mathbf{D},$$

we obtain

$$\boldsymbol{\tau}^{(p)} + \lambda\frac{\delta}{\delta t}\boldsymbol{\tau}^{(p)} = 2\eta_p\mathbf{D}, \tag{7.66}$$

where

$$\eta_p = G\lambda = \frac{\nu\zeta Nb^2}{12} = \frac{1}{2}\pi\nu aNb^2\eta_s \tag{7.67}$$

is the polymer-contributed viscosity. The model (7.62), or (7.66), is called the Upper Convected Maxwell (UCM) model in honour of Maxwell, who introduced the linear version in his kinetic theory of gases in 1867. Since $\mathbf{S}^{(p)}$ is proportional to $\langle \mathbf{RR} \rangle$, it is positive definite, whereas $\boldsymbol{\tau}^{(p)}$ is not. In some numerical applications, (7.62) may be preferred to (7.66), because the lack of positive definiteness in $\mathbf{S}^{(p)}$ can be conveniently tested numerically; this lack can be used as an indication of impending numerical divergence.

7.5.5 Oldroyd-B Model

When the solvent and the polymer-contributed stresses are combined, c.f. (7.61),

$$\mathbf{S} = \mathbf{S}^{(s)} + \boldsymbol{\tau}^{(p)} = 2\eta_s\mathbf{D} + \boldsymbol{\tau}^{(p)}, \tag{7.68}$$

one has

$$(\mathbf{S} - 2\eta_s \mathbf{D}) + \lambda \frac{\delta}{\delta t} (\mathbf{S} - 2\eta_s \mathbf{D}) = 2\eta_p \mathbf{D}$$

$$\mathbf{S} + \lambda_1 \frac{\delta \mathbf{S}}{\delta t} = 2\eta \left(\mathbf{D} + \lambda_2 \frac{\delta \mathbf{D}}{\delta t} \right), \tag{7.69}$$

where $\lambda_1 = \lambda$ is the relaxation time, $\eta = \eta_s + \eta_p$ is the total viscosity, $\lambda_2 = \lambda \eta_s / \eta$ is the retardation time. The model (7.69) is called the Oldroyd fluid B, or Oldroyd-B model.

The Oldroyd-B constitutive equation qualitatively describes many features of the so-called Boger fluids.[8] In a steady state simple shear flow, this constitutive equation predicts a constant viscosity, a first normal stress difference which is quadratic in the shear rate, and a zero second normal stress difference. In an unsteady state shear flow, the stresses increase monotonically in time to their steady values, without stress overshoots which are sometimes observed with some dilute polymer solutions. In an elongational flow, the elongational viscosity becomes infinite at a finite elongation rate of $1/(2\lambda)$ – these will be explored in a series of problems.

7.6 Main Features of the Oldroyd-B Model

Recall the relative strain tensor

$$\mathbf{C}_t (t - s) = \mathbf{F}_t (s)^T \mathbf{F}_t (s), \ \mathbf{C}_t (s)^{-1} = \mathbf{F}_t (s)^{-1} \mathbf{F}_t (s)^{-T},$$

$$\mathbf{F}_t (s) = \mathbf{F} (t - s) \mathbf{F} (t)^{-1}, \ \mathbf{F}_t (t - s)^{-1} = \mathbf{F} (t) \mathbf{F} (t - s)^{-1},$$

$$\dot{\mathbf{F}} (t) = \mathbf{L} (t) \mathbf{F} (t), \ \dot{\mathbf{F}}_t (s)^{-1} = \mathbf{L} (t) \mathbf{F}_t (s), \ \dot{\mathbf{F}}_t (s)^{-T} = \mathbf{F}_t (s) \mathbf{L} (t)^T,$$

$$\dot{\mathbf{C}}_t (s)^{-1} = \mathbf{L} (t) \mathbf{C}_t (s)^{-1} + \mathbf{C}_t (s)^{-1} \mathbf{L} (t)^T. \tag{7.70}$$

From the results (7.70), it can be shown that the UCM model is solved by (Problem 7.5)

$$\mathbf{S}^{(p)} (t) = \frac{G}{\lambda} \int_{-\infty}^{t} e^{(s-t)/\lambda} \mathbf{C}_t (s)^{-1} \, ds = G\mathbf{I} + \tau^{(p)}. \tag{7.71}$$

This integral version of the UCM model is called the Lodge rubber-like liquid model [53]. It was derived from a network of polymer strands, a model meant for concentrated polymer solutions and melts. It is remarkable that two models for two distinct microstructures, a dilute suspension of dumbbells and a concentrated network of polymer strands, share a common constitutive framework.

---

[8]Dilute solutions of polymers in highly viscous solvents [9].

7.6.1   *Simple Flows*

In a simple shear flow, with a time-dependent shear rate $\dot{\gamma}(t)$, the stress components of the UCM model (7.66) obey

$$
\begin{aligned}
&\tau_{11}^{(p)} + \lambda \left\{ \dot{\tau}_{11}^{(p)} - 2\dot{\gamma}\tau_{12}^{(p)} \right\} = 0, \\
&\tau_{22}^{(p)} + \lambda \dot{\tau}_{22}^{(p)} = 0, \\
&\tau_{33}^{(p)} + \lambda \dot{\tau}_{33}^{(p)} = 0, \\
&\tau_{12}^{(p)} + \lambda \left\{ \dot{\tau}_{12}^{(p)} - \dot{\gamma}\tau_{22}^{(p)} \right\} = \eta_p \dot{\gamma}.
\end{aligned}
\tag{7.72}
$$

If the stresses start from zero initial states, then $\tau_{22}^{(p)} = \tau_{33}^{(p)} = 0$ for all time, and the only two non-trivial components are $\tau_{11}^{(p)}$ and $\tau_{12}^{(p)}$.

Start-Up Shear Flow. In a start-up of a shear flow, with a constant shear rate $\dot{\gamma}$, the solution to (7.72) is

$$
\tau_{12}^{(p)} = \eta_p \dot{\gamma} \left(1 - e^{-t/\lambda} \right),
\tag{7.73}
$$

$$
\tau_{11}^{(p)} = 2\eta_p \lambda \dot{\gamma}^2 \left(1 - e^{-t/\lambda} \right) - 2\eta_p t \dot{\gamma}^2 e^{-t/\lambda}.
\tag{7.74}
$$

At steady state, $\tau_{12}^{(p)} = \eta_p \dot{\gamma}$, $\tau_{11}^{(p)} = 2\eta_p \lambda \dot{\gamma}^2$ and thus the viscometric functions are

$$
\eta = \eta_s + \eta_p, \quad N_1 = 2\eta_p \lambda \dot{\gamma}^2, \quad N_2 = 0.
\tag{7.75}
$$

Oscillatory Shear Flow. In an oscillatory flow with shear rate $\dot{\gamma} = \dot{\gamma}_0 \Re \left(e^{i\omega t} \right)$, where $\Re$ denotes the real part, (7.72) becomes

$$
\tau_{11}^{(p)} + \lambda \dot{\tau}_{11}^{(p)} = 2\dot{\gamma}_0 \Re \left(e^{i\omega t} \right) \tau_{12}^{(p)}, \quad \tau_{12}^{(p)} + \lambda \dot{\tau}_{12}^{(p)} = \eta_p \dot{\gamma}_0 \Re \left(e^{i\omega t} \right).
\tag{7.76}
$$

We search for the steady solution

$$
\tau_{12}^{(p)} = \Re \left(S_0 e^{i\omega t} \right), \quad \tau_{11}^{(p)} = \Re \left(N_0 e^{2i\omega t} \right).
\tag{7.77}
$$

When these are substituted into (7.72), we find that

$$
S_0 = \frac{\eta_p \dot{\gamma}_0}{1 + i\lambda\omega}, \quad N_0 = \frac{2\eta_p \lambda \dot{\gamma}_0^2}{(1 + i\lambda\omega)(1 + 2i\lambda\omega)}.
\tag{7.78}
$$

Consequently, the dynamic properties of the Oldroyd-fluid are

$$
\eta^* = \eta_s + \frac{\eta_p}{1 + i\lambda\omega},
\tag{7.79}
$$

$$\eta' = \eta_s + \frac{\eta_p}{1 + \lambda^2\omega^2}, \quad \eta'' = \frac{\eta_p\lambda\omega}{1 + \lambda^2\omega^2}, \tag{7.80}$$

$$G' = \frac{G\lambda^2\omega^2}{1 + \lambda^2\omega^2}, \quad G'' = \eta_s\omega + \frac{G\lambda\omega}{1 + \lambda^2\omega^2}. \tag{7.81}$$

Elongational Flow. In the start-up of an uniaxial elongational flow, where the velocity gradient is

$$[\mathbf{L}] = diag\left(\dot{\varepsilon}, -\dot{\varepsilon}/2, -\dot{\varepsilon}/2\right),$$

the stress components of the UCM model (7.66) obey (note that $\tau_{22}^{(p)} = \tau_{33}^{(p)}$)

$$\tau_{11}^{(p)} + \lambda\left(\dot{\tau}_{11}^{(p)} - 2\dot{\varepsilon}\tau_{11}^{(p)}\right) = 2\eta_p\dot{\varepsilon}, \tag{7.82}$$

$$\tau_{22}^{(p)} + \lambda\left(\dot{\tau}_{22}^{(p)} + \dot{\varepsilon}\tau_{22}^{(p)}\right) = -\eta_p\dot{\varepsilon}.$$

The solution is

$$\tau_{11}^{(p)} = \frac{2\eta_p\dot{\varepsilon}}{1 - 2\lambda\dot{\varepsilon}}\left(1 - e^{-(1-2\lambda\dot{\varepsilon})t/\lambda}\right), \tag{7.83}$$

$$\tau_{22}^{(p)} = \tau_{33}^{(p)} = -\frac{\eta_p\dot{\varepsilon}}{1 + \lambda\dot{\varepsilon}}\left(1 - e^{-(1+\lambda\dot{\varepsilon})t/\lambda}\right).$$

Thus, if either $\lambda\dot{\varepsilon} \geq 1/2$ or $\lambda\dot{\varepsilon} \leq -1$, then at least one component of the stress grows unboundedly. This reflects the linear spring in the model that allows the end-to-end vector of the dumbbell to grow without bound in a strong flow. For $-1 < \lambda\dot{\varepsilon} < 1/2$ and at steady state, the elongational viscosity of the Oldroyd-B model can be shown to be, using (7.83),

$$\eta_E = \frac{S_{11} - S_{22}}{\dot{\varepsilon}} = 3\eta_s + \frac{3\eta_p}{(1 - 2\lambda\dot{\varepsilon})(1 + \lambda\dot{\varepsilon})}. \tag{7.84}$$

The Trouton ratio η_E/η increases from the Newtonian value of 3, when $\dot{\varepsilon} = 0$ and becomes unbounded at $\lambda\dot{\varepsilon}$ approaches either -1 of $1/2$.

The prediction of an infinite stress at a finite elongational rate is not physically realistic. It is due to the linear dumbbell model being allowed to stretch infinitely. Constraining the dumbbell to a maximum allowable length will fix this problem (e.g., FENE dumbbell, Phan-Thien/Tanner model [6, 39, 85]).

The linear elastic dumbbell model is also inadequate in oscillatory flow: it predicts a shear stress proportional to the amplitude of the shear strain, irrespective of the latter magnitude. This is unrealistic: in practice this proportionality is only found when the shear strain is small (typically less than about 10% for polymer solutions and melts).

7.6.2 Multiple Relaxation Time UCM Model

The frequency response of the dumbbell model is also inadequate, due to only one relaxation time in the model. With multiple relaxation times, the Rouse model, Fig. 7.1a, results in

$$\tau^{(p)} = \sum_{j=1}^{N} \tau^{(j)}, \tag{7.85}$$

$$\tau^{(j)} + \lambda_j \frac{\delta \tau^{(j)}}{\delta t} = 2\eta_j \mathbf{D},$$

where $\{\lambda_j, \eta_j\}$ is the discrete relaxation spectrum. The dynamic properties are now much improved:

$$\eta' = \eta_s + \sum_{j=1}^{N} \frac{\eta_j}{1 + \lambda_j^2 \omega^2}, \quad \eta'' = \sum_{j=1}^{N} \frac{\eta_j \lambda_j \omega}{1 + \lambda_j^2 \omega^2}, \tag{7.86}$$

$$G' = \sum_{j=1}^{N} \frac{G_j \lambda_j^2 \omega^2}{1 + \lambda_j^2 \omega^2}, \quad G'' = \eta_s \omega + \sum_{j=1}^{N} \frac{G_j \lambda_j \omega}{1 + \lambda_j^2 \omega^2}.$$

In a steady shear flow, the model still predicts a constant viscosity, a quadratic first normal stress difference in the shear rate, and a zero second normal stress difference. The Boger fluids show little shear thinning over a large range of shear rates, but this is no doubt due to the high solvent viscosity that completely masks the contribution from the polymer viscosity; any amount of shear-thinning from the polymer contribution would hardly show up on the total fluid viscosity. In general, dilute polymer solutions usually show some degree of shear thinning. The fix is to adopt a more realistic force law for the chain. One such model is the FENE (Finitely Extendable Nonlinear Elastic) model [6].

Problems

Problem 7.1 Use the solution (7.24) in (7.27) to show that

$$\mathbf{D} = kT\zeta^{-1}.$$

This is the *Stokes–Einstein relation*, relating the diffusivity to the mobility of a Brownian particle.

Problem 7.2 Starting from the Langevin equation in configuration space, in the limit $\mathbf{m} \to \mathbf{0}$,

$$\dot{x} = -\zeta^{-1} \cdot \mathbf{K}\mathbf{x} + \zeta^{-1}\mathbf{F}^{(b)}\left(t\right), \tag{7.87}$$

show that

$$\Delta\mathbf{x}\left(t\right) = -\zeta^{-1} \cdot \mathbf{K}\mathbf{x}\Delta t + \int_{t}^{t+\Delta t} \zeta^{-1}\mathbf{F}^{(b)}\left(t'\right)dt'. \tag{7.88}$$

From this, show that

$$\langle \Delta\mathbf{x} \rangle = -\zeta^{-1} \cdot \mathbf{K}\mathbf{x}, \quad \langle \Delta\mathbf{x}\left(t\right)\Delta\mathbf{x}\left(t\right) \rangle = 2kT\zeta^{-1}\Delta t \tag{7.89}$$

and conclude that the Fokker–Planck equation is

$$\frac{\partial\phi}{\partial t} = \frac{\partial}{\partial\mathbf{x}} \cdot \left[kT\zeta^{-1}\frac{\partial\phi}{\partial\mathbf{x}} + \zeta^{-1} \cdot \mathbf{K}\mathbf{x}\phi\right].$$

Problem 7.3 Investigate the migration problem in a plane Poiseuille flow.

Problem 7.4 Show that the solution to (7.56) is

$$\langle \mathbf{R}\left(t\right) \rangle = e^{-t/2\lambda}e^{\mathbf{L}t}\mathbf{R}_0. \tag{7.90}$$

Thus conclude that the flow is strong if

$$\text{eigen}\left(\mathbf{L}\right) \geq 1/2\lambda,$$

where eigen $(\mathbf{L})$ is the maximum eigenvalue of $\mathbf{L}$.

Problem 7.5 Using the result (7.70), show that the following solves the Maxwell equation (7.62):

$$\mathbf{S}^{(p)}\left(t\right) = \frac{G}{\lambda}\int_{-\infty}^{t} e^{(s-t)/\lambda}\mathbf{C}_t\left(s\right)^{-1}ds = G\mathbf{I} + \tau^{(p)}.$$

Chapter 8
Suspensions

Particulates

Suspension is a term used to describe an effective fluid made up of particles suspended in a liquid; examples of such liquids abound in natural and man-made materials: blood, milk, paints, inks. The concept of a suspension is meaningful only when there are two widely different length scales in the problem: l is a typical dimension of a suspended particle, L is a typical size of the apparatus, and $l \ll L$. When this is not met, we simply have a collection of discrete individual particles suspended in a liquid. Most progress has been made with Newtonian suspensions, i.e., suspensions of particles in a Newtonian liquid. The review paper by Metzner [57] contained most of the relevant information on the subject.

If the particles are small enough (less than $10\,\mu$m in size), then they will undergo Brownian motion, their micromechanics are described by a set of stochastic differential equations, together with some relevant fluctuation-dissipation theorems, and the full solution of the relevant equations can only be obtained by specifying the probability distribution of the system.

The relative importance of Brownian motion is characterised by a Péclet number, such as $Pe = O(\eta_s \dot{\gamma} l^3 / kT)$, the ratio of viscous stress to stress induced by thermal excitation, where $\dot{\gamma}$ is a typical strain rate, η_s is the solvent viscosity, and kT is the Boltzmann temperature. At low Péclet numbers, Brownian motion is strong, and the particles' orientation tends to be randomised, leading to a larger dissipation (i.e., higher effective viscosity) than when the Péclet number is large, the Brownian motion is weak, and the particles tend to align with the flow most of the time. Thus, we expect shear-thinning with the inclusion of Brownian motion (increasing shear rate leads to an increase in the Péclet number). We will focus on non-Brownian flow regime, where the particles are large enough (typically of ~$10\,\mu$m in size), but yet orders of magnitudes smaller than L.

With $l \ll L$, the microscale Reynolds number is small. Thus, the relevant equations governing the micromechanics are the Stokes equations,

$$\nabla \cdot \mathbf{u} = 0, \quad -\nabla p + \eta \nabla^2 \mathbf{u} = \mathbf{0}. \tag{8.1}$$

© Springer International Publishing AG 2017
N. Phan-Thien and N. Mai-Duy, *Understanding Viscoelasticity*,
Graduate Texts in Physics, DOI 10.1007/978-3-319-62000-8_8

Stokes equations are linear and instantaneous in the driving boundary data. Consequently the microdynamics are also linear and instantaneous in the driving forces; only the present boundary data are important, not their past history. This does not imply that the overall response will have no memory, nor does it imply that the macroscaled Reynolds number is small. In most studies, the particle's inertia is neglected, its inclusion may lead to a non-objective constitutive model, since the stress contributed by micro inertia may not be objective (Ryskin and Rallison [79]).

The linearity of the micromechanics implies that the particle-contributed stress will be linear in the strain rate; in particular all the rheological properties (shear stress, first and second normal stress differences, and elongational stresses) will be linear in the strain rate. Several investigators have indeed found Newtonian behaviour in shear for suspensions up to a large volume fraction. However, experiments with some concentrated suspensions usually show shear-thinning behaviour, but the particles in these experiments are in the μm range, where Brownian motion would be important. Shear-thickening behaviour, and indeed, yield stress and discontinuous behaviour in the viscosity-shear-rate relation have been observed, e.g., Metzner [57]. This behaviour cannot be accommodated within the framework of hydrodynamic interaction alone; for a structure to be formed, we need forces and torques of a non-hydrodynamic origin.

8.1 Bulk Suspension Properties

Consider now a volume V which is large enough to contain many particles but small enough so that macroscopic variables hardly change on the scale $V^{1/3}$, i.e., $l \ll V^{1/3} \ll L$. The effective stress tensor seen from a macroscopic level is simply the volume-averaged stress [4, 47],

$$\langle \sigma_{ij} \rangle = \frac{1}{V} \int_V \sigma_{ij} \, dV = \frac{1}{V} \int_{V_f} \sigma_{ij} \, dV + \frac{1}{V} \int_{V_p} \sigma_{ij} \, dV,$$

where V_f is the volume occupied by the solvent, V_p is the volume of the particles in V, and the angle brackets denote a volume-averaged quantity. If the solvent is Newtonian, we have

$$\sigma_{ij}(\mathbf{x}) = -p\delta_{ij} + 2\eta_s D_{ij} \equiv \sigma_{ij}^{(f)}, \quad \mathbf{x} \in V_f.$$

Thus

$$\frac{1}{V} \int_{V_f} \sigma_{ij} dV = -\langle p \rangle \, \delta_{ij} + 2\eta_s \langle D_{ij} \rangle - \frac{1}{V} \int_{V_p} \sigma_{ij}^{(f)} dV.$$

Furthermore, from the equations of motion in the absence of inertia and the body force,

$$\frac{\partial}{\partial x_k} \left(x_i \sigma_{kj} \right) = \sigma_{ij},$$

and we find that

$$\frac{1}{V} \int_{V_p} \sigma_{ij} \, dV = \frac{1}{V} \int_{V_p} \frac{\partial}{\partial x_k} \left(x_i \sigma_{kj} \right) \, dV = \frac{1}{V} \int_{S_p} x_i t_j \, dS,$$

where $t_j = \sigma_{kj} n_k$ is the traction vector and S_p is the bounding surface of all the particles. In addition,

$$\frac{1}{V} \int_{V_p} \sigma_{ij}^{(f)} \, dV = \frac{1}{V} \int_{V_p} \left[-p \delta_{ij} + \eta_s \left(\frac{\partial u_i}{\partial x_j} + \frac{\partial u_j}{\partial x_i} \right) \right] \, dV$$

$$= \frac{1}{V} \int_{S_p} \left[\eta_s \left(u_i n_j + u_j n_i \right) \right] \, dS$$

to within an isotropic tensor which can be lumped into a generic hydrostatic pressure P, which is determined through the balance of momentum and the incompressibility constraint. The average stress is thus given by

$$\langle \sigma_{ij} \rangle = -p' \delta_{ij} + \underbrace{2 \eta_s \langle D_{ij} \rangle}_{\text{solvent}} + \underbrace{\frac{1}{V} \int_{S_p} \left\{ x_i t_j - \eta_s \left(u_i n_j + u_j n_i \right) \right\} \, dS}_{\text{particles}}, \qquad (8.2)$$

consisting of a solvent contribution, and a particle contribution; p' is just a scalar pressure (the prime will be dropped from hereon). The particle contribution can be decomposed into a symmetric part, and an antisymmetric part. The symmetric part is in fact the sum of the stresslets $S_{ij}^{(p)}$ defined by

$$S_{ij} = \frac{1}{2} \int_{S_p} \left\{ x_i t_j + x_j t_i - 2 \eta_s \left(u_i n_j + u_j n_i \right) \right\} \, dS = \sum_p S_{ij}^{(p)}, \qquad (8.3)$$

and the antisymmetric part leads to the rotlet:

$$R_{ij} = \frac{1}{2} \int_{S_p} \left(x_i t_j - x_j t_i \right) \, dS = \frac{1}{2} \sum_p \epsilon_{ijk} T_k^{(p)}, \qquad (8.4)$$

where $T_k^{(p)}$ is the torque exerted on the particle p, and the summation is over all particles in the volume V. The particle-contributed stress is therefore given by

$$\sigma_{ij}^{(p)} = \frac{1}{V} \sum_p \left(S_{ij}^{(p)} + \frac{1}{2} \epsilon_{ijk} T_k^{(p)} \right). \qquad (8.5)$$

The total rate of energy dissipation can be calculated by consider a large enough volume V to contain all the particles - the rate of energy dissipation in V is thus

$$\Phi = \int_V \sigma_{ij} D_{ij} \, dV = \int_V \frac{\partial}{\partial x_j} \left(\sigma_{ij} u_i \right) \, dV = \sum_p \int_{S_p} \sigma_{ij} u_i n_j \, dS. \tag{8.6}$$

The second equality comes from the balance of momentum, and the third one from an application of the divergence theorem, assuming that the condition at infinity is quiescent (the bounding surface of V consists of particles' surfaces and surface at infinity). For a system of rigid particles, the boundary condition on the surface of a particle p is that

$$\mathbf{u} = \mathbf{U}^{(p)} + \Omega^{(p)} \times \mathbf{x}, \tag{8.7}$$

where $\mathbf{U}^{(p)}$ and $\Omega^{(p)}$ are the translational and rotation velocities of the particle, which can be taken outside the integral in (8.6). The terms remaining can be identified with the force $\mathbf{F}^{(p)}$, and the torque $\mathbf{T}^{(p)}$ imparted by the particle p to the fluid. Thus the total rate of energy dissipation is

$$\Phi = \sum_p \left(\mathbf{U}^{(p)} \cdot \mathbf{F}^{(p)} + \Omega^{(p)} \cdot \mathbf{T}^{(p)} \right). \tag{8.8}$$

Note also that for a system of rigid particles, the integral

$$\int_{S_p} (\mathbf{u}\mathbf{n} + \mathbf{n}\mathbf{u}) \, dS = \mathbf{0},$$

since

$$\int_{S_p} \mathbf{U}^{(p)} \mathbf{n} dS = \mathbf{0}, \quad \int_{S_p} \left(\Omega^{(p)} \times \mathbf{x}\mathbf{n} + \mathbf{n}\Omega^{(p)} \times \mathbf{x} \right) dS = \mathbf{0},$$

by applications of the divergence theorem.

8.2　Dilute Suspension of Spheroids

We consider now a dilute suspension of force- and torque-free monodispersed spheres in a general homogeneous deformation. The dilute assumption means the volume fraction

$$\phi = \nu \frac{4\pi a^3}{3} \ll 1, \tag{8.9}$$

where ν is the number density of the spheres, of radius a each. In this case, in a representative volume V we expect to find only one sphere. Thus, the microscale problem consists of a single sphere in an effectively unbounded fluid; the superscript p on the generic particle can be omitted, and the coordinate system can be conveniently placed at the origin of the sphere. The boundary conditions for this microscale problem are

$$\mathbf{u} = \mathbf{U} + (\mathbf{D} + \mathbf{W}) \cdot \mathbf{x}, \quad \text{far from the particle, } |\mathbf{x}| \to \infty, \tag{8.10}$$

and

$$\mathbf{u} = \mathbf{V} + \mathbf{w} \cdot \mathbf{x}, \quad \text{on the particle's surface, } |\mathbf{x}| = a, \tag{8.11}$$

where $\mathbf{L} = \mathbf{D} + \mathbf{W}$ is the far-field velocity gradient tensor; $\mathbf{D}$ is the strain rate tensor, $\mathbf{W}$ is the vorticity tensor, $\mathbf{w}$ is the skew-symmetric tensor such that $w_{ij} = -\epsilon_{ijk}\Omega_k$, with Ω being the angular velocity of the particle. The far-field boundary condition must be interpreted to be far away from the particle under consideration, but not far enough so that another sphere can be expected. The solution to this unbounded flow problem is well known, [35]

$$\mathbf{u} = \mathbf{U} + \mathbf{L} \cdot \mathbf{x} + \frac{a^3}{x^3}(\mathbf{w} - \mathbf{W}) \cdot \mathbf{x} + \left(\frac{3a}{4x} + \frac{a^3}{4x^3}\right)(\mathbf{V} - \mathbf{U}) - \frac{a^5}{x^5}\mathbf{D} \cdot \mathbf{x} \tag{8.12}$$

$$+ \frac{3(\mathbf{V} - \mathbf{U}) \cdot \mathbf{x}}{x^2}\left(\frac{a}{x} - \frac{a^3}{x^3}\right)\mathbf{x} - \frac{5\mathbf{D} : \mathbf{xx}}{2x^2}\left(\frac{a^3}{x^3} - \frac{a^5}{x^5}\right)\mathbf{x}, \tag{8.13}$$

and

$$p = \frac{3}{2}\eta_s a \frac{(\mathbf{V} - \mathbf{U}) \cdot \mathbf{x}}{x^3} - 5\eta_s a^3 \frac{\mathbf{D} : \mathbf{xx}}{x^5}. \tag{8.14}$$

The traction on the surface of the sphere is

$$\mathbf{t} = \sigma \cdot \mathbf{n}|_{x=a} = -\frac{3\eta_s}{2a}(\mathbf{V} - \mathbf{U}) - \frac{3\eta_s}{a}(\mathbf{w} - \mathbf{W}) \cdot \mathbf{x} + 5\frac{\eta_s}{a}\mathbf{D} \cdot \mathbf{x}. \tag{8.15}$$

The force and the torque on the particle can be evaluated:

$$\mathbf{F} = \int_S \sigma \cdot \mathbf{n} \, dS = -6\pi\eta_s a(\mathbf{V} - \mathbf{U}) \tag{8.16}$$

and

$$\mathbf{T} = \int_S \mathbf{x} \times \sigma \cdot \mathbf{n} dS = -8\pi\eta_s a^3(\Omega - \boldsymbol{\omega}), \tag{8.17}$$

where $\omega_i = \frac{1}{2}\epsilon_{ijk}W_{jk}$ is the local vorticity vector. Thus, if the particle is force-free and torque-free, then it will translate with $\mathbf{U}$ and spin with an angular velocity of $\boldsymbol{\omega}$.

Returning now to the particle-contributed stress, (8.5),

$$\left\langle \sigma_{ij}^{(p)} \right\rangle = \frac{1}{V} \sum_p \mathcal{S}_{ij} = \nu \mathcal{S}_{ij},$$

where the stresslet is given in (8.3). From (8.12), and noting that

$$\int_S \mathbf{x}\, dS = \mathbf{0}, \quad \int_S \mathbf{xx}\, dS = \frac{4\pi a^4}{3}\mathbf{1},$$

we find

$$\mathcal{S}_{ij} = \frac{1}{2} \int_S \left(x_i t_j + x_j t_i \right) dS = 5\frac{\eta_s}{a} \int_S D_{ik} x_k x_j\, dS = 5\eta_s \left(\frac{4\pi a^3}{3} \right) D_{ij}.$$

Recall that the volume fraction of particles is $\phi = 4\pi a^3 \nu / 3$, the effective stress will now become

$$\langle \sigma \rangle = -p\mathbf{1} + 2\eta_s \left(1 + \frac{5}{2}\phi \right) \mathbf{D}. \tag{8.18}$$

This is the celebrated Einstein's result [20], who arrived at the conclusion from the equality of the dissipation at the microscale and the dissipation at the macroscale as described by an effective Newtonian viscosity.

A similar theory has been worked out for a dilute suspension of spheroids by Leal and Hinch [51], when the spheroids may be under the influence of Brownian motion, using the solution for flow around a spheroid due to Jeffery [41]. Here, if $\mathbf{p}$ denotes a unit vector directed along the major axis of the spheroid, then Jeffery's solution states that

$$\dot{\mathbf{p}} = \mathbf{W} \cdot \mathbf{p} + \frac{R^2 - 1}{R^2 + 1} \left(\mathbf{D} \cdot \mathbf{p} - \mathbf{D} : \mathbf{ppp} \right), \tag{8.19}$$

where R is the aspect ratio of the particle (major to minor diameter ratio). The particle-contributed stress may be shown to be

$$\sigma^{(p)} = 2\eta_s \phi \{ A\mathbf{D} : \langle \mathbf{pppp} \rangle + B \left(\mathbf{D} \cdot \langle \mathbf{pp} \rangle + \langle \mathbf{pp} \rangle \cdot \mathbf{D} \right) \tag{8.20}$$
$$+ C\mathbf{D} + d_R F \langle \mathbf{pp} \rangle \},$$

where the angular brackets denote the ensemble average with respect to the distribution function of $\mathbf{p}$; A, B, C and F are some shape factors, and d_R is the rotational diffusivity. If the particles are large enough so that Brownian motion can be ignored, then the last term, as well as the angular brackets, can be omitted in (8.20). The asymptotic values of the shape factors are given in Table 8.1.

Table 8.1 Asymptotic values of the shape factors

Asymptotic limit	$R \rightarrow \infty$ (rod-like)	$R = 1 + \delta,\ \delta \ll 1$ (near-sphere)	$R \rightarrow 0$ (disk-like)
A	$\dfrac{R^2}{2(\ln 2R - 1.5)}$	$\dfrac{395}{147}\delta^2$	$\dfrac{10}{3\pi R} + \dfrac{208}{9\pi^2} - 2$
B	$\dfrac{6\ln 2R - 11}{R^2}$	$\dfrac{15}{14}\delta - \dfrac{395}{588}\delta^2$	$-\dfrac{8}{3\pi R} + 1 - \dfrac{128}{9\pi^2}$
C	2	$\dfrac{5}{2}\left(1 - \dfrac{2}{7}\delta + \dfrac{1}{3}\delta^2\right)$	$\dfrac{8}{3\pi R}$
F	$\dfrac{3R^2}{\ln 2R - 1/2}$	9δ	$-\dfrac{12}{\pi R}$

The rheological predictions of this constitutive equation have also been considered by Hinch and Leal [37]. In essence, the viscosity is shear-thinning, the first normal stress difference is positive while the second normal stress difference is negative, but of a smaller magnitude. The precise values depend on the aspect ratio and the strength of the Brownian motion. The predictions of the constitutive equation (8.20) are considered in Problems 8.2–8.4. In particular, the reduced elongational viscosity of the suspension can be shown to be

$$\frac{N_1 - 3\eta_s\dot\gamma}{\eta_s\dot\gamma} = 2\,(A + 2B + C)\,\phi \approx \frac{\phi R^2}{\ln 2R - 1.5}. \qquad (8.21)$$

Strictly speaking, the dilute assumption means that the volume fraction is low enough, so that a particle can rotate freely without any hindrance from its nearby neighbours. The distance Δ between any two particles must therefore satisfy $l < \Delta$, so that a volume of l^3 contains only one particle, where l is the length of the particle and d is its diameter. The volume fraction therefore satisfies

$$\phi \sim \frac{d^2 l}{\Delta^3}, \quad \phi R^2 < 1.$$

Thus, the reduced elongational viscosity is only $O(1)$ in the dilute limit, not $O(R^2)$ as suggested by the formula (8.21). As the concentration increases, we get subsequently into the semi-dilute regime, the isotropic concentrated solution, and the liquid crystalline solution. The reader is referred to Doi and Edwards [18] for more details. Here, we simply note that the concentration region $1 < \phi R^2 < R$ is called semiconcentrated. Finally, the suspension with $\phi R > 1$ is called concentrated, where the average distance between fibres is less than a fibre diameter, and therefore fibres cannot rotate independently except around their symmetry axes. Any motion of the fibre must necessarily involve a cooperative motion of surrounding fibres.

Problems

Problem 8.1 Use the instantaneous nature of the micromechanics to explain the shear reversal experiments of Gadala-Maria and Acrivos [30]. They found that if shearing is stopped after a steady state has been reached in a Couette device, the torque is reduced to zero instantaneously. If shearing is resumed in the same direction after a period of rest, then the torque would attain its final value that corresponds to the resumed shear rate almost instantaneously. However, if shearing is resumed in the opposite direction, then the torque attains an intermediate value and gradually settles down to a steady state. How would you classify the memory of the liquid, zero, fading or infinite?

Problem 8.2 Show that (8.19) is solved by

$$\mathbf{p} = \frac{\mathbf{Q}}{Q},\tag{8.22}$$

where

$$\dot{\mathbf{Q}} = \mathcal{L} \cdot \mathbf{Q}, \quad \mathcal{L} = \mathbf{L} - \frac{2}{R^2 + 1}\mathbf{D}.\tag{8.23}$$

The effective velocity gradient tensor is $\mathcal{L} = \mathbf{L} - \zeta\mathbf{D}$, where $\zeta = 2/(R^2 + 1)$ is a 'non-affine' parameter.

Problem 8.3 In the start-up of a simple shear flow, where the shear rate is $\dot{\gamma}$, show that

$$Q_1 = Q_{10}\cos\omega t + \sqrt{\frac{2-\zeta}{\zeta}}Q_{20}\sin\omega t,$$

$$Q_2 = Q_{20}\cos\omega t - \sqrt{\frac{\zeta}{2-\zeta}}Q_{10}\sin\omega t,$$

and $Q_3 = Q_{30}$, where $\{Q_{10}, Q_{20}, Q_{30}\}$ are the initial components of $\mathbf{Q}$, and the frequency of the oscillation is

$$\omega = \frac{1}{2}\dot{\gamma}\sqrt{\zeta(2-\zeta)} = \frac{\dot{\gamma}R}{R^2 + 1}.$$

From these results, obtain the particle-contributed stress and the viscometric functions as

The reduced viscosity:

$$\frac{\langle\sigma_{12}\rangle - \eta_s\dot{\gamma}}{\eta_s\dot{\gamma}\phi} = 2Ap_1^2p_2^2 + B\left(p_1^2 + p_2^2\right) + C,\tag{8.24}$$

The reduced first normal stress difference:

$$\frac{N_1}{\eta_s \dot{\gamma} \phi} = 2A p_1 p_2 \left(p_1^2 - p_2^2 \right), \tag{8.25}$$

and the reduced second normal stress difference:

$$\frac{N_2}{\eta_s \dot{\gamma} \phi} = 2 p_1 p_2 \left(A p_2^2 + B \right). \tag{8.26}$$

Thus, the particles tumble along with the flow, with a period of $T = 2\pi (R^2 + 1)/\dot{\gamma} R$, spending most of their time aligned with the flow.

Problem 8.4 In the start-up of an elongational flow with a positive elongational rate $\dot{\gamma}$, show that

$$Q_1 = Q_{10} \exp \left\{ (1 - \varsigma) \dot{\gamma} t \right\},$$
$$Q_2 = Q_{20} \exp \left\{ -\frac{1}{2} (1 - \varsigma) \dot{\gamma} t \right\},$$
$$Q_3 = Q_{30} \exp \left\{ -\frac{1}{2} (1 - \varsigma) \dot{\gamma} t \right\},$$

so that the particle is quickly aligned with the flow in a time scale $O(\dot{\gamma}^{-1})$. At a steady state, show that the reduced elongational viscosity is given by

$$\frac{N_1 - 3\eta_s \dot{\gamma}}{\eta_s \dot{\gamma} \phi} = 2A + 4B + 3C \approx \frac{R^2}{\ln 2R - 1.5}. \tag{8.27}$$

Chapter 9
Dissipative Particle Dynamics (DPD)

A Particle-Based Method

We have discussed constitutive modelling techniques for some simple micro-structure models in previous chapters, including dilute suspensions of dumbbells (as a model for polymer solutions) and spheroids (as a model for suspension of rigid particles). Clearly close form solutions for more complex models may be very limited and are difficult to find. In these cases, a more suitable method is required that can handle both the constitutive modelling and the flow problems. In this chapter, we discuss a particle-based method called the Dissipative Particle Dynamics (DPD) method. The method, originally conceived as a mesoscale technique by Hoogerbrugge and Koelman [38], has its basis in statistical mechanics (Español and Warren [22], Marsh [56]). In the method, the fluid is modelled by a system of particles (called DPD particles) in their Newton 2nd law motions. In the original technique, these DPD particles are regarded as clusters of molecules, undergoing a soft pairwise repulsion, in addition to other dissipative and random forces designed to conserve mass and momentum in the mean. The microstructure of the DPD fluid can be made as complicated as we like: DPD particles can be connected to form strings to model polymer solutions (Kong et al. [44]), to form rigid particles to model suspensions (Boek et al. [8]), to form immiscible droplets to model multiphase fluids (Novik and Coveney [63]), to form vapour phase to model liquid/vapour interaction (Arienti et al. [1]). The DPD system exists in continuous space, rather than on a lattice as the Lattice-Gas Automata (LGA) (e.g., Frisch et al. [28]), and hence it removes some of the isotropy and Galilean invariance problems facing LGA, but it still retains the computational efficiency of the LGA. Another popular numerical method for simulating complex-structure fluids is the Brownian Dynamics Simulation (BDS) (e.g., Fan et al. [25]). In BDS, the bulk flow field kinematics are specified a-priori; then the effects of the fluid microstructure evolution on the flow field are taken into account by coupling the BDS to the kinematics in an iterative manner. Furthermore, BDS conserves particles (mass), but not momentum. In contrast, DPD method conserves both the number of particles and also the total momentum of the system; its transport equations are of the familiar form of mass and momentum conservation. Thus, both the flow kinematics,

© Springer International Publishing AG 2017
N. Phan-Thien and N. Mai-Duy, *Understanding Viscoelasticity*,
Graduate Texts in Physics, DOI 10.1007/978-3-319-62000-8_9

and the stress tensor (constitutive equation) can be found as a part of the solution procedure. In this point of view, DPD can be regarded as a particle-based solver for continuum problems – DPD particles are regarded as fictitious constructs to satisfy conservation laws. We are especially attracted to the DPD method because of the ease and flexibility of its modelling of a complex-structure fluid. The method conserves mass and momentum in the mean, and therefore is not only restricted to mesoscale problems – it is also applicable to problems of arbitrary scales and therefore it may be regarded as a particle-based method for solving continuum flow problems. We review the technique here in details, together with some test problems of interest, to complete our adventure in the constitutive modelling of complex-structure fluids.

9.1   1-D Model

Although one-dimensional models do not make sense in fluid systems, they provide considerable pedagogic insights into the DPD method, and therefore we first consider a 1-D system under the action of a "conservative" force $F_C(r)$, a function of the particle's position r, a dissipative force $F_D = \gamma w_D(r) v$ proportional to the particle's velocity $v = \dot{r}$, with strength $\gamma w_D(r)$, and a random force $F_R = \sigma w_R(r)\theta(t)$, with strength $\sigma w_R(r)$, and $\theta(t)$ is a Gaussian white noise, with zero mean and unit variance:

$$\frac{dr}{dt} = v, \quad m\frac{dv}{dt} = F_c - \gamma w_D v + \sigma w_R \theta(t), \quad r(0) = r_0, \; v(0) = v_0, \qquad (9.1)$$
$$\langle \theta(t) \rangle = 0, \quad \langle \theta(t)\theta(t+\tau) \rangle = \delta(\tau).$$

The zone of influence of the dissipative and random forces may be prescribed by specifying the weighting functions $w_D(r)$ and $w_R(r)$ - they are defined to be dimensionless and are zero outside a certain cutoff radius r_c, for $r \geq r_c$. The angular brackets denote an ensemble average with respect to the distribution function of the quantity concerned. Of course, there is no need to separate the dissipative and random forces strength into a scalar and a configuration-dependent weighting function in the manner indicated – this is done only to conform with existing DPD notation.

Langevin Equation Since the random force has only well-defined statistical properties, the so-called Langevin equation (9.1) is understood as a stochastic differential equation, its complete solution is specified by the joint probability distribution function $f(r, v, t)$ of the process $\{r, v\}$, defined so that $f(r, v, t)\,dr\,dv$ is the probability of finding the process at the states between $\{r, v\}$ and $\{r + dr, v + dv\}$ at time t. Specifying an initial state, as done in (9.1), is equivalent to prescribing an initial delta probability function:

$$f(r, v, 0) = \delta(r - r_0)\,\delta(v - v_0). \qquad (9.2)$$

A specification of $f(r, v, t)$ leads to the phase-space description of the stochastic system (9.1). The velocity space v, or the configuration space r, may be integrated out of $f(r, v, t)$, in which case we have a configuration-space, or a velocity-space description of the stochastic system (9.1). The equation governing the probability is sometimes known as the Fokker–Planck or Smoluchowski, or simply the diffusion equation of the process.

Fluctuation-Dissipation Theorem There are three time scales in the stochastic differential equation (9.1):

1. a fluctuation time scale τ_R of the random force, which is arbitrarily small;
2. an inertial time scale $\tau_I = O\left(m\gamma^{-1}\right)$;
3. and a relaxation time scale $\tau = O\left(\gamma H^{-1}\right)$ where H is the stiffness of the system, $H = O\left(|\partial_r F_c|\right)$.

In a typical physical system with small inertia, we have a natural separation between the time scales: $\tau_R \ll \tau_I \ll \tau$. The random force fluctuations on the time scale τ_I inject kinetic energies into the system and raise its temperature, defined as its Boltzmann temperature $k_B T$,

$$\frac{1}{2}k_B T = \frac{1}{2}m \langle v^2(t) \rangle. \tag{9.3}$$

In order to focus on events on the time scale τ_I we can regard the restoring force F_c as constant, so that it can be absorbed in a re-definition of the system state system (by re-defining r and v appropriately) to obtain (in the new state space),

$$\frac{dv}{dt} = -m^{-1}\gamma w_D v + m^{-1}\sigma w_R \theta(t). \tag{9.4}$$

A formal solution for this is

$$v(t) = \int_0^t e^{m^{-1}\gamma w_D(t'-t)} m^{-1}\sigma w_R \theta(t')\, dt'. \tag{9.5}$$

With this result, the existence of the temperature assumption (9.3) demands that (Problem 9.1)

$$\sigma^2 w_R^2 \gamma^{-1} w_D^{-1} = 2k_B T. \tag{9.6}$$

Equation (9.6) is a consequence of the temperature assumption (9.3), sometimes known as Equi-Partition Principle, one of the many fluctuation-dissipation theorems in Statistical Fluid Mechanics ([47]).

It is customary in the DPD literature to choose,

$$w^D(r) = \left[w^R(r)\right]^2 \text{ and } \gamma = \frac{\sigma^2}{2k_B T}. \tag{9.7}$$

This choice guarantees the thermodynamic temperature as defined by (9.3). Note that the requirement (9.6), or (9.7), implies that the cutoff radius for the random and the dissipative forces must be the same.

Phase-Space Description: Fokker–Planck Equation The phase-space description of (9.1) is given by the Fokker–Planck equation, sometimes known as Liouville equation ([14]):

$$\left(\frac{\partial}{\partial t} + v\frac{\partial}{\partial r}\right) W(r, v, t) = \lim_{\Delta t \to 0} \frac{\partial}{\partial v}\left[\frac{\langle \Delta v \Delta v \rangle}{2\Delta t}\frac{\partial}{\partial v} - \frac{\langle \Delta v \rangle}{\Delta t}\right] W(r, v, t). \quad (9.8)$$

Its derivation is based on the assumption of the Markovian nature of the process (9.1), namely, what is going to happen at any given instant t depends on the current state of the system at time t, not on what has already happened preceding time t ([14]). For the stochastic system (9.1),

$$\Delta v = m^{-1}(-\gamma w_D v + F_c)\Delta t + \int_0^{\Delta t} m^{-1}\sigma w_R \theta(t') dt'. \quad (9.9)$$

Consequently,

$$\langle \Delta v \rangle = m^{-1}(-\gamma w_D v + F_c)\Delta t,$$

$$\langle \Delta v \Delta v \rangle = O(\Delta t^2) + \int_0^{\Delta t}\int_0^{\Delta t} m^{-2}\sigma^2 w_R^2 \langle \theta(t')\theta(t)\rangle dt' dt$$

$$= O(\Delta t^2) + \int_0^{\Delta t} m^{-2}\sigma^2 w_R^2 dt',$$

$$= O(\Delta t^2) + m^{-2}\sigma^2 w_R^2 \Delta t,$$

and the phase-space diffusion equation for the stochastic system (9.1) is

$$\left(\frac{\partial}{\partial t} + v\frac{\partial}{\partial r}\right) W = \frac{\partial}{\partial v}\left[\frac{\sigma^2 w_R^2}{2m^2}\frac{\partial}{\partial v} - m^{-1}(F_c - \gamma w_D v)\right] W, \quad (9.10)$$

subjected to some relevant initial condition, for example (9.2). From the phase space description (9.10), one can derive balance, or conservation equations for the system, but it is not meaningful to do so for this simplistic 1D system.

Configuration Space: Fokker-Planck-Smoluchowski Equation Sometimes, it is more convenient to deal directly with the configuration-space distribution; the velocity-space can be integrated out from (9.10) and the resulting equation is usually called the Fokker–Planck equation in configuration space, or the Smoluchowski equation. The Fokker–Planck equation in configuration space can be shown to be ([14]):

$$\frac{\partial}{\partial t} W(r, t) = \lim_{\Delta t \to 0} \frac{\partial}{\partial r}\left[\left(\frac{\langle \Delta r \Delta r \rangle}{2\Delta t}\frac{\partial}{\partial r} - \frac{\langle \Delta r \rangle}{\Delta t}\right) W(r, t)\right]. \quad (9.11)$$

In (9.11) $\lim_{t\to\infty} \langle \Delta r/\Delta t \rangle$ is called the drift velocity, and $\lim_{t\to\infty} \langle \Delta r \Delta r/2\Delta t \rangle$ is called the diffusivity. In the limit of small inertia, the stochastic system (9.1) becomes

$$\frac{dr}{dt} = \gamma^{-1} w_D^{-1} F_c + \gamma^{-1} w_D^{-1} \sigma w_R \theta(t). \tag{9.12}$$

Problem 9.2 shows how to derive the following Fokker–Planck equation in configuration space:

$$\frac{\partial}{\partial t} W(r,t) = \frac{\partial}{\partial r} \left[\left(\frac{k_B T}{\gamma w_D} \frac{\partial}{\partial r} - \frac{F_c}{\gamma w_D} \right) W(r,t) \right]. \tag{9.13}$$

9.2 DPD Fluid

9.2.1 Langevin Equations

Most of the essential ideas in DPD are contained in the 1-D theory discussed in 9.1. We continue our discussion of DPD in a 3D setting. We define a *DPD fluid* as the ensemble of N particles, call DPD particles, each of mass m_i, $i = 1, \ldots, N$, located at position $\mathbf{r}_i$, with velocity $\mathbf{v}_i$. Furthermore, we assume identical mass $m_i = m$, without much loss of generality (Fig. 9.1). In the original viewpoint, a DPD particle may be thought of as a cluster of fluid molecules; alternatively, it may be regarded as a fictitious construct to solve a flow problem of a complex-structure fluid – we will return to these different viewpoints later. The DPD particles interact with each other in their Newton's second law motions:

$$\frac{d\mathbf{r}_i}{dt} = \mathbf{v}_i, \quad m\frac{d\mathbf{v}_i}{dt} = \sum_j \mathbf{F}_{ij} + \mathbf{F}_e. \tag{9.14}$$

Here, $\mathbf{F}_e$ is an external force on particle i (for example, gravity to simulate the effect of a pressure gradient), and $\mathbf{F}_{ij}$ is the pairwise additive interparticle force by particle j on particle i; this force consists of three parts, a conservative force $\mathbf{F}_{ij}^C$, a dissipative force $\mathbf{F}_{ij}^D$, and a random force $\mathbf{F}_{ij}^R$:

$$\mathbf{F}_{ij} = \mathbf{F}_{ij}^C + \mathbf{F}_{ij}^D + \mathbf{F}_{ij}^R. \tag{9.15}$$

In (9.14), the sum runs over all other particles except i (note, by definition $\mathbf{F}_{ii} = \mathbf{0}$). Outside a certain cutoff radius r_c, the interactions are zero. Here we may allow the cutoff radius to be different for different type of forces. There are some key ideas in the DPD theory, which we will go through in details.

Key Idea 1: DPD is an MD-like Method The first key idea is that DPD looks very much like molecular dynamics simulation in the absence of dissipative and random

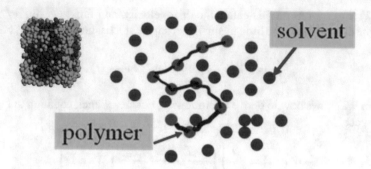

Fig. 9.1 A DPD fluid is made up of DPD particles of various connectivities

forces. In fact, the system (9.14), with $\mathbf{F}_{ij}^D = \mathbf{F}_{ij}^R = \mathbf{0}$, is called the *associate system*. When the conservative forces are chosen appropriately, the associate system is the basis for molecular dynamics (MD) simulation.

Key Idea 2: Conservative Force is Soft Repulsion The second key idea is that the conservative force is chosen to be a soft repulsion. This allows a much larger time step to be employed (easily by a factor of 10), as compared to a much smaller time step, when a standard MD molecular potential, such as Lennard-Jones potential, is used (Keaveny et al. [42]). Here the standard DPD soft conservative force, cutoff outside a critical normalised radius $r_c = 1$ is used:

$$\mathbf{F}_{ij}^C = -\frac{\partial \varphi\left(r_{ij}\right)}{\partial \mathbf{r}_i} = \begin{cases} a_{ij}\left(1 - r_{ij}\right)\mathbf{e}_{ij}, & r_{ij} < 1, \\ 0, & r_{ij} \geq 1, \end{cases} \tag{9.16}$$

where a_{ij} is a coefficient of interaction, r_{ij} is the distance between particles i and j, and $\mathbf{e}_{ij}$ is the unit vector directed from particles j to i:

$$\mathbf{r}_{ij} = \mathbf{r}_i - \mathbf{r}_j, \ r_{ij} = \left|\mathbf{r}_{ij}\right|, \ \mathbf{e}_{ij} = \mathbf{r}_{ij}/r_{ij}. \tag{9.17}$$

The DPD particles can be of a complex type, for examples, interconnecting particles to form (polymer) chains, in which case there may be connector forces pertaining to the nature of the connectors, or forces arisen due to some rigidity constraints to form a rigid body. We may consider lumping these connector forces in the conservative forces $\mathbf{F}_{ij}^C$ as a matter of convenience. The conservative forces (together with any connector forces) determine the rheology of the system.

Key 3: Dissipative Force is Centre-to-Centre The dissipative force slows down the particle, extracts parts of the kinetic energy injected by the random force. It is directed from centre-to-centre and is given by

$$\mathbf{F}_{ij}^D = -\gamma w_D\left(r_{ij}\right)\left(\mathbf{e}_{ij} \cdot \mathbf{v}_{ij}\right)\mathbf{e}_{ij}, \ \mathbf{v}_{ij} = \mathbf{v}_i - \mathbf{v}_j, \tag{9.18}$$

Its strength is controlled by the parameter γ, and its zone of influence is governed by the weighting function $w_D\left(r_{ij}\right)$.

Key Idea 4: Random Force is Centre-to-Centre The random force compensates for the loss of kinetic energy from dissipation; it is directed from centre-to-centre, its strength and domain of influenced are governed by the parameter σ, and the weighting function $w_R\left(r_{ij}\right)$:

$$\mathbf{F}_{ij}^{R} = \sigma w_R\left(r_{ij}\right)\theta_{ij}\hat{\mathbf{r}}_{ij}, \tag{9.19}$$

where $\theta_{ij}\left(t\right) = \theta_{ji}\left(t\right)$ is a Gaussian white noise with the properties

$$\langle\theta_{ij}\left(t\right)\rangle = 0,$$
$$\langle\theta_{ik}\left(t\right)\theta_{jl}\left(t'\right)\rangle = \left(\delta_{ij}\delta_{kl} + \delta_{il}\delta_{jk}\right)\delta\left(t - t'\right),\ i \neq k,\ j \neq l. \tag{9.20}$$

Note that the random force is central, and it is anticipated that the strengths and the weighting functions of the random and the dissipative forces cannot be prescribed independently: they act together in a precise manner to keep the Boltzmann temperature of the system constant. Furthermore, because of the central nature of the random forces,

$$\sum_{i,j\neq i}\mathbf{F}_{ij}^{R} = 0. \tag{9.21}$$

Key Idea 5: Fluctuation-Dissipation Theorem Links the Dissipative and Random Forces Again, there are three time scales in the system: the fluctuation time scale of the random force, τ_R, which is arbitrarily small, the inertial time scale, $\tau_I = O\left(m/\gamma^{-1}w_D^{-1}\right)$, and the relaxation time scale, $\tau = O\left(\gamma w_D H^{-1}\right)$, where $H = O\left(|\partial_r\mathbf{F}^C|\right)$ is the stiffness of the system. Since the particles' mass is typically small for a physical system, we have the natural separation of the time scales, $\tau_R \ll \tau_I \ll \tau$. The fluctuations of the random force inject kinetic energies into and heat up the system, which is then taken away and cooled down by the dissipative force. This balance allows a thermodynamic temperature to be defined. In order to see this detailed balance, one needs only to consider the motion at time scale τ_I. In this time scale, the conservative and external forces can be regarded as constant, and can be scaled out of the equations of motion (by re-define the state variables, in the same manner as the 1D model discussed previously),

$$\frac{d\mathbf{v}_i}{dt} + \sum_j m^{-1}\gamma w_D\mathbf{e}_{ij}\mathbf{e}_{ij}\cdot\mathbf{v}_{ij} = \sum_j m^{-1}\sigma w_R\theta_{ij}\mathbf{e}_{ij}. \tag{9.22}$$

Collecting the velocity vectors into a state variable equation (9.22) then results into a linear system:

$$\dot{\mathbf{v}}_\alpha + \sum_\beta m^{-1}\mathbf{A}_{\alpha\beta} \cdot \mathbf{v}_\beta = m^{-1}\mathbf{B}_\alpha, \tag{9.23}$$

$$\mathbf{A}_{\alpha\beta} = \delta_{\alpha\beta} \sum_j \gamma w_D \mathbf{e}_{\alpha j}\mathbf{e}_{\alpha j} - \gamma w_D \mathbf{e}_{\alpha\beta}\mathbf{e}_{\alpha\beta},$$

$$\mathbf{B}_\alpha = \sum_j \sigma w_R \theta_{\alpha j}\mathbf{e}_{\alpha j}, \quad \alpha, \beta = 1, \ldots, N.$$

Here, we reserve Greek indices to refer to state vectors and matrices; the Roman indices are used to enumerate the particles as usual – summations on them will be indicated explicitly to avoid confusion. It is to be noted that $\mathbf{A}$ is an $[N, N]$ matrix, with second-order tensors as elements. It is also a symmetric matrix of zero row and column sums; it has zero as one eigenvalue, with corresponding unit eigen-vector. Likewise, $\mathbf{B}$ is an $[N, 1]$ matrix, with three-dimensional vectors as elements. Note also that, because (9.21), the centre of gravity of the system, $\mathbf{r}_C = \frac{1}{N}\sum_i \mathbf{r}_i$ moves deterministically.

A formal solution of the linear system (9.23) can be derived, noting the communicative property of $\mathbf{A}$ and its exponent function $\exp(\mathbf{A}t)$,

$$\mathbf{v}_\alpha = \sum_\beta \int_0^t \exp\left[m^{-1}\mathbf{A}_{\alpha\beta}\left(t' - t\right)\right] \cdot m^{-1}\mathbf{B}_\beta dt'. \tag{9.24}$$

Thus, the specific kinetic energy is

$$\langle \mathbf{v}_\alpha \mathbf{v}_\nu \rangle = m^{-2} \sum_{\beta,\gamma} \int_0^t \int_0^t \exp\left[m^{-1}\mathbf{A}_{\alpha\beta}\left(t' - t\right)\right] \cdot \langle \mathbf{B}_\beta\left(t'\right)\mathbf{B}_\gamma\left(t''\right)\rangle \cdot$$

$$\exp\left[m^{-1}\mathbf{A}_{\nu\gamma}\left(t'' - t\right)\right] dt'dt''. \tag{9.25}$$

But, from the property of the random force (9.20),

$$\langle \mathbf{B}_\beta\left(t'\right)\mathbf{B}_\gamma\left(t''\right)\rangle = \sum_j \sum_k \sigma_{\beta j}\sigma_{\gamma k}\mathbf{e}_{\beta j}\mathbf{e}_{\gamma k}\left(\delta_{\beta\gamma}\delta_{jk} + \delta_{\beta k}\delta_{\gamma j}\right)\delta\left(t' - t''\right),$$

where we have used the short-hand notation σ_{ij} to denote the scalar $\sigma w_R\left(r_{ij}\right)$. This, when used in (9.25), yields

$$\langle \mathbf{v}_\alpha \mathbf{v}_\nu \rangle = m^{-2} \sum_{\beta,\gamma} \int_0^t \exp\left[m^{-1}\mathbf{A}_{\alpha\beta}\left(t' - t\right)\right] \cdot \tag{9.26}$$

$$\left\{\delta_{\beta\gamma} \sum_k \sigma_{\beta k}\sigma_{\gamma k}\mathbf{e}_{\beta k}\mathbf{e}_{\gamma k} + \sigma_{\beta\gamma}\sigma_{\gamma\beta}\mathbf{e}_{\beta\gamma}\mathbf{e}_{\gamma\beta}\right\} \cdot \exp\left[m^{-1}\mathbf{A}_{\nu\gamma}\left(t' - t\right)\right] dt'.$$

If the existence of a thermodynamic temperature is assumed, then the left hand is simply $m^{-1}k_B T \mathbf{I}_{\alpha\nu}$, where $k_B T$ is the Boltzmann temperature of the system, and $\mathbf{I}_{\alpha\nu}$ is the $[N, N]$ unit tensor (having diagonal elements as second-order unit tensors, and zero entries elsewhere). Thus

$$k_B T \mathbf{A}_{\eta\nu} = \sum_{\beta,\gamma} \int_0^t m^{-1} \mathbf{A}_{\eta\alpha} \exp\left[m^{-1} \mathbf{A}_{\alpha\beta} \left(t' - t\right)\right] \cdot \tag{9.27}$$

$$\left\{ \delta_{\beta\gamma} \sum_k \sigma_{\beta k} \sigma_{\gamma k} \mathbf{e}_{\beta k} \mathbf{e}_{\gamma k} + \sigma_{\beta\gamma} \sigma_{\gamma\beta} \mathbf{e}_{\beta\gamma} \mathbf{e}_{\gamma\beta} \right\} \cdot \exp\left[m^{-1} \mathbf{A}_{\nu\gamma} \left(t' - t\right)\right] dt'.$$

Similar to 1-D case, an integration by parts is done, recognising $\mathbf{A} \exp(\mathbf{A}t)$ as the time derivative of $\exp(\mathbf{A}t)$ and take the limit of large time (compared to the inertial time scale),

$$k_B T \mathbf{A}_{\eta\nu} = \lim_{t \to \infty} \sum_{\beta,\gamma} \exp\left[m^{-1} \mathbf{A}_{\eta\beta} \left(t' - t\right)\right] \cdot$$

$$\left\{ \delta_{\beta\gamma} \sum_k \sigma_{\beta k} \sigma_{\gamma k} \mathbf{e}_{\beta k} \mathbf{e}_{\gamma k} + \sigma_{\beta\gamma} \sigma_{\gamma\beta} \mathbf{e}_{\beta\gamma} \mathbf{e}_{\gamma\beta} \right\} \cdot \exp\left[m^{-1} \mathbf{A}_{\nu\gamma} \left(t' - t\right)\right] \Bigg]_{t'=0}^{t'=t}$$

$$- \sum_{\beta,\gamma} \int_0^t dt' \exp\left[m^{-1} \mathbf{A}_{\eta\beta} \left(t' - t\right)\right] \cdot$$

$$\left\{ \delta_{\beta\gamma} \sum_k \sigma_{\beta k} \sigma_{\gamma k} \mathbf{e}_{\beta k} \mathbf{e}_{\gamma k} + \sigma_{\beta\gamma} \sigma_{\gamma\beta} \mathbf{e}_{\beta\gamma} \mathbf{e}_{\gamma\beta} \right\} \cdot \exp\left[m^{-1} \mathbf{A}_{\delta\gamma} \left(t' - t\right)\right] \mathbf{A}_{\nu\delta}.$$

The last integral on the right side of the preceding equation is recognised as the right side of (9.27), and the limit at large time (compared to the inertial time scale) can be taken to yield

$$2k_B T \mathbf{A}_{\eta\nu} = \mathbf{I}_{\eta\beta} \cdot \left\{ \delta_{\beta\gamma} \sum_k \sigma_{\beta k} \sigma_{\gamma k} \mathbf{e}_{\beta k} \mathbf{e}_{\gamma k} + \sigma_{\beta\gamma} \sigma_{\gamma\beta} \mathbf{e}_{\beta\gamma} \mathbf{e}_{\gamma\beta} \right\} \cdot \mathbf{I}_{\nu\gamma},$$

or

$$2k_B T \left(\delta_{\eta\nu} \sum_j \gamma w_D \mathbf{e}_{\eta j} \mathbf{e}_{\eta j} - \gamma w_D \mathbf{e}_{\eta\nu} \mathbf{e}_{\eta\nu} \right) = \delta_{\eta\nu} \sum_k \sigma_{\eta k} \sigma_{\eta k} \mathbf{e}_{\eta k} \mathbf{e}_{\eta k}$$

$$- \sigma_{\eta\nu} \sigma_{\eta\nu} \mathbf{e}_{\eta\nu} \mathbf{e}_{\eta\nu}. \tag{9.28}$$

Equation (9.28) is the consequence of the existence of the temperature, and may be regarded as the fluctuation-dissipation theorem for our DPD system. From the form of $\mathbf{A}$, it is sufficient that

$$2k_B T \gamma w_D (r) = \sigma^2 w_R^2 (r) \qquad (9.29)$$

for the equality (9.28) to hold true – and this is the final result for the existence of temperature. It is customary in DPD to take

$$w_D (r) = w_R^2(r) \text{ and } \gamma = \frac{\sigma^2}{2k_B T} \qquad (9.30)$$

to ensure a constant Boltzmann temperature $k_B T$, or strictly speaking, a constant specific kinetic energy of the system. The requirement (9.29), or (9.30) implies that the cutoff radii for the random and the dissipative forces must be the same.

As far as thermal energy is concerned, the random two-particle force, $\mathbf{F}_{ij}^R$, representing the results of thermal motion of all molecules contained in particles i and j, "heats up" the system. The dissipative force, $\mathbf{F}_{ij}^D$, reduces the relative velocity of two particles and removes kinetic energy from their mass centre to "cool down" the system. When the detailed balance, (9.29), is satisfied, the system temperature will approach the given value. The dissipative and random forces act like a thermostat in the conventional molecular dynamics (MD) system.

Clearly, the addition of the dissipative and random forces to a conservative system may appear to be artificial, but the versatility of DPD lies in its ability to satisfy conservation laws in the mean (to be shown later). It is possible to think of the DPD system as a coarse-grained model of a physical model. Therefore we could construct models of complex-structure fluids by endowing the simple DPD particles with features in a manner similar to modelling, for example, polymeric solution by a suspension of Rouse chains.

9.2.2 Phase-Space Description: Fokker–Planck Equation

The complete solution of the stochastic system (9.22) would require a specification of the probability density distribution $f (\mathbf{r}_1, \mathbf{v}_1, \ldots, \mathbf{r}_N, \mathbf{v}_N, t)$ of the state space $\chi = \{\mathbf{r}_i, \mathbf{v}_i, t\}, i = 1, \ldots, N$; $f (\chi, t) \, d\chi$ is the probability of finding the process at the state between χ and $\chi + d\chi$ at time t. With the assumption of the Markovian nature of the process, Chandrasekhar [14] showed that this distribution function obeys the following Fokker–Planck (or Liouville) equation,

$$\frac{\partial f}{\partial t} + \sum_i \frac{\partial}{\partial \mathbf{r}_i} \cdot (\mathbf{v}_i f) + \sum_i \frac{\partial}{\partial \mathbf{v}_i} \cdot \left(\left\langle \frac{\Delta \mathbf{v}_i}{\Delta t} \right\rangle f \right) \qquad (9.31)$$

$$= \sum_{i,j} \frac{\partial}{\partial \mathbf{v}_i} \cdot \left(\left\langle \frac{\Delta \mathbf{v}_i \Delta \mathbf{v}_j}{2\Delta t} \right\rangle \cdot \frac{\partial f}{\partial \mathbf{v}_j} \right),$$

where the limit $\Delta t \to 0$ is implied in the ensemble averages.

From the Langevin equation (9.14), the drift and the diffusion terms can be found (Problem 9.5), and the Fokker–Planck equation for the process is

$$\frac{\partial f}{\partial t} + \sum_i \frac{\partial}{\partial \mathbf{r}_i} \cdot (\mathbf{v}_i f) + \sum_{i,j} \frac{\partial}{\partial \mathbf{p}_i} \cdot (\mathbf{F}_{ij}^C f) = \gamma \sum_{i,j} w_{ij}^D \mathbf{e}_{ij} \frac{\partial}{\partial \mathbf{p}_i} \cdot (\mathbf{e}_{ij} \cdot \mathbf{v}_{ij} f)$$

$$+ \gamma k_B T \sum_{i,j} w_{ij}^D \mathbf{e}_{ij} \cdot \frac{\partial}{\partial \mathbf{p}_i} \cdot \left(\mathbf{e}_{ij} \cdot \left(\frac{\partial f}{\partial \mathbf{p}_i} - \frac{\partial f}{\partial \mathbf{p}_j} \right) \right),\tag{9.32}$$

where $\mathbf{p}_i = m\mathbf{v}_i$ is the linear momentum of particle i.

9.2.3 Distribution Functions

Consider a function $Q = Q(\mathbf{X})$ of the state $\mathbf{X} = \{\mathbf{r}_1, \mathbf{v}_1, \ldots, \mathbf{r}_N, \mathbf{v}_N\}$. Its ensemble average is defined to be

$$\langle Q \rangle = \int \ldots \int Q f(\mathbf{r}_1, \mathbf{v}_1, \ldots, \mathbf{r}_N, \mathbf{v}_N) \, d\mathbf{r}_1 d\mathbf{v}_1 \ldots d\mathbf{r}_N d\mathbf{v}_N \tag{9.33}$$

$$= \int Q(\mathbf{X}) f(\mathbf{X}, t) \, d\mathbf{X}.$$

Sometimes, a full statistical description $f(\mathbf{X}, t)$ is not required, since the quantities of interest only involve a subset of the state variable, the rest of the state variable can be integrated out. In particular, we are interested in the one-particle, $f_1(\chi, t) = f_1(\mathbf{r}, \mathbf{v}, t)$, $\chi = \{\mathbf{r}, \mathbf{v}\}$, or the two-particles distribution function, $f_2(\chi, \chi', t) = f_2(\mathbf{r}, \mathbf{v}, \mathbf{r}', \mathbf{v}', t)$, $\chi = \{\mathbf{r}, \mathbf{v}\}$, $\chi' = \{\mathbf{r}', \mathbf{v}'\}$, defined as:

$$f_1(\chi, t) = \left\langle \sum_i \delta(\chi - \chi_i) \right\rangle = \left\langle \sum_i \delta(\mathbf{r} - \mathbf{r}_i) \delta(\mathbf{v} - \mathbf{v}_i) \right\rangle, \tag{9.34}$$

$$f_2(\chi, \chi', t) = \left\langle \sum_{i,j \neq i} \delta(\chi - \chi_i) \delta(\chi' - \chi_j) \right\rangle \tag{9.35}$$

$$= \left\langle \sum_{i,j \neq i} \delta(\mathbf{r} - \mathbf{r}_i) \delta(\mathbf{v} - \mathbf{v}_i) \delta(\mathbf{r}' - \mathbf{r}_j) \delta(\mathbf{v}' - \mathbf{v}_j) \right\rangle.$$

The one-particle distribution function, $f_1(\chi, t)$, is the probability distribution function of finding a particle at the state $\chi = \{\mathbf{r}, \mathbf{v}\}$, at time t. The two-particles distribution function, $f_2(\chi, \chi', t)$, is the probability distribution function of finding two particles simultaneously at state $\chi = \{\mathbf{r}, \mathbf{v}\}$, of position $\mathbf{r}$ and velocity $\mathbf{v}$, and at state

$\chi' = \{\mathbf{r}', \mathbf{v}'\}$, of position $\mathbf{r}'$ and velocity $\mathbf{v}'$ at time t. The velocities can be integrated out of (9.34) and (9.35), in which case one has the configuration-dependent one-particle distribution function

$$\bar{f}_1 (\mathbf{r}, t) = \int f_1 (\mathbf{r}, \mathbf{v}, t) \, d\mathbf{v}, \tag{9.36}$$

and the configuration-dependent two-particle, distribution function

$$\bar{f}_2 (\mathbf{r}, \mathbf{r}', t) = \int d\mathbf{v} \int d\mathbf{v}' f_2 (\mathbf{r}, \mathbf{r}', \mathbf{v}, \mathbf{v}', t). \tag{9.37}$$

9.2.4   Equation of Change

The equation of change for any dynamical process $Q(\mathbf{X})$ can be derived by pre-multiplying Q with (9.32), and integrate the resulting equation, using Green's theorem whenever necessary. First, we re-write (9.32) as

$$\frac{\partial f}{\partial t} + \mathcal{L} f = 0, \tag{9.38}$$

where the operator $\mathcal{L}$ is defined by

$$\mathcal{L} f = \sum_i \frac{\partial}{\partial \mathbf{r}_i} \cdot (\mathbf{v}_i f) + \sum_{i,j} \frac{\partial}{\partial \mathbf{p}_i} \cdot \left((\mathbf{F}_{ij}^C + \mathbf{F}_{ij}^D) \, f \right)$$
$$- \gamma k_B T \sum_{i,j} w_{ij}^D \mathbf{e}_{ij} \cdot \frac{\partial}{\partial \mathbf{p}_i} \left(\mathbf{e}_{ij} \cdot \left(\frac{\partial}{\partial \mathbf{p}_i} - \frac{\partial}{\partial \mathbf{p}_j} \right) f \right).$$

Next, multiply Q with (9.38) and integrate over the state space to obtain

$$\frac{\partial}{\partial t} \langle Q \rangle = - \int Q (\mathbf{X}) \mathcal{L} f (\mathbf{X}, t) \, d\mathbf{X} = \int f (\mathbf{X}, t) \mathcal{L}' Q \, d\mathbf{X} = \langle \mathcal{L}' Q \rangle,$$

where the operator $\mathcal{L}'$ is defined from applications of Green's theorem (the surface integrals resulted from these applications vanish, because the probability distribution function is assumed to vanish on these surfaces). In full,

$$\frac{\partial}{\partial t} \langle Q \rangle = \langle \mathcal{L}' Q \rangle = \sum_i \left\langle \mathbf{v}_i \cdot \frac{\partial}{\partial \mathbf{r}_i} Q \right\rangle + \sum_{i,j} \left\langle \mathbf{F}_{ij}^C \cdot \frac{\partial}{\partial \mathbf{p}_i} Q \right\rangle \tag{9.39}$$
$$- \gamma \sum_{i,j} \left\langle w_{ij}^D \mathbf{v}_{ij} \cdot \mathbf{e}_{ij} \mathbf{e}_{ij} \cdot \frac{\partial}{\partial \mathbf{p}_i} Q \right\rangle$$

$$+ \gamma k_B T \sum_{i,j} \left\langle w_{ij}^D \mathbf{e}_{ij} \cdot \frac{\partial}{\partial \mathbf{p}_i} \left(\mathbf{e}_{ij} \cdot \left(\frac{\partial Q}{\partial \mathbf{p}_i} - \frac{\partial Q}{\partial \mathbf{p}_j} \right) \right) \right\rangle.$$

9.2.5 Conservation of Mass

If we take, as our dynamical variable

$$Q = \sum_i \delta (\mathbf{r} - \mathbf{r}_i), \tag{9.40}$$

then the element of this sum is zero, unless a DPD particle is at the position $\mathbf{r}$. Thus, the average of this variable is the number density, which is precisely the one-particle configuration-dependent distribution function:

$$n (\mathbf{r}, t) = \left\langle \sum_i \delta (\mathbf{r} - \mathbf{r}_i) \right\rangle = \int f_1 (\mathbf{r}, \mathbf{v}, t) \, d\mathbf{v} = \bar{f}_1 (\mathbf{r}, t). \tag{9.41}$$

Now, if we take

$$Q = \sum_i m\delta (\mathbf{r} - \mathbf{r}_i), \tag{9.42}$$

then its average is the fluid density:

$$\rho (\mathbf{r}, t) = \left\langle \sum_i m\delta (\mathbf{r} - \mathbf{r}_i) \right\rangle = \int mf_1 (\mathbf{r}, \mathbf{v}, t) \, d\mathbf{v} = mn (\mathbf{r}, t). \tag{9.43}$$

Now, noting that $\partial_{\mathbf{r}_i} f (\mathbf{r} - \mathbf{r}_i) = -\partial_{\mathbf{r}} f (\mathbf{r} - \mathbf{r}_i)$, we find, for our chosen dynamical variable (9.42),

$$\langle \mathcal{L}' Q \rangle = \sum_{i,j} \left\langle m\mathbf{v}_j \cdot \frac{\partial}{\partial \mathbf{r}_j} \delta (\mathbf{r} - \mathbf{r}_i) \right\rangle = -\frac{\partial}{\partial \mathbf{r}} \cdot \left\langle \sum_j m\mathbf{v}_j \delta (\mathbf{r} - \mathbf{r}_j) \right\rangle = -\frac{\partial}{\partial \mathbf{r}} \cdot (\rho \mathbf{u}).$$

The term

$$\rho (\mathbf{r}, t) \mathbf{u} (\mathbf{r}, t) = \left\langle \sum_j m\mathbf{v}_j \delta (\mathbf{r} - \mathbf{r}_j) \right\rangle = \int m\mathbf{v} f_1 (\mathbf{r}, \mathbf{v}, t) \, d\mathbf{v} \tag{9.44}$$

can be identified as the linear momentum density. We thus obtain the usual equation for the conservation of mass:

$$\frac{\partial}{\partial t}\rho\left(\mathbf{r}, t\right) + \nabla \cdot \left(\rho\left(\mathbf{r}, t\right)\mathbf{u}\left(\mathbf{r}, t\right)\right) = 0, \qquad \nabla = \partial/\partial\mathbf{r}. \tag{9.45}$$

This is merely a statement of conservation of the mass probability (note that the probability distribution is normalised to one, and (9.45) is simply a Reynolds transport theorem).

9.2.6 Conservation of Linear Momentum

Next, taking our dynamical variable Q the linear momentum (9.44),

$$Q = \sum_j m\mathbf{v}_j\delta\left(\mathbf{r} - \mathbf{r}_j\right),$$

and note that

$$\mathbf{v}_i \cdot \frac{\partial}{\partial\mathbf{r}_i}Q = \mathbf{v}_i \cdot \frac{\partial}{\partial\mathbf{r}_i}\sum_\alpha m\mathbf{v}_\alpha\delta\left(\mathbf{r} - \mathbf{r}_\alpha\right) = -\nabla \cdot \left(m\mathbf{v}_i\mathbf{v}_i\delta\left(\mathbf{r} - \mathbf{r}_i\right)\right),$$

$$\frac{\partial}{\partial\mathbf{p}_i}Q = \frac{\partial}{\partial\mathbf{p}_i}\sum_\alpha m\mathbf{v}_\alpha\delta\left(\mathbf{r} - \mathbf{r}_\alpha\right) = \sum_\alpha \mathbf{I}_{\alpha i}\delta\left(\mathbf{r} - \mathbf{r}_i\right),$$

and thus,

$$\frac{\partial}{\partial t}\langle Q\rangle = \langle\mathcal{L}'Q\rangle = -\nabla \cdot \sum_i \langle m\mathbf{v}_i\mathbf{v}_i\delta\left(\mathbf{r} - \mathbf{r}_i\right)\rangle + \sum_{i,j} \langle\mathbf{F}_{ij}^C\delta\left(\mathbf{r} - \mathbf{r}_i\right)\rangle$$
$$+ \sum_{i,j} \langle\mathbf{F}_{ij}^D\delta\left(\mathbf{r} - \mathbf{r}_i\right)\rangle. \tag{9.46}$$

There are three terms on the right side of (9.46), each leads to a stress tensor, after recasting them into suitable divergence forms (the first term is already in a divergence form). Although the second and the third terms are similar in form, the conservative force in second term involves only the configuration whereas the dissipative force in the third term involves both the configuration and the velocity; different treatments will be necessary.

Kinetic Pressure Tensor Although the first term on the right of (9.46) is already in a divergence form, it can be simplified further. We first define the peculiar velocity as

$$\mathbf{V}_i = \mathbf{v}_i - \mathbf{u} \tag{9.47}$$

i.e., the velocity fluctuation of particle i with respect to the mean field velocity defined by (9.44). The first term on the right of (9.46) is then re-written as

$$-\nabla \cdot \sum_i \langle m\mathbf{v}_i\mathbf{v}_i\delta\left(\mathbf{r}-\mathbf{r}_i\right)\rangle = -\nabla \cdot \sum_i \langle m\delta\left(\mathbf{r}-\mathbf{r}_i\right)\left(\mathbf{u}\mathbf{u}+\mathbf{V}_i\mathbf{V}_i\right)\rangle$$

$$= -\nabla \cdot \left(\rho\mathbf{u}\mathbf{u}\right) - \nabla \cdot \mathbf{P}_K, \qquad (9.48)$$

where the terms linear in the fluctuations are averaged to zero, because of their definitions, and the kinetic pressure tensor is defined as

$$\mathbf{P}_K\left(\mathbf{r}, t\right) = \sum_i \langle m\mathbf{V}_i\mathbf{V}_i\delta\left(\mathbf{r}-\mathbf{r}_i\right)\rangle = \int m\left(\mathbf{v}-\mathbf{u}\right)\left(\mathbf{v}-\mathbf{u}\right)f_1\left(\mathbf{r},\mathbf{v},t\right)d\mathbf{v}. \quad (9.49)$$

Stress Tensors from Interaction Forces The second and third terms on the right of (9.46) can be written as

$$\sum_{i,j} \langle\left(\mathbf{F}_{ij}^C+\mathbf{F}_{ij}^D\right)\delta\left(\mathbf{r}-\mathbf{r}_i\right)\rangle = \frac{1}{2}\sum_{i,j} \langle\left(\mathbf{F}_{ij}^C+\mathbf{F}_{ij}^D\right)\left(\delta\left(\mathbf{r}-\mathbf{r}_i\right)-\delta\left(\mathbf{r}-\mathbf{r}_j\right)\right)\rangle,$$

$$(9.50)$$

where we have used the anti-symmetric nature of the interaction forces between particles i and j. Irving and Kirkwood [40] showed how to recast the right side of (9.50) into a divergence form and we summarise their method here.

First, by expressing

$$\mathbf{r}-\mathbf{r}_i = \mathbf{r}-\mathbf{r}_j-\mathbf{r}_{ij},$$

and expanding $\delta\left(\mathbf{r}-\mathbf{r}_i\right)$ in a Taylor's series expansion about $\mathbf{r}-\mathbf{r}_j$

$$\delta\left(\mathbf{r}-\mathbf{r}_i\right) = \delta\left(\mathbf{r}-\mathbf{r}_j\right) - \nabla \cdot \left[\mathbf{r}_{ij}\left(1-\frac{1}{2}\mathbf{r}_{ij}\cdot\nabla+\cdots\right)\delta\left(\mathbf{r}-\mathbf{r}_j\right)\right], \quad (9.51)$$

it is seen that, from (9.50) and (9.51),

$$\sum_{i,j} \langle\left(\mathbf{F}_{ij}^C+\mathbf{F}_{ij}^D\right)\delta\left(\mathbf{r}-\mathbf{r}_i\right)\rangle = -\frac{1}{2}\nabla \cdot \sum_{i,j}\left(\mathbf{F}_{ij}^C+\mathbf{F}_{ij}^D\right)$$

$$\mathbf{r}_{ij}\left(1-\frac{1}{2}\mathbf{r}_{ij}\cdot\nabla+\cdots\right)\delta\left(\mathbf{r}-\mathbf{r}_j\right)$$

$$= \nabla \cdot \mathbf{S} = \nabla \cdot \left(\mathbf{S}_C+\mathbf{S}_D\right), \qquad (9.52)$$

where the particle-interaction stress is given by

$$\mathbf{S}\left(\mathbf{r}, t\right) = -\frac{1}{2}\sum_{i,j} \left\langle\left(\mathbf{F}_{ij}^C+\mathbf{F}_{ij}^D\right)\mathbf{r}_{ij}\left(1-\frac{1}{2}\mathbf{r}_{ij}\cdot\nabla+\cdots\right)\delta\left(\mathbf{r}-\mathbf{r}_j\right)\right\rangle$$

$$= \mathbf{S}_C\left(\mathbf{r}, t\right) + \mathbf{S}_D\left(\mathbf{r}, t\right). \qquad (9.53)$$

Here, the stress contributed from the conservative forces is defined as

$$
\mathbf{S}_C\left(\mathbf{r}, t\right) = -\frac{1}{2} \sum_{i,j} \left\langle \mathbf{F}_{ij}^C \mathbf{r}_{ij} \left(1 - \frac{1}{2}\mathbf{r}_{ij} \cdot \nabla + \cdots\right) \delta\left(\mathbf{r} - \mathbf{r}_j\right)\right\rangle, \tag{9.54}
$$

and the stress from the dissipative forces is given by

$$
\mathbf{S}_D\left(\mathbf{r}, t\right) = -\frac{1}{2} \sum_{i,j} \left\langle \mathbf{F}_{ij}^D \mathbf{r}_{ij} \left(1 - \frac{1}{2}\mathbf{r}_{ij} \cdot \nabla + \cdots\right) \delta\left(\mathbf{r} - \mathbf{r}_j\right)\right\rangle. \tag{9.55}
$$

Thus, finally the conservation of linear momentum takes the usual familiar form

$$
\frac{\partial}{\partial t}\left(\rho \mathbf{u}\right) + \nabla \cdot \left(\rho \mathbf{u}\mathbf{u}\right) = \nabla \cdot \mathbf{T}, \tag{9.56}
$$

where the total stress is

$$
\mathbf{T}\left(\mathbf{r}, t\right) = -\mathbf{P}_K\left(\mathbf{r}, t\right) + \mathbf{S}_C\left(\mathbf{r}, t\right) + \mathbf{S}_D\left(\mathbf{r}, t\right). \tag{9.57}
$$

Key Idea 6: DPD Method Conserves Mass and Linear Momentum Herein lies the power of DPD method: from our seemingly artificial construct (9.14) that bears a great deal of resemblance to MD, it has been shown that its mean quantities satisfy both the conservation of mass and momentum, with the stress derived directly from the microstructure through (9.49) and (9.53). Conversely, flow problems for a complex-structure fluid may be solved by this particle-based method; the stress (i.e., constitutive law) needs not be specified a-priori, but can be found by a post-processing application.

Of the total stress (9.57), the kinetic pressure has been recast into an average with respect to the one-particle distribution function (9.49) – likewise, the particle-interaction stress can be recast into an average with respect to the two-particle distribution function, in the manner described by Irving and Kirkwood [40], and we outline the development here below.

Stress from Conservative Forces We first re-write the conservative-force-contributed stress (9.54) as

$$
\mathbf{S}_C\left(\mathbf{r}, t\right) = -\frac{1}{2} \sum_{i,j} \left\langle \int \delta\left(\mathbf{R} - \mathbf{r}_{ij}\right) \mathbf{F}^C \mathbf{R} \left(1 - \frac{1}{2}\mathbf{R} \cdot \nabla + \cdots\right) \delta\left(\mathbf{r} - \mathbf{r}_j\right) d\mathbf{R}\right\rangle, \tag{9.58}
$$

where any dependence on $\mathbf{r}_{ij}$ of $\mathbf{F}^C$ has been replaced by the dependence on $\mathbf{R}$, because the pre-multiplied delta function $\delta\left(\mathbf{R} - \mathbf{r}_{ij}\right)$ in the integration shifts to role of $\mathbf{r}_{ij}$ to $\mathbf{R}$. Next the sum can be taken inside the integral:

$$\mathbf{S}_C\left(\mathbf{r}, t\right) = -\frac{1}{2} \int d\mathbf{R} \mathbf{F}^C \mathbf{R} \left(1 - \frac{1}{2}\mathbf{R} \cdot \nabla + \cdots\right) \sum_{i,j} \langle \delta\left(\mathbf{R} - \mathbf{r}_{ij}\right) \delta\left(\mathbf{r} - \mathbf{r}_j\right)\rangle.$$

Furthermore, the product $\delta\left(\mathbf{R} - \mathbf{r}_{ij}\right) \delta\left(\mathbf{r} - \mathbf{r}_j\right)$ is recognised as equivalent to $\delta\left(\mathbf{R} + \mathbf{r} - \mathbf{r}_i\right) \delta\left(\mathbf{r} - \mathbf{r}_j\right)$, because of the property of delta functions, resulting in

$$\mathbf{S}_C\left(\mathbf{r}, t\right) = -\frac{1}{2} \int d\mathbf{R} \mathbf{F}^C \mathbf{R} \left\{1 - \frac{1}{2}\mathbf{R} \cdot \nabla + \cdots\right\} \sum_{i,j} \langle \delta\left(\mathbf{R} + \mathbf{r} - \mathbf{r}_i\right) \delta\left(\mathbf{r} - \mathbf{r}_j\right)\rangle.$$
(9.59)

The sum in (9.59) is recognised as the two-point probability distribution function,

$$\sum_{i,j} \langle \delta\left(\mathbf{R} + \mathbf{r} - \mathbf{r}_i\right) \delta\left(\mathbf{r} - \mathbf{r}_j\right)\rangle = \bar{f}_2\left(\mathbf{r} + \mathbf{R}, \mathbf{r}, t\right),$$
(9.60)

and thus we obtain for the stress contributed from the conservative forces,

$$\mathbf{S}_C\left(\mathbf{r}, t\right) = -\frac{1}{2} \int d\mathbf{R} \mathbf{F}^C \mathbf{R} \left\{1 - \frac{1}{2}\mathbf{R} \cdot \nabla + \cdots\right\} \bar{f}_2\left(\mathbf{r} + \mathbf{R}, \mathbf{r}, t\right).$$
(9.61)

Stress from Dissipative Forces The dissipative-force-contributed stress can be treated in the same manner,

$$\mathbf{S}_D\left(\mathbf{r}, t\right) = -\frac{1}{2} \sum_{i,j} \left\langle \int \delta\left(\mathbf{R} - \mathbf{r}_{ij}\right) \mathbf{F}^D\left(\mathbf{R}, \mathbf{v} - \mathbf{v}'\right) \mathbf{R} \right.$$
$$\left. \left\{1 - \frac{1}{2}\mathbf{R} \cdot \nabla + \cdots\right\} \delta\left(\mathbf{r} - \mathbf{r}_j\right) d\mathbf{R}\right\rangle,$$

where we have used the notation for convenience

$$\mathbf{F}^D\left(\mathbf{R}, \mathbf{v} - \mathbf{v}'\right) = \gamma w^D\left(R\right) \hat{\mathbf{R}}\hat{\mathbf{R}} \cdot \left(\mathbf{v} - \mathbf{v}'\right), \quad \hat{\mathbf{R}} = \mathbf{R}/R, \quad R = |\mathbf{R}|.$$
(9.62)

The extra complication here is due to the explicit appearance of the velocities. As before, we note that

$$\mathbf{S}_D\left(\mathbf{r}, t\right) = -\frac{1}{2} \int d\mathbf{R} \int d\mathbf{v} \int d\mathbf{v}' \mathbf{F}^D\left(\mathbf{R}, \mathbf{v} - \mathbf{v}'\right) \mathbf{R} \left\{1 - \frac{1}{2}\mathbf{R} \cdot \nabla + \cdots\right\}$$
$$\cdot \sum_{i,j} \langle \delta\left(\mathbf{R} - \mathbf{r}_{ij}\right) \delta\left(\mathbf{r} - \mathbf{r}_j\right) \delta\left(\mathbf{v} - \mathbf{v}_i\right) \delta\left(\mathbf{v}' - \mathbf{v}_j\right)\rangle$$

$$= -\frac{1}{2} \int d\mathbf{R} \int d\mathbf{v} \int d\mathbf{v}' \mathbf{F}^D\left(\mathbf{R}, \mathbf{v} - \mathbf{v}'\right) \mathbf{R} \left\{1 - \frac{1}{2}\mathbf{R} \cdot \nabla + \cdots\right\}$$
$$\cdot \sum_{i,j} \langle \delta\left(\mathbf{R} + \mathbf{r} - \mathbf{r}_i\right) \delta\left(\mathbf{r} - \mathbf{r}_j\right) \delta\left(\mathbf{v} - \mathbf{v}_i\right) \delta\left(\mathbf{v}' - \mathbf{v}_j\right)\rangle$$

The sum is recognised as the two-point probability distribution function; and thus we obtain,

$$\mathbf{S}_D(\mathbf{r}, t) = -\frac{1}{2} \int d\mathbf{R} \int d\mathbf{v} \int d\mathbf{v}' \mathbf{F}^D(\mathbf{R}, \mathbf{v} - \mathbf{v}') \mathbf{R} \left\{ 1 - \frac{1}{2} \mathbf{R} \cdot \nabla + \cdots \right\}$$
$$\cdot f_2(\mathbf{r} + \mathbf{R}, \mathbf{r}, \mathbf{v}, \mathbf{v}', t). \tag{9.63}$$

Irving and Kirkwood pointed out that all the terms inside the curly brackets in (9.61) and (9.63) beyond the first may be neglected, due to the fact that they are of higher orders (in $O(\mathbf{R})$).

Marsh's Equivalent Results for the Stresses Marsh [56] used a different technique in deriving the stresses, which involves expressing the delta function (9.51) as an integral. Problem 9.7 explores this. From this, it can be shown that the kinetic pressure tensor, and the stresses from the interaction forces can be expressed as,

$$\mathbf{P}_K(\mathbf{r}, t) = \sum_i \langle m \mathbf{V}_i \mathbf{V}_i \delta(\mathbf{r} - \mathbf{r}_i) \rangle = \int m(\mathbf{v} - \mathbf{u})(\mathbf{v} - \mathbf{u}) f_1(\mathbf{r}, \mathbf{v}, t) d\mathbf{v}, \tag{9.64}$$

$$\mathbf{S}_C(\mathbf{r}, t) = -\frac{1}{2} \int d\mathbf{v}' \int d\mathbf{v}'' \int \mathbf{F}^C \mathbf{R} \bar{W}_2(\chi', \chi'', t; \mathbf{r}) d\mathbf{R}, \tag{9.65}$$

$$\mathbf{S}_D(\mathbf{r}, t) = -\frac{1}{2} \int d\mathbf{v}' \int d\mathbf{v}'' \int \mathbf{F}^D(\mathbf{R}, \mathbf{v}' - \mathbf{v}'') \mathbf{R} \bar{W}_2(\chi', \chi'', t; \mathbf{r}), \tag{9.66}$$

where

$$\bar{W}_2(\mathbf{r}', \mathbf{v}', \mathbf{r}'', \mathbf{v}'', t; \mathbf{r}) = \int_0^1 f_2(\mathbf{r} + \lambda \mathbf{R}, \mathbf{r} - (1 - \lambda)\mathbf{R}, \mathbf{v}', \mathbf{v}'', t) d\lambda. \tag{9.67}$$

9.2.7   Energy Equation

Energy is not a conserved quantity, but one can write a statement of balance for the specific energy $e(\mathbf{r}, t)$, defined as [56]

$$e(\mathbf{r}, t) = \left\langle \sum_i e_i \delta(\mathbf{r} - \mathbf{r}_i) \right\rangle, \quad e_i = \frac{1}{2} m_i \mathbf{v}_i \cdot \mathbf{v}_i + \frac{1}{2} \sum_j \varphi(r_{ij}), \tag{9.68}$$

where e_i is the energy per particle. First we define the kinetic energy

$$e_K = \left\langle \sum_i \frac{1}{2} m_i \mathbf{v}_i \cdot \mathbf{v}_i \delta(\mathbf{r} - \mathbf{r}_i) \right\rangle = \langle KE \rangle. \tag{9.69}$$

Using $Q = KE$ in the equation of change (9.39) yields

$$\frac{\partial}{\partial t} e_K = -\nabla \cdot \left\langle \sum_i \frac{1}{2} m \mathbf{v} \mathbf{v}_i \cdot \mathbf{v}_i \delta \left(\mathbf{r} - \mathbf{r}_i \right) \right\rangle + \left\langle \sum_{i,j} \mathbf{F}_{ij}^C \cdot \mathbf{v}_i \delta \left(\mathbf{r} - \mathbf{r}_i \right) \right\rangle \qquad (9.70)$$

$$- \gamma \left\langle \sum_{i,j} w_{ij}^D \mathbf{v}_{ij} \cdot \mathbf{e}_{ij} \mathbf{e}_{ij} \cdot \mathbf{v}_i \delta \left(\mathbf{r} - \mathbf{r}_i \right) \right\rangle + \gamma k_B T \left\langle \sum_{i,j} \frac{1}{m_i} w_{ij}^D \delta \left(\mathbf{r} - \mathbf{r}_i \right) \right\rangle .$$

The same can be done for the potential energy,

$$e_P = \left\langle \sum \frac{1}{2} \varphi \left(r_{ij} \right) \delta \left(\mathbf{r} - \mathbf{r}_i \right) \right\rangle = \langle PE \rangle . \qquad (9.71)$$

Again, by using $Q = PE$ in the equation of change (9.39):

$$\frac{\partial}{\partial t} e_P = \left\langle \sum_{i,j,k} \frac{1}{2} \mathbf{v}_k \cdot \left[\frac{\partial \varphi \left(r_{ij} \right)}{\partial \mathbf{r}_k} \delta \left(\mathbf{r} - \mathbf{r}_i \right) + \varphi \left(r_{ij} \right) \frac{\partial}{\partial \mathbf{r}_k} \delta \left(\mathbf{r} - \mathbf{r}_i \right) \right] \right\rangle \qquad (9.72)$$

$$= \left\langle \sum_{i,j} \frac{1}{2} \left\{ -\mathbf{F}_{ij}^C \cdot \mathbf{v}_i \delta \left(\mathbf{r} - \mathbf{r}_i \right) + \mathbf{F}_{ij}^C \cdot \mathbf{v}_j \delta \left(\mathbf{r} - \mathbf{r}_j \right) - \nabla \cdot \varphi \left(r_{ij} \right) \delta \left(\mathbf{r} - \mathbf{r}_i \right) \right\} \right\rangle .$$

Thus,

$$\frac{\partial}{\partial t} e = -\nabla \cdot \left\langle \sum_i \mathbf{v}_i e_i \delta \left(\mathbf{r} - \mathbf{r}_i \right) \right\rangle + \left\langle \sum_{i,j} \frac{1}{2} \mathbf{F}_{ij}^C \cdot \left(\mathbf{v}_i + \mathbf{v}_j \right) \delta \left(\mathbf{r} - \mathbf{r}_i \right) \right\rangle \qquad (9.73)$$

$$- \gamma \left\langle \sum_{i,j} w_{ij}^D \mathbf{v}_{ij} \cdot \mathbf{e}_{ij} \mathbf{e}_{ij} \cdot \mathbf{v}_i \delta \left(\mathbf{r} - \mathbf{r}_i \right) \right\rangle + \gamma k_B T \left\langle \sum_{i,j} \frac{1}{m_i} w_{ij}^D \delta \left(\mathbf{r} - \mathbf{r}_i \right) \right\rangle .$$

The first term on the right of (9.73) is already in the form of a divergence of a vector $-\nabla \cdot \mathbf{q}_K$, where

$$\mathbf{q}_K \left(\mathbf{r}, t \right) = \sum_i \mathbf{v}_i e_i \delta \left(\mathbf{r} - \mathbf{r}_i \right)$$

$$= \int d\mathbf{v} \int d\mathbf{v}' \int d\mathbf{R} \left(\frac{m}{2} v^2 + \frac{1}{2} \varphi \left(R \right) \right) \mathbf{v} f_2 \left(\chi, \chi', t \right) . \qquad (9.74)$$

The second term on the right of (9.73) can be converted into a divergence of a vector by the use of (9.51):

$$
\left\langle \sum_{i,j} \frac{1}{2} \mathbf{F}_{ij}^C \cdot \left(\mathbf{v}_i + \mathbf{v}_j \right) \delta \left(\mathbf{r} - \mathbf{r}_i \right) \right\rangle = -\nabla \cdot \sum_{i,j} \left\langle \frac{1}{4} \mathbf{F}_{ij}^C \cdot \left(\mathbf{v}_i + \mathbf{v}_j \right) \mathbf{r}_{ij} \right.
$$

$$
\left. \left(1 - \frac{1}{2} \mathbf{r}_{ij} \cdot \nabla + \cdots \right) \delta \left(\mathbf{r} - \mathbf{r}_j \right) \right\rangle
$$

$$
= -\nabla \cdot \mathbf{q}_C, \tag{9.75}
$$

where

$$
\mathbf{q}_C = \left\langle \sum_{i,j} \frac{1}{4} \mathbf{F}_{ij}^C \cdot \left(\mathbf{v}_i + \mathbf{v}_j \right) \mathbf{r}_{ij} \left(1 - \frac{1}{2} \mathbf{r}_{ij} \cdot \nabla + \cdots \right) \delta \left(\mathbf{r} - \mathbf{r}_j \right) \right\rangle. \tag{9.76}
$$

This can be converted into average with respect to two-particle distribution function as before (Problem 9.8).

The third term on the right of (9.73) can be re-cast into a source/sink term and a flux, using the fact that $\mathbf{v}_i = \mathbf{v}_{ij}/2 + \left(\mathbf{v}_i + \mathbf{v}_j \right)/2$:

$$
\gamma \left\langle \sum_{i,j} w_{ij}^D \mathbf{v}_{ij} \cdot \mathbf{e}_{ij} \mathbf{e}_{ij} \cdot \mathbf{v}_i \delta \left(\mathbf{r} - \mathbf{r}_i \right) \right\rangle = \frac{\gamma}{2} \left\langle \sum_{i,j} w_{ij}^D \left(\mathbf{v}_{ij} \cdot \mathbf{e}_{ij} \right)^2 \delta \left(\mathbf{r} - \mathbf{r}_i \right) \right\rangle
$$

$$
+ \frac{\gamma}{2} \left\langle \sum_{i,j} w_{ij}^D \mathbf{v}_{ij} \cdot \mathbf{e}_{ij} \mathbf{e}_{ij} \cdot \left(\mathbf{v}_i + \mathbf{v}_j \right) \delta \left(\mathbf{r} - \mathbf{r}_i \right) \right\rangle
$$

$$
= \Lambda_D + \nabla \cdot \mathbf{q}_D,
$$

with

$$
\Lambda_D \left(\mathbf{r}, t \right) = \frac{\gamma}{2} \left\langle \sum_{i,j} w_{ij}^D \left(\mathbf{v}_{ij} \cdot \mathbf{e}_{ij} \right)^2 \delta \left(\mathbf{r} - \mathbf{r}_i \right) \right\rangle \tag{9.77}
$$

$$
= -\frac{\gamma}{2} \int d\mathbf{v} \int d\mathbf{v}' \int d\mathbf{R} w^D \left(R \right) \left[\hat{\mathbf{R}} \cdot \left(\mathbf{v} - \mathbf{v}' \right) \right]^2 f_2 \left(\chi, \chi', t \right),
$$

and

$$
\mathbf{q}_D \left(\mathbf{r}, t \right) = -\frac{\gamma}{4} \sum_{i,j} \left\langle w_{ij}^D \mathbf{v}_{ij} \cdot \mathbf{e}_{ij} \mathbf{e}_{ij} \cdot \left(\mathbf{v}_i + \mathbf{v}_j \right) \mathbf{r}_{ij} \right.
$$

$$
\left. \left(1 - \frac{1}{2} \mathbf{r}_{ij} \cdot \nabla + \cdots \right) \delta \left(\mathbf{r} - \mathbf{r}_j \right) \right\rangle
$$

$$
= \frac{\gamma}{4} \int d\mathbf{R} \int d\mathbf{v} \int d\mathbf{v}' w^D \left(R \right) \left(\mathbf{v} - \mathbf{v}' \right) \cdot \hat{\mathbf{R}} \hat{\mathbf{R}} \cdot \left(\mathbf{v} + \mathbf{v}' \right) \mathbf{R}
$$

$$
\cdot \left\{ 1 - \frac{1}{2} \mathbf{R} \cdot \nabla + \cdots \right\} f_2 \left(\mathbf{r} + \mathbf{R}, \mathbf{r}, \mathbf{v}, \mathbf{v}', t \right). \tag{9.78}
$$

Finally, the last term on the right of (9.73) is recognised as a source term due to injection of kinetic energy via the random forces:

$$\Lambda_R(\mathbf{r}, t) = \gamma k_B T \left\langle \sum_{i,j} \frac{1}{m_i} w_{ij}^D \delta(\mathbf{r} - \mathbf{r}_i) \right\rangle \tag{9.79}$$

$$= \frac{\gamma k_B T}{m} \int d\mathbf{v} \int d\mathbf{v}' \int d\mathbf{R} w^D(R) f_2(\chi, \chi', t).$$

The energy flow balance equation then takes the form

$$\frac{\partial e}{\partial t} = -\nabla \cdot (\mathbf{q}_K + \mathbf{q}_C + \mathbf{q}_D) + \Lambda_R - \Lambda_D. \tag{9.80}$$

The two source/sink terms will cancel each other when the temperature reaches the value dictated by the fluctuation-dissipation theorem, as can be recognised by their forms:

$$\Lambda_R - \Lambda_D = \gamma \left\langle \sum_{i,j} \left\{ \frac{k_B T}{m_i} - \frac{1}{2} (\mathbf{v}_{ij} \cdot \mathbf{e}_{ij})^2 \right\} w_{ij}^D \delta(\mathbf{r} - \mathbf{r}_i) \right\rangle. \tag{9.81}$$

9.3 Some Approximate Results

9.3.1 High Damping Limit

Exclusion of Conservative Forces Marsh [56] derived some important results in the limit of high γ, where the conservatives forces are zero, $\mathbf{F}_{ij}^C = 0$. This limit is thought to be relevant to the general DPD case. In this limit, he showed that a compressible Newtonian fluid is the solution to the resulting linearised Fokker-Planck-Boltzmann equations,

$$\mathbf{T}(\mathbf{r}, t) = -nk_B T \mathbf{I} + (\zeta_K + \zeta_D) \nabla \cdot \mathbf{u} \mathbf{I} \tag{9.82}$$

$$+ (\eta_K + \eta_D) \left(\nabla \mathbf{u} + \nabla \mathbf{u}^T - \frac{2}{3} \nabla \cdot \mathbf{u} \mathbf{I} \right),$$

with shear and bulk viscosities given by

$$\eta = \eta_K + \eta_D = \frac{3mk_B T}{2\gamma [w^D]_R} + \frac{\gamma n^2 [R^2 w^D]_R}{30}$$

$$= \frac{\rho D}{2} + \frac{\gamma n^2 [R^2 w^D]_R}{30}, \tag{9.83}$$

$$\zeta = \zeta_K + \zeta_D = \frac{mk_BT}{\gamma \left[w^D\right]_R} + \frac{\gamma n^2 \left[R^2 w^D\right]_R}{18}, \tag{9.84}$$

where the diffusivity of a "tagged" DPD particle is

$$D = \frac{3k_BT}{n\gamma \left[w^D\right]_R}, \tag{9.85}$$

and the square brackets define the following operation

$$\left[w^D\right]_R = \int d\mathbf{R} w^D(R), \quad \left[R^2 w^D\right]_R = \int d\mathbf{R} R^2 w^D(R). \tag{9.86}$$

This "compressible" Newtonian fluid is the analogue to an elastic solid with Lamé moduli η and ζ. We could define its "Poisson's ratio" ν as,

$$\nu = \frac{\zeta - \frac{2}{3}\eta}{2\left(\zeta + \frac{1}{3}\eta\right)} = \frac{X}{1+4X}, \quad X = \frac{\gamma^2 n^2 \left[w^D\right]_R \left[R^2 w^D\right]_R}{90 mk_BT}. \tag{9.87}$$

As this parameter X varies from zero to infinity, the fluid equivalent Poisson's ratio is less than 0.25.

With the standard weighting function for DPD,

$$w^D(r) = \begin{cases} (1 - r/r_c)^2, & r < r_c, \\ 0, & r \geq r_c, \end{cases} \tag{9.88}$$

the integrations indicated in (9.86) can be performed, resulting in

$$\left[w^D\right]_R = \frac{2\pi}{15}r_c^3, \quad \left[R^2 w^D\right]_R = \frac{4\pi}{105}r_c^5, \tag{9.89}$$

which then leads to

$$\eta = \frac{45mkT}{4\pi\gamma r_c^3} + \frac{2\pi\gamma n^2 r_c^5}{1575}, \quad \zeta = \frac{15mkT}{2\pi\gamma r_c^3} + \frac{2\pi\gamma n^2 r_c^5}{945}, \quad D = \frac{45k_BT}{2\pi\gamma n r_c^3}. \tag{9.90}$$

Inclusion of Conservative Forces In the limit of $m/\gamma^2 \to 0$ and no external force, (9.14) reduces to

$$0 = \sum_{j=1, j\neq i}^{N} a_{ij} w_C \mathbf{e}_{ij} - \sum_{j=1, j\neq i}^{N} \gamma w_D \left(\mathbf{e}_{ij}\mathbf{e}_{ij} \cdot \mathbf{v}_{ij}\right) + \sum_{j=1, j\neq i}^{N} \sigma w_R \theta_{ij} \mathbf{e}_{ij}, \tag{9.91}$$

where we note that N is the total number of DPD particles, and when necessary, we use the superscript k on relevant variables to denote their values at the time level

$t = t^k$. This limit is not equivalent to a steady-state flow assumption - it only guarantees that the inertial terms (as represented by Reynolds number) in the momentum equations are zero, but any other time-dependent behaviour inherited, for example from the boundary conditions, has not been eliminated.

Taking the dissipative force to the left side, (9.91) becomes

$$\sum_{j=1, j\neq i}^{N} \gamma w_D e_{ij} e_{ij} \cdot (\mathbf{v}_i - \mathbf{v}_j) = \sum_{j=1, j\neq i}^{N} a_{ij} w_C e_{ij} + \sum_{j=1, j\neq i}^{N} \sigma w_R \theta_{ij} e_{ij}. \quad (9.92)$$

The DPD system is obtained by letting i taking values from 1 to N in (9.92), in the matrix-vector notation,

$$\mathcal{A}\mathbf{v} = \mathbf{f}, \quad (9.93)$$

where $\mathcal{A}$ is the system matrix, whose elements are second-order tensors, $\mathbf{v}$ the column vector of unknown velocity vectors and $\mathbf{f}$ the column vector of right hand force vectors:

$$\mathcal{A} = \begin{bmatrix} \sum_{j=2}^{N} \gamma w_D e_{1j} e_{1j} & -\gamma w_D e_{12} e_{12} & \cdots & -\gamma w_D e_{1N} e_{1N} \\ -\gamma w_D e_{21} e_{21} & \sum_{j=1, j\neq 2}^{N} \gamma w_D e_{2j} e_{2j} & \cdots & -\gamma w_D e_{2N} e_{2N} \\ \vdots & \vdots & \ddots & \vdots \\ -\gamma w_D e_{N1} e_{N1} & -\gamma w_D e_{N2} e_{N2} & \cdots & \sum_{j=1}^{N-1} \gamma w_D e_{Nj} e_{Nj} \end{bmatrix}, \quad (9.94)$$

$$\mathbf{v} = \begin{pmatrix} \mathbf{v}_1^k \\ \mathbf{v}_2^k \\ \vdots \\ \mathbf{v}_N^k \end{pmatrix}, \quad \mathbf{f} = \begin{pmatrix} \sum_{j=2}^{N} a_{1j} w_C e_{1j} + \sum_{j=2}^{N} \sigma w_R \theta_{1j} e_{1j} \\ \sum_{j=1, j\neq 2}^{N} a_{2j} w_C e_{2j} + \sum_{j=1, j\neq 2}^{N} \sigma w_R \theta_{2j} e_{2j} \\ \vdots \\ \sum_{j=1}^{N-1} a_{Nj} w_C e_{Nj} + \sum_{j=1}^{N-1} \sigma w_R \theta_{Nj} e_{Nj} \end{pmatrix}. \quad (9.95)$$

This system is singular since the row (and column) sum of $\mathcal{A}$ is zero. The eigenvector of the system, corresponding to an eigenvalue of 0, is $\{\mathbf{U}, \ldots, \mathbf{U}\}^T$, where $\mathbf{U}$ is an arbitrary vector. Below are two schemes proposed to deal with this singularity.

Scheme 1

We first record a physical constraint for the DPD system. The centre of mass (for $m_i = m$),

$$\mathbf{R}_c = \frac{1}{Nm} \sum_{i=1}^{N} m_i \mathbf{r}_i, \quad (9.96)$$

satisfies

$$Nm \frac{d^2 \mathbf{R}_c}{dt^2} = \mathbf{0} = \sum_{i=1}^{N} m_i \frac{d^2 \mathbf{r}_i}{dt^2}, \quad (9.97)$$

due to pair-wise force interactions, and thus the centre of mass of the DPD system moves deterministically with constant velocity $\mathbf{U}_c$. Without much loss to generality, one can take this constant as zero,

$$\mathbf{U}_c = \frac{1}{N}\sum_{r=1}^{N}\mathbf{v}_r = \mathbf{0}. \tag{9.98}$$

By adding the term $\sum_{j=1, j\neq i}^{N}\gamma w_D\mathbf{e}_{ij}\mathbf{e}_{ij}\left(\sum_{r=1, r\neq j}^{N}\mathbf{v}_r\right)$ to both sides of equation (9.92), we obtain, in the matrix-vectorial notation,

$$\sum_{j=1, j\neq i}^{N}\gamma w_D\mathbf{e}_{ij}\mathbf{e}_{ij}\left(\mathbf{v}_i - \mathbf{v}_j\right) + \sum_{j=1, j\neq i}^{N}\gamma w_D\mathbf{e}_{ij}\mathbf{e}_{ij}\left(\sum_{r=1, r\neq j}^{N}\mathbf{v}_r\right) =$$
$$\sum_{j=1, j\neq i}^{N}a_{ij}w_C\mathbf{e}_{ij} + \sum_{j=1, j\neq i}^{N}\gamma w_D\mathbf{e}_{ij}\mathbf{e}_{ij}\left(\sum_{r=1, r\neq j}^{N}\mathbf{v}_r\right) + \sum_{j=1, j\neq i}^{N}\sigma w_R\theta_{ij}\mathbf{e}_{ij}. \tag{9.99}$$

It can be rearranged as

$$\sum_{j=1, j\neq i}^{N}\gamma w_D\mathbf{e}_{ij}\mathbf{e}_{ij}\left(\mathbf{v}_i + \sum_{r=1, r\neq j}^{N}\mathbf{v}_r\right) = \sum_{j=1, j\neq i}^{N}a_{ij}w_C\mathbf{e}_{ij} +$$
$$\sum_{j=1, j\neq i}^{N}\gamma w_D\mathbf{e}_{ij}\mathbf{e}_{ij}\left(\sum_{r=1}^{N}\mathbf{v}_r\right) + \sum_{j=1, j\neq i}^{N}\sigma w_R\theta_{ij}\mathbf{e}_{ij}. \tag{9.100}$$

Substitution of (9.98) into (9.100) yields

$$\sum_{j=1, j\neq i}^{N}\gamma w_D\mathbf{e}_{ij}\mathbf{e}_{ij}\left(\mathbf{v}_i + \sum_{r=1, r\neq j}^{N}\mathbf{v}_r\right) = \sum_{j=1, j\neq i}^{N}a_{ij}w_C\mathbf{e}_{ij} + \sum_{j=1, j\neq i}^{N}\sigma w_R\theta_{ij}\mathbf{e}_{ij}. \tag{9.101}$$

The DPD system matrix corresponding to (9.101) is

$$\mathcal{A} = \begin{bmatrix} 2\sum_{j=2}^{N}\gamma w_D\mathbf{e}_{1j}\mathbf{e}_{1j} & \sum_{j=3}^{N}\gamma w_D\mathbf{e}_{1j}\mathbf{e}_{1j} & \cdots & \sum_{j=2}^{N-1}\gamma w_D\mathbf{e}_{1j}\mathbf{e}_{1j} \\ \sum_{j=3}^{N}\gamma w_D\mathbf{e}_{2j}\mathbf{e}_{2j} & 2\sum_{j=1, j\neq 2}^{N}\gamma w_D\mathbf{e}_{2j}\mathbf{e}_{2j} & \cdots & \sum_{j=1, j\neq 2}^{N-1}\gamma w_D\mathbf{e}_{2j}\mathbf{e}_{2j} \\ \vdots & \vdots & \ddots & \vdots \\ \sum_{j=2}^{N-1}\gamma w_D\mathbf{e}_{Nj}\mathbf{e}_{Nj} & \sum_{j=1, j\neq 2}^{N-1}\gamma w_D\mathbf{e}_{Nj}\mathbf{e}_{Nj} & \cdots & 2\sum_{j=1}^{N-1}\gamma w_D\mathbf{e}_{Nj}\mathbf{e}_{Nj} \end{bmatrix}, \tag{9.102}$$

which is invertible, but fully populated and not diagonally dominant. The present DPD system can be solved by a direct solver such as the one based on the LU factorisation with partial pivoting, with a high computational cost at large number of particles.

Scheme 2

The system matrix $\mathcal{A}$ in (9.93) can be decomposed into a diagonal block $\mathcal{I}$ and the remainder $\mathcal{K}$:

$$\mathcal{I}\mathbf{v} + \mathcal{K}\mathbf{v} = \mathbf{f}, \tag{9.103}$$

where

$$\mathcal{I} = \begin{bmatrix} \sum_{j=2}^{N} \gamma w_D \mathbf{e}_{1j}\mathbf{e}_{1j} & \mathbf{0} & \cdots & \mathbf{0} \\ \mathbf{0} & \sum_{j=1, j\neq 2}^{N} \gamma w_D \mathbf{e}_{2j}\mathbf{e}_{2j} & \cdots & \mathbf{0} \\ \vdots & \vdots & \ddots & \vdots \\ \mathbf{0} & \mathbf{0} & \cdots & \sum_{j=1}^{N-1} \gamma w_D \mathbf{e}_{Nj}\mathbf{e}_{Nj} \end{bmatrix}, \tag{9.104}$$

and

$$\mathcal{K} = \begin{bmatrix} \mathbf{0} & -\gamma w_D \mathbf{e}_{12}\mathbf{e}_{12} & \cdots & -\gamma w_D \mathbf{e}_{1N}\mathbf{e}_{1N} \\ -\gamma w_D \mathbf{e}_{21}\mathbf{e}_{21} & \mathbf{0} & \cdots & -\gamma w_D \mathbf{e}_{2N}\mathbf{e}_{2N} \\ \vdots & \vdots & \ddots & \vdots \\ -\gamma w_D \mathbf{e}_{N1}\mathbf{e}_{N1} & -\gamma w_D \mathbf{e}_{N2}\mathbf{e}_{N2} & \cdots & \mathbf{0} \end{bmatrix}. \tag{9.105}$$

Now, since $\mathcal{I}$ is diagonal block, its inverse is

$$\mathcal{I}^{-1} = \begin{bmatrix} \left(\sum_{j=2}^{N} \gamma w_D \mathbf{e}_{1j}\mathbf{e}_{1j} \right)^{-1} & \cdots & \mathbf{0} \\ \vdots & \ddots & \vdots \\ \mathbf{0} & \cdots & \left(\sum_{j=1}^{N-1} \gamma w_D \mathbf{e}_{Nj}\mathbf{e}_{Nj} \right)^{-1} \end{bmatrix}, \tag{9.106}$$

and we can rewrite (9.103) as

$$\mathbf{v} + \mathcal{H}\mathbf{v} = \mathbf{b}, \tag{9.107}$$

where $\mathcal{H} = \mathcal{I}^{-1}\mathcal{K}$ and $\mathbf{b} = \mathcal{I}^{-1}\mathbf{f}$.

Because of the zero row (and column) sum of the operator $\mathcal{I} + \mathcal{K}$, an eigenvector for $\mathcal{H}$ is

$$\epsilon = (\mathbf{U}, \mathbf{U}, ..., \mathbf{U})^T / \sqrt{N}U, \quad U^2 = \mathbf{U} \cdot \mathbf{U}, \tag{9.108}$$

where $\mathbf{U}$ is any constant vector - this eigenvector corresponds to a normalised -1 eigenvalue. Therefore solution of (9.107), if exist, is non-unique - any solution plus a multiple of the eigenvector (9.108) is also another solution.

Following the Wielandt's deflation approach ([7, 68]), we can render a unique solution to (9.107), and at the same time, deflate the eigenvalue at -1 to zero, by considering the equivalent system,

$$\mathbf{v} + \mathcal{H}\mathbf{v} + \epsilon_x \left[\mathbf{v}, \epsilon_x\right] + \epsilon_y \left[\mathbf{v}, \epsilon_y\right] + \epsilon_z \left[\mathbf{v}, \epsilon_z\right] = \mathbf{b}, \tag{9.109}$$

where $[\cdot, \cdot]$ denotes the natural inner product of two vector elements in the underlying space, and

$$\epsilon_x = (\mathbf{U}_x, \mathbf{U}_x, ..., \mathbf{U}_x)^T / \sqrt{N} U_x, \quad \mathbf{U}_x = (1, 0, 0)^T, \quad U_x^2 = \mathbf{U}_x \cdot \mathbf{U}_x, \tag{9.110}$$

$$\epsilon_y = (\mathbf{U}_y, \mathbf{U}_y, ..., \mathbf{U}_y)^T / \sqrt{N} U_y, \quad \mathbf{U}_y = (0, 1, 0)^T, \quad U_y^2 = \mathbf{U}_y \cdot \mathbf{U}_y, \tag{9.111}$$

$$\epsilon_z = (\mathbf{U}_z, \mathbf{U}_z, ..., \mathbf{U}_z)^T / \sqrt{N} U_z, \quad \mathbf{U}_z = (0, 0, 1)^T, \quad U_z^2 = \mathbf{U}_z \cdot \mathbf{U}_z. \tag{9.112}$$

This is seen by expanding the additional inner-product terms

$$\epsilon_x \left[\mathbf{v}, \epsilon_x\right] + \epsilon_y \left[\mathbf{v}, \epsilon_y\right] + \epsilon_z \left[\mathbf{v}, \epsilon_z\right] = \frac{1}{NU_x^2} (\mathbf{U}_x, \mathbf{U}_x, \ldots, \mathbf{U}_x)^T \mathbf{U}_x \cdot \sum_{i=1}^{N} \mathbf{v}_{ix} +$$

$$\frac{1}{NU_y^2} (\mathbf{U}_y, \mathbf{U}_y, \ldots, \mathbf{U}_y)^T \mathbf{U}_y \cdot \sum_{i=1}^{N} \mathbf{v}_{iy} +$$

$$\frac{1}{NU_z^2} (\mathbf{U}_z, \mathbf{U}_z, \ldots, \mathbf{U}_z)^T \mathbf{U}_z \cdot \sum_{i=1}^{N} \mathbf{v}_{iz} = \mathbf{0}, \tag{9.113}$$

and therefore they do not contribute to the governing equation; they represent the constraint (9.98). However, by writing this in this fashion, we have (i) mapped the eigenvalue -1 to zero (ϵ is still the eigenvector, but with corresponding eigenvalue of zero); and (ii) rendered a unique solution to (9.103).

If the previous spectral radius (before deflation) is one, it is now less than one and iterative numerical implementation to (9.103) may work well; a simplest one is the Picard's iteration:

$$\mathbf{v}_i^l = -\mathcal{H}_i \mathbf{v}^{l-1} - \epsilon_x \langle \mathbf{v}^{l-1}, \epsilon_x \rangle - \epsilon_y \langle \mathbf{v}^{l-1}, \epsilon_y \rangle - \epsilon_z \langle \mathbf{v}^{l-1}, \epsilon_z \rangle + \mathbf{b}_i, \tag{9.114}$$

where $i = (1, 2, \ldots, N)$,

$$\mathcal{H}_i = \left(\sum_{j=1, j \neq i}^{N} \gamma w_D \mathbf{e}_{ij} \mathbf{e}_{ij} \right)^{-1}$$

$$\left[-\gamma w_D \mathbf{e}_{i1} \mathbf{e}_{i1}, \ldots, -\gamma w_D \mathbf{e}_{i(i-1)} \mathbf{e}_{i(i-1)}, \mathbf{0}, \ldots, -\gamma w_D \mathbf{e}_{iN} \mathbf{e}_{iN} \right], \tag{9.115}$$

and

$$\mathbf{b}_i = \left(\sum_{j=1, j \neq i}^{N} \gamma w_D \mathbf{e}_{ij} \mathbf{e}_{ij} \right)^{-1} \left(\sum_{j=1, j \neq i}^{N} a_{ij} w_C \mathbf{e}_{ij} + \sum_{j=1, j \neq i}^{N} \sigma w_R \theta_{ij} \mathbf{e}_{ij} \right). \quad (9.116)$$

The iterative process stops when the difference of the solutions between two successive iterations is less than a specified tolerance - a possible tolerance may be

$$\text{norm}(\mathbf{v}^l - \mathbf{v}^{l-1}) < 10^{-7} \text{norm}(\mathbf{v}^l). \quad (9.117)$$

It can be seen that $\mathcal{H}_i$ is a sparse matrix because (i) the construction of this matrix involves only particles within the interaction zone whose size is defined by the cutoff radius r_c; and (ii) the value of r_c is generally much smaller than a typical linear dimension of the flow domain. Furthermore, it is straightforward to implement the iterative scheme in parallel as (9.114) is applied to each particle, independent of the rest of the particles.

For both Schemes 1 and 2, the Euler algorithm is employed to advance the position of the particles

$$\mathbf{r}_i^{k+1} = \mathbf{r}_i^k + \mathbf{v}_i^k \Delta t, \quad (9.118)$$

where $\Delta t = t^{k+1} - t^k$.

9.3.2 Standard DPD Parameters

It is a convenient point to mention the "standard" choice of DPD parameters. The standard weighting function (9.88) has been mentioned. Mass and cutoff radius are commonly normalised to unity ($m = 1 = r_c$), then the density depends on the initial arrangement of DPD particles – for an initial face centered cubic (FCC) arrangement, for example, the density is 4 (four DPD particles per unit cell, $r_c \times r_c \times r_c$). Boltzmann temperature is also normalised to unity, $k_B T$. For a liquid with a compressibility like that of water, Groot and Warren [33] recommended

$$a_{ij} = \frac{75 k_B T}{n r_c^4}, \quad (9.119)$$

in 3D, and in 2D, the repulsive coefficients are given by

$$a_{ij} = \frac{57.23 k_B T}{n r_c^3}. \quad (9.120)$$

They further recommended a random force strength of $\sigma = 3$; consequently, from the fluctuation-dissipation theorem, $\gamma = 4.5$. Too high a value of σ may render the system unstable. We refer to the foregoing choice of DPD parameters as the "standard" choice.

9.3.3   Effective Size of a DPD Particle

A DPD particle is a point mass in its Newton's 2nd law of motion, under the action of inter-particle forces. However, it is endowed with a soft repulsive potential, and therefore has an "effective" size, which is the exclusion zone of the particle. One could provide an estimate for this effective size of a DPD particle by various means. In what follows, we analyse roles of the DPD forces on the particle's exclusion size.

Role of the Conservative Force The inclusion of the conservative force (repulsive force) into the DPD formulation is to provide an independent mean of controlling the speed of sound (compressibility) to the number density and the temperature of the DPD system [56]. Keeping the dissipative and random forces unchanged, an increase in the repulsion strength a_{ij} will promote incompressibility of the DPD fluid [65]. However, at large values of a_{ij}, the mean squared particle displacement $\langle(\mathbf{r}(t) - \mathbf{r}(0))^2\rangle$ is observed to be no longer linear in time (a characteristic of a Brownian particle) with crystallisation occurring and the DPD system has a solid-like structure. Here, we limit our attention to the case where the compressibility of the system is matched to that of the water at room temperature [33].

If the weighting function w_C is fixed at its linear form, the size of solvent particles induced by the conservative forces will be controlled by means of the repulsion parameter a_{ij}. A larger value of a_{ij} results in a larger size of the particle and vice versa. From expressions (9.120) and (9.119), one can reduce the particle size by increasing n, increasing r_c or reducing $k_B T$. These expressions also reveal that the larger the DPD particle's size (corresponding to larger a_{ij}), the coarser level (smaller n) the DPD system will be (this comes from the scaling property of the DPD system [29]).

Role of the Dissipative and Random Forces Consider a generic "tagged" DPD particle in a sea of other DPD particles undergoing a diffusion process. Its size a_{eff} may be estimated by the Stokes–Einstein relation

$$a_{eff} = \frac{k_B T}{6\pi D\eta},\tag{9.121}$$

where D is the diffusion coefficient of the tagged particle subject to Brownian motion in an unbounded domain and η the shear viscosity of the surrounding fluid. This can be equated to the diffusivity in (9.90) with $\mathbf{F}_{ij}^C = 0$ to yield

$$a_{eff} = \frac{\rho\gamma r_c^3}{135\eta} = \frac{630 \times 2\pi\gamma^2 n^2 r_c^6}{27\left(1575 \times 45\rho k_B T + 8\pi^2\gamma^2 n^3 r_c^8\right)}.\tag{9.122}$$

For our standard DPD parameters $a_{eff} \approx 0.12$. Note that expression (9.122) is established assuming the Stokes–Einstein relation, which is valid for a dispersion of mesoscopic particles in a continuous solvent. The DPD particles representing the solvent phase are assumed not to be clustered, and are of a size considerably less than that

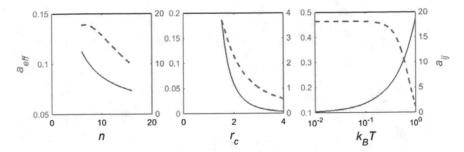

Fig. 9.2 DPD system, standard parameters: effects of the number density (*left*), cutoff radius (*middle*) and temperature (*right*) on the particle size (the *dash line* representing the zone size caused by the dissipative and random forces (9.122) and the *solid line* by the conservative force (9.119))

of the tagged particle that has no inertia. This latter condition is of course not satisfied here and the expression (9.122) can only be considered as best an estimate. However, the solvent particle size approaches zero as any one of the number density n, the cutoff radius r_c or the thermodynamic temperature $k_B T$ approaches infinity.

Although expression (9.122) is only an approximate result, but it serves as a mean to gauge the effect of different parameters. Figure 9.2 (left to right) show respectively the effects of n, r_c and $k_B T$ on the solvent particle size. Figure 9.2 (left and middle) clearly indicate that the exclusion zones caused by the conservative force and by the dissipative and random forces are both smaller with increasing either n or r_c. However, in Fig. 9.2 (right), the solvent particle size is an increasing function of $k_B T$, for the conservative force, and a decreasing function of $k_B T$, for the dissipative and random forces. Special care is thus needed if one tries to control the solvent particle size via $k_B T$. It is noted that (i) reducing $k_B T$ makes a_{ij} smaller which can result in the clustering of particles; and (ii) the particle size, defined in (9.122), is inversely proportional to r_c^2 and n, and the value of a_{ij}, defined in (9.119), is inversely proportional to r_c^4 and n. These observations imply that (i) controlling particles size via n and r_c is clearly more effective than via $k_B T$; and (ii) increasing r_c results in a faster decrease in the particle size than increasing n.

From this approximate analysis, we can see that there are many possible combinations of n and r_c that reduces a_{eff}. One can employ $r_c = 1$ with a large value of n (finer coarse-graining level with a standard cutoff radius). Or one can employ, for example $n = 3$ with a larger value of $r_c > 1$ (upper coarse graining level with a larger cutoff radius).

The determination of the exclusion size of a DPD particle can be accomplished with the concept of radial distribution function [76]

$$g(q) = \frac{1}{N/V} \frac{\langle s \rangle}{4\pi q^2 \Delta q}, \tag{9.123}$$

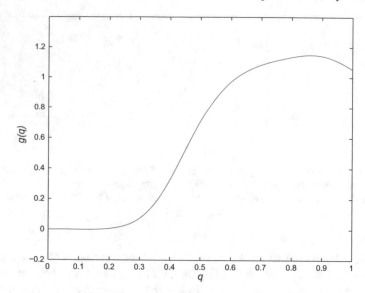

Fig. 9.3 Radial distribution function for the DPD particles

where V is the volume of the domain of interest containing N particles, and $\langle s \rangle$ is the average number of particles in a spherical shell of width Δq at a distance q from a particle in the fluid. Figure 9.3 shows the variation of $g(q)$, where Δq is chosen as 0.05, for the standard DPD fluid. Let $\bar{q}$ be the value of q at which $g(q) > 0.05$. The effective radius of the solvent particles can be estimated as $a_{eff} = \bar{q}/2 = 0.32/2 = 0.16$, a result close to Pan et al.'s [67].

9.4 Modification of the Weighting Function

The dissipative force directly affects the rate of momentum transfer of the system and thus the dynamic properties of the system; a measure of this response is the Schmidt number, defined as the rate of the speed of momentum transfer to the speed of particle's diffusion

$$S_c = \frac{\eta}{\rho D}. \tag{9.124}$$

Because the soft interaction between DPD particles, the speed of momentum transfer is slow, of the same order as particle diffusion. Therefore the Schmidt number is about unity. For a real fluid of physical properties like those of water, the Schmidt number is $O(10^3)$, and therefore there is the need to improve on the dynamic behaviour of the DPD system.

A simple way to do this has been proposed (Fan et al. [24]): the standard weighting function (9.88) for the dissipative force is to be replaced by

$$w^D(r) = \left[w^R(r)\right]^2 = \begin{cases} (1 - r/r_c)^s, & r < r_c, \\ 0, & r \geq r_c. \end{cases} \qquad (9.125)$$

When $s = 2$, the weight function for the conventional DPD formulation is recovered. From (9.86), and using the new weighting function (9.125), it can be shown that

$$\left[w^D\right]_R = 4\pi r_c^3 \left[\frac{1}{1+s} - \frac{2}{2+s} + \frac{1}{3+s}\right], \qquad (9.126)$$

$$\left[R^2 w^D\right]_R = 4\pi r_c^5 \left[\frac{1}{1+s} - \frac{4}{2+s} + \frac{6}{3+s} - \frac{4}{4+s} + \frac{1}{5+s}\right],$$

leading to the following diffusivity and viscosity,

$$D = \frac{3k_B T}{n\gamma \left[w^D\right]_R} = \frac{3k_B T}{4\pi n\gamma r_c^3} \left[\frac{1}{1+s} - \frac{2}{2+s} + \frac{1}{3+s}\right]^{-1}, \qquad (9.127)$$

$$\eta = \frac{3mk_B T}{8\pi\gamma r_c^3} \left[\frac{1}{1+s} - \frac{2}{2+s} + \frac{1}{3+s}\right]^{-1} \qquad (9.128)$$
$$+ \frac{2\pi r_c^5 \gamma n^2}{15} \left[\frac{1}{1+s} - \frac{4}{2+s} + \frac{6}{3+s} - \frac{4}{4+s} + \frac{1}{5+s}\right],$$

Consequently, Schmidt number is

$$S_c = \frac{1}{2} + \frac{8\pi^2 \gamma^2 n^2 r_c^8}{45mk_B T} \left[\frac{1}{1+s} - \frac{4}{2+s} + \frac{6}{3+s} - \frac{4}{4+s} + \frac{1}{5+s}\right].$$
$$\cdot \left[\frac{1}{1+s} - \frac{2}{2+s} + \frac{1}{3+s}\right] \qquad (9.129)$$

The physical properties of two DPD systems, corresponding to the standard weighting function ($s = 2$), and a modified weighting function ($s = \frac{1}{2}$) are tabulated in Table 9.1. With the standard choice of DPD parameters, a simple change

Table 9.1 Properties of the two DPD systems

Formulation	Conventional ($s = 2$)	Modified ($s = 1/2$)
Diffusivity, D	$\dfrac{45k_B T}{2\pi\gamma nr_c^3}$	$\dfrac{315k_B T}{64\pi\gamma nr_c^3}$
Viscosity, η	$\dfrac{45mk_B T}{4\pi\gamma r_c^3} + \dfrac{2\pi\gamma n^2 r_c^5}{1575}$	$\dfrac{315mk_B T}{128\pi\gamma r_c^3} + \dfrac{512\pi\gamma n^2 r_c^5}{51975}$
Bulk viscosity, ζ	$\dfrac{15mk_B T}{2\pi\gamma r_c^3} + \dfrac{2\pi\gamma n^2 r_c^5}{945}$	$\dfrac{105mk_B T}{436\pi\gamma r_c^3} + \dfrac{512\pi\gamma n^2 r_c^5}{31185}$
Schmidt number, Sc	$\dfrac{1}{2} + \dfrac{\left(2\pi\gamma nr_c^4\right)^2}{70875mk_B T}$	$\dfrac{1}{2} + \dfrac{32\,768\left(\pi\gamma nr_c^4\right)^2}{16\,372\,125mk_B T}$

from $s = 2$ to $s = \frac{1}{2}$ increases the Sc number by a factor of 10, and thereby improve on the dynamic response of the DPD fluid. Since $Sc = O\left(\gamma^2, r_c^8\right)$, a modest increase in either γ or r_c results in a large increase in Sc. Increasing γ strengthens the random forces, resulting in larger fluctuations of thermal energy injected into the system. Clearly, the most efficient way to increase Sc is to increase r_c. However, the computation demand of calculating the dissipative and random forces is $O\left(r_c^3\right)$, and can be significantly increased with r_c. We prefer to combine the modified weighting function with a moderate increase in the cutoff radius for dissipative weighting function, so that a physical level of Sc can be reached with a moderate increase in the computation cost. For $s = \frac{1}{2}, r_c = 2$, and standard choice of the remaining DPD parameters we find that $Sc = 1.6 \times 10^3$.

9.5 Imposition of Physical Parameters

Equations (9.45) and (9.56) show that the motion of a DPD fluid is governed by the Navier–Stokes equation, with its physical properties, such as viscosity, Schmidt number, isothermal compressibility, relaxation and inertia time scales, etc., in fact its whole rheology, resulted from the choice of the DPD model parameters. In this section, we will explore the response of a DPD fluid with respect to its parameter space, where the model input parameters can be chosen in advance so that (i) the ratio between the relaxation and inertia time scales is fixed; (ii) the isothermal compressibility of water at room temperature is enforced; and (iii) the viscosity and Schmidt number can be specified as inputs. These impositions are possible with some extra degrees of freedom in the weighting functions for the conservative and dissipative forces.

9.5.1 Time Scales

Let us focus on a tagged, but otherwise arbitrary DPD particle in the system, and let α be the ratio of its relaxation time scale to inertia time scale

$$\alpha = \frac{\tau}{\tau_I} = O\left(\frac{\gamma^2 r_c}{m a_{ij}}\right), \tag{9.130}$$

where $\tau = O(\gamma H^{-1})$, $H = O\left(|\partial_r F_C|\right) = O\left(a_{ij} r_c^{-1}\right)$ and $\tau_I = O\left(m\gamma^{-1}\right)$.
   Substitution of (9.119) and (9.120) into (9.130) yield, respectively,

$$\alpha = \frac{\tau}{\tau_I} \sim \frac{\gamma^2 r_c^5 n}{71.54 m k_B T}, \quad \text{for 3D space,} \tag{9.131}$$

$$\alpha = \frac{\tau}{\tau_I} \sim \frac{\gamma^2 r_c^4 n}{57.23 m k_B T}, \quad \text{for 2D space,} \tag{9.132}$$

which reveal the dependence of the dimensionless quantity α on m, r_c, n, γ and $k_B T$.

For the case of small m (small τ_I, large α), the particles' inertia can be neglected leading to a fast response. Small time steps are required for a numerical simulation, but particles are well distributed because of a low Mach number. The low-mass DPD system will be further discussed in Sect. 9.6.2.

We now keep the mass and number density fixed ($m = 1$ and $n = 4$) and vary the cutoff radius ($r_c \geq 1$). With standard DPD input values ($m = 1$, $k_B T = 1$, $n = 4$, $\sigma = 3$ ($\gamma = 4.5$), $r_c = 1$), where the DPD system is observed to well behave, the ratio between the two time scales can be estimated as

$$\alpha = \frac{\tau}{\tau_I} \sim 1.1322 = O(1), \quad \text{for 3D space,} \tag{9.133}$$

$$\alpha = \frac{\tau}{\tau_I} \sim 1.4153 = O(1), \quad \text{for 2D space.} \tag{9.134}$$

Expressions (9.131) and (9.132) indicate a very strong influence of r_c on the time-scale ratio. For example, with $r_c = 2.5$, one has $\alpha = O(10^2)$. Also, as r_c increases from 1 to 2.5 (a_{ij}, from (9.120), is reduced from 14.30 to 0.91), numerical experiments reveal that DPD particles tend to form local clusters. It appears that a large value of α, due to large r_c, may adversely affect the stability of the DPD system through clustering. Our recommendation here is to make α constant (independently of the input parameters) and having an appropriate value to avoid the clustering of particles. The optimal value of α can be determined numerically.

9.5.2 Imposition of Dimensionless Compressibility and Time-Scale Ratio

Our goal here is to create a new form of the conservative force that can control both the isothermal compressibility and time scale ratio. To do so, apart from a_{ij}, there is a need for having another free parameter. A conservative force is proposed to be

$$F_{ij,C} = a_{ij} \left(1 - \frac{r}{r_c}\right)^{\bar{s}}, \quad \bar{s} > 0, \tag{9.135}$$

whose average gradient over $0 \leq r \leq r_c$ is also a_{ij}/r_c. These two free parameters, a_{ij} and $\bar{s}$, can be designed to satisfy

$$\alpha = \frac{\tau}{\tau_I} = \frac{\gamma^2 r_c}{m a_{ij}}, \tag{9.136}$$

$$\kappa^{-1} = \frac{1}{k_B T} \frac{\partial p}{\partial n}, \tag{9.137}$$

where p and κ are the pressure and isothermal compressibility, respectively (α and κ are given constants, e.g., for water, $\kappa = 1/15.98$).

From the virial theorem, the pressure is computed as

$$p = nk_B T + \frac{n^2}{2d} \int d\mathbf{R}r \, F_{ij,C}(r)g(r), \tag{9.138}$$

where d is the flow dimensionality and $g(r)$ the radial distribution function - a possible function may be $g(r) = 1$ corresponding to an infinite number of DPD particles.

Expression (9.138) results in

$$p = nk_B T + \frac{4\pi a_{ij} n^2 r_c^4}{(\bar{s} + 1)(\bar{s} + 2)(\bar{s} + 3)(\bar{s} + 4)}, \tag{9.139}$$

$$\frac{\partial p}{\partial n} = k_B T + \frac{8\pi a_{ij} n r_c^4}{(\bar{s} + 1)(\bar{s} + 2)(\bar{s} + 3)(\bar{s} + 4)}, \tag{9.140}$$

for 3D case, and

$$p = nk_B T + \frac{\pi a_{ij} n^2 r_c^3}{(\bar{s} + 1)(\bar{s} + 2)(\bar{s} + 3)}, \tag{9.141}$$

$$\frac{\partial p}{\partial n} = k_B T + \frac{2\pi a_{ij} n r_c^3}{(\bar{s} + 1)(\bar{s} + 2)(\bar{s} + 3)}, \tag{9.142}$$

for 2D case. Equations (9.140) and (9.142) for the variable $\bar{s}$ can be solved analytically and we are interested in only physical positive values of $\bar{s}$.

Analytic solution to (9.136) and (9.137) can thus be found as

$$a_{ij} = \frac{1}{\alpha} \frac{\gamma^2 r_c}{m}, \tag{9.143}$$

$$\bar{s} = \frac{\sqrt{5 + 4\sqrt{C + 1}} - 5}{2}, \quad C = \frac{8\pi a_{ij} n r_c^4}{(\kappa^{-1} - 1)k_B T}, \tag{9.144}$$

for 3D space, and

$$a_{ij} = \frac{1}{\alpha} \frac{\gamma^2 r_c}{m}, \tag{9.145}$$

$$\bar{s} = \frac{1}{3B} + B - 2, \quad B = \left(\frac{C}{2} + \sqrt{\frac{C^2}{4} - \frac{1}{27}} \right)^{1/3}, \quad C = \frac{2\pi a_{ij} n r_c^3}{(\kappa^{-1} - 1)k_B T}, \tag{9.146}$$

for 2D space. For (9.144) and (9.146) to have a physical value (i.e. $\bar{s} > 0$), it requires $C > 24$ and $C > 6$, respectively, which can be easily satisfied.

Alternatively, the condition of isothermal compressibility can be replaced with the speed of sound

$$c_s^2 = \frac{\partial p}{\partial \rho},$$ (9.147)

where $\rho = mn$ is the density. The corresponding solutions a_{ij} and $\bar{s}$ have the same forms as (9.143)–(9.144) and (9.145)–(9.146), except that values of C in (9.144) and (9.146) are replaced with $8\pi a_{ij}nr_c^4/(mc_s^2 - k_BT)$ and $2\pi a_{ij}nr_c^3/(mc_s^2 - k_BT)$, respectively.

9.5.3 Imposition of Viscosity and Dynamic Response

Following [24], we employ

$$w_D = \left(1 - \frac{r}{r_c}\right)^s, \quad s > 0.$$ (9.148)

The dissipative force now involves 2 free parameters, γ and s, and we utilise them to match the two thermodynamic properties: flow resistance and dynamic response.

As shown in (9.83), there are two contributions to the viscosity, the kinetic part η_K (gaseous contribution) and the dissipative part η_D (liquid contribution). We are interested in the case where the dissipative contribution is a dominant part, i.e. $\eta_D \gg \eta_K$ (liquid-like behaviour) under which the two following constraints are approximately satisfied

$$\eta_D = \bar{\eta},$$ (9.149)

$$\frac{\eta_D}{\eta_K} = 2\bar{S}_c,$$ (9.150)

or

$$\frac{\gamma n^2 [R^2 w_D]_R}{2d(d+2)} = \bar{\eta},$$ (9.151)

$$\frac{2\gamma \bar{\eta}[w_D]_R}{dmk_BT} = 2\bar{S}_c,$$ (9.152)

where $\bar{\eta}$ and $\bar{S}_c$ are the specified viscosity and Schmidt number, respectively. This system can be solved analytically for the two variables γ and s. In 3D space, its solution is

$$s = \frac{-9 + \sqrt{1 + 4C}}{2}, \quad C = \frac{6\overline{S}_c m k_B T n^2 r_c^2}{5\overline{\eta}^2}, \tag{9.153}$$

$$\gamma = \frac{5\overline{\eta}(s + 1)(s + 2)(s + 3)(s + 4)(s + 5)}{16\pi n^2 r_c^5}. \tag{9.154}$$

Since $s > 0$, it requires

$$\overline{\eta} < \sqrt{\frac{3\overline{S}_c m k_B T n^2 r_c^2}{50}} \quad \text{for a given } \overline{S}_c, \tag{9.155}$$

$$\overline{S}_c > \frac{50\overline{\eta}^2}{3 m k_B T n^2 r_c^2} \quad \text{for a given } \overline{\eta}. \tag{9.156}$$

In 2D, the solution to the system takes the form

$$s = \frac{-7 + \sqrt{1 + 4C}}{2}, \quad C = \frac{3\overline{S}_c m k_B T n^2 r c^2}{4\overline{\eta}^2}, \tag{9.157}$$

$$\gamma = \frac{4\overline{\eta}(s + 1)(s + 2)(s + 3)(s + 4)}{3\pi n^2 r_c^4}. \tag{9.158}$$

Since $s > 0$, it requires

$$\overline{\eta} < \sqrt{\frac{\overline{S}_c m k_B T n^2 r_c^2}{16}} \quad \text{for a given } \overline{S}_c, \tag{9.159}$$

$$\overline{S}_c > \frac{16\overline{\eta}^2}{m k_B T n^2 r_c^2} \quad \text{for a given } \overline{\eta}. \tag{9.160}$$

Verification The four parameters, namely isothermal compressibility, time-scale ratio, dynamic response and viscosity, can be prescribed as inputs to the DPD equations. They define a particular fluid; but unlike conventional DPDs, a change in r_c should not affect the fluid characteristics. For a given shear rate $\dot{\gamma}$, shear stresses in Couette flow can be computed as $\tau_{xy} = \overline{\eta}\dot{\gamma}$ and we use these "exact values" to assess the accuracy of the present DPD.

For a typical set of input parameters ($\overline{\eta} = 30$, $\overline{S}_c = 400$, $r_c = 2.5$, $m = 1$, $k_B T = 1$, $n = 4$, $\dot{\gamma} = 0.2$), we plot the standard deviations of n and τ_{xy} on the cross section and percentage errors of τ_{xy} against the time-scale ratio α (Fig. 9.4), which show that their minimum values occur within a range $\alpha = 10^{-1} - 10^1$. For simplicity, α can be taken as 1.

To examine the effects of enforcing a fixed time scale ratio, the obtained results with $\alpha = 1$ are compared with those by the DPD using conventional conservative forces (i.e. variable α). As shown in Table 9.2, the former is more stable and accurate

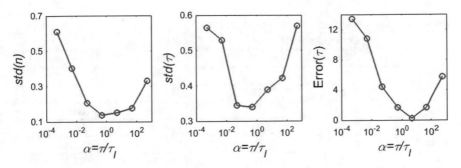

Fig. 9.4 Couette flow: Standard deviations of the number density n and shear stress τ_{xy} on the cross section and percentage errors of τ_{xy} against time-scale ratio $\alpha = \tau/\tau_I$

Table 9.2 Couette flow: mean and standard deviation of the number density n and shear stress τ_{xy} on the cross section

	Mean (n)	Std(n)	Mean (τ_{xy})	Error (τ_{xy}) (%)	Std (τ_{xy})
$r_c = 2$					
Conventional F_C	4.0000	0.2140	5.7873	3.5450	0.4293
Modified F_C	4.0000	0.1043	6.0956	1.5933	0.4208
$r_c = 2.5$					
Conventional F_C	4.0000	0.2672	5.7110	4.8167	0.4688
Modified F_C	4.0000	0.1701	6.1176	1.9600	0.4022
$r_c = 3.0$					
Conventional F_C	4.0000	0.2964	5.6590	5.6833	0.4119
Modified F_C	4.0000	0.1396	6.0286	0.4767	0.3177
$r_c = 3.5$					
Conventional F_C	4.0000	0.3958	5.6559	5.7350	0.6375
Modified F_C	4.0000	0.1246	6.0648	1.0800	0.3372
$r_c = 4$					
Conventional F_C	4.0000	0.3859	5.5519	7.4683	0.4635
Modified F_C	4.0000	0.1464	6.0403	0.6717	0.4576

than the latter for all values of r_c employed. In Table 9.3, several viscosities corresponding to s between 0 and 2 (standard value) are imposed, and the case of using conservative forces with a fixed α is seen to yield a better agreement.

Table 9.3 Couette flow: imposed and computed viscosities.

	Conventional F_C		Modified F_C	
Imposed η	Computed η	Error (%)	Computed η	Error (%)
$r_c = 2.5$				
39	38.3277	1.72	38.8942	0.27
35	34.1332	2.47	35.0943	0.26
31	29.7771	3.94	31.4797	1.54
$r_c = 3.5$				
58	57.5698	0.74	58.0389	0.06
50	49.3714	1.25	49.8442	0.31
42	41.1593	2.00	42.0830	0.19

9.6   Numerical Implementation

A typical DPD simulation is done in a similar manner to that of an MD: an initial configuration of fluid and wall particles are generated separately by a pre-processing program and read in as input data. The total number of particles depends on the size and geometry of the flow domain and the densities of the fluid and wall materials. Typically, fluid particles are initially located at conveniently points, for example, at the sites of a face-centred cubic (FCC) lattice. The initial velocities of the particles are set randomly according to a given temperature but the wall particles are frozen. At the beginning of the simulation the particles are allowed to move without applying the external force until a thermodynamic equilibrium is reached. Then the external force field is applied to fluid particles and the non-equilibrium simulation starts. There are many schemes to integrate the equations of motion for a DPD system in time, including Verlet, leap-frog, predictor-corrector algorithms (Nikunena et al. [60]), but the standard one is the velocity Verlet scheme, in a modified form - it is reviewed here below.

9.6.1   Velocity Verlet Scheme

The standard velocity Verlet algorithm follows directly from a Taylor's series expansion, omitting the subscript,

$$\mathbf{r}(t + \Delta t) = \mathbf{r}(t) + \Delta t \dot{\mathbf{r}}(t) + \frac{1}{2}\Delta t^2 \ddot{\mathbf{r}}(t) + \frac{1}{6}\Delta t^3 \dddot{\mathbf{r}}(t) + O\left(\Delta t^4\right), \quad (9.161)$$

where the acceleration in $\mathbf{r}$ is a known function, from the total force calculation. Similarly,

$$\mathbf{r}(t - \Delta t) = \mathbf{r}(t) - \Delta t \dot{\mathbf{r}}(t) + \frac{1}{2}\Delta t^2 \ddot{\mathbf{r}}(t) - \frac{1}{6}\Delta t^3 \dddot{\mathbf{r}}(t) + O\left(\Delta t^4\right). \quad (9.162)$$

From these, the basic Verlet updating scheme is derived,

$$\mathbf{r}(t + \Delta t) = 2\mathbf{r}(t) - \mathbf{r}(t - \Delta t) + \Delta t^2 \ddot{\mathbf{r}}(t) + O\left(\Delta t^4\right). \quad (9.163)$$

Since the evaluation of the acceleration (the forces) requires the current value of the velocity field, a slightly modified version of (9.163) is required:

$$\mathbf{r}(t + \Delta t) = \mathbf{r}(t) + \Delta t \mathbf{v}(t) + \frac{1}{2}\Delta t^2 \dot{\mathbf{v}}(t),$$
$$\tilde{\mathbf{v}}(t + \Delta t) = \mathbf{v}(t) + \lambda \Delta t \dot{\mathbf{v}}(t),$$
$$\dot{\mathbf{v}}(t + \Delta t) = \dot{\mathbf{v}}(\mathbf{r}(t + \Delta t), \tilde{\mathbf{v}}(t + \Delta t)),$$
$$\mathbf{v}(t + \Delta t) = \mathbf{v}(t) + \frac{1}{2}\Delta t [\dot{\mathbf{v}}(t) + \dot{\mathbf{v}}(t + \Delta t)], \quad (9.164)$$

where the acceleration $\dot{\mathbf{v}}(t)$ (i.e. the total net force) is a known function, given the position and velocity at instant t, $\tilde{\mathbf{v}}(t + \Delta t)$ is the velocity prediction at the instant $t + \Delta t$, λ is an empirically introduced parameter, which account for some additional effects of the stochastic interactions. If the total force is velocity independent, the standard velocity-Verlet algorithm is recovered at $\lambda = 1/2$. Groot and Warren [33] found $\lambda = 0.65$ to be optimum. At this value, the time step can be increased to $\Delta t \sim 0.05$ without a loss of temperature control.

9.6.2 Exponential Time Differencing Scheme

DPD fluid in its standard form is compressible in nature and its dynamic response is rather slow - the Schmidt number (Sc) is about unity because of the soft interaction between particles in the DPD system. On the other hand, many practical applications involve incompressible flows that exhibit strongly viscous behaviour at low fluid inertia, (i.e. low Reynolds number (Re) flows). Incompressibility is a good approximation in many practical flows at low Mach numbers ($M < 0.3$). One effective way to induce an incompressible slow viscous flow in a DPD fluid and simultaneously enhance its dynamic response is to reduce the mass of the particles (m). In the limit of $m \to 0$, the DPD system is singular with a corresponding loss of the highest order derivative and a strongly overdamped system results, with $Re \to 0$, $M \to 0$ and $Sc \to \infty$.

At low mass, $m \to 0$, the system (9.14) is stiff, and a different numerical integration scheme designed for stiff systems, such as the Exponential Time Differencing (ETD) method, may be a better alternative. To illustrate different ETD schemes, it is sufficient to consider the scalar system

$$\frac{d\phi(t)}{dt} + c\phi(t) = A + B\theta(t), \tag{9.165}$$

where the parameter c is either large and positive or large and imaginary, with A representing the deterministic part, and $B\theta(t)$, the stochastic part of the forcing terms, with $\theta(t)$ a white noise of unit strength. In our DPD case, when the mass is significantly small, $c = O(m^{-1})$ is positively large, and the system (9.165) is stiff. The solution in this case contains very different time scales occurring simultaneously (not counting the time scales inherent in $\theta(t)$), and standard integration schemes, such as the velocity-Verlet, are not accurate enough [81].

We multiply (9.165) by the integrating factor e^{ct}, and then integrate the equation over a single time step from t to $t + \Delta t$ to get

$$e^{c(t+\Delta t)}\phi(t + \Delta t) = e^{ct}\phi(t) + e^{ct}\int_0^{\Delta t} e^{c\tau}(A(t+\tau) + B(t+\tau)\theta(t+\tau))\,d\tau,$$

or

$$\phi(t + \Delta t) = e^{-c\Delta t}\phi(t) + \Delta A + \Delta B, \tag{9.166}$$

where

$$\Delta A(t; \Delta t) = e^{-c\Delta t}\int_0^{\Delta t} e^{c\tau}A(t+\tau)\,d\tau, \tag{9.167}$$

and

$$\Delta B(t; \Delta t) = e^{-c\Delta t}\int_0^{\Delta t} e^{c\tau}B(t+\tau)\theta(t+\tau)\,d\tau. \tag{9.168}$$

The basis of different ETD schemes lies in the approximations to the integrals (9.167) and (9.168).

First-Order Stochastic ETD Scheme In the simplest approximation, where A and B are treated as constants within this time step, one arrives at the first-order ETD scheme. In this case

$$\Delta A = \Delta A_1 = e^{-c\Delta t}\int_0^{\Delta t} e^{c\tau}A(t+\tau)\,d\tau = \frac{A(t)}{c}\left(1 - e^{-c\Delta t}\right), \tag{9.169}$$

and

$$\Delta B(t; \Delta t) = e^{-c\Delta t}B(t)\int_0^{\Delta t} e^{c\tau}\theta(t+\tau)\,d\tau$$
$$= B(t)\,\Delta W_1, \tag{9.170}$$

where ΔW_1 is the Wiener process

$$\Delta W_1(\Delta t) = e^{-c\Delta t}\int_0^{\Delta t} e^{c\tau}\theta(t+\tau)\,d\tau, \tag{9.171}$$

with zero mean,

$$\langle \Delta W_1 (\Delta t) \rangle = 0, \tag{9.172}$$

and auto-correlation function

$$\langle \Delta W_1 (\Delta t) \, \Delta W_1 (\Delta t) \rangle = e^{-2c\Delta t} \int_0^{\Delta t} dt' \int_0^{\Delta t} dt'' e^{ct'} \langle \theta (t + t') \, \theta (t + t'') \rangle e^{ct''}$$

$$= e^{-2c\Delta t} \int_0^{\Delta t} e^{2ct'} dt'$$

$$= \frac{1}{2c} \left(1 - e^{-2c\Delta t} \right). \tag{9.173}$$

We call the resulting scheme (9.166), with (9.169) and (9.170), the first-order stochastic exponential time differencing scheme. In summary, the scheme is defined as

$$\phi (t + \Delta t) = e^{-c\Delta t} \phi (t) + \Delta A_1 (t; \Delta t) + B (t) \, \Delta W_1. \tag{9.174}$$

Second-Order Stochastic EDT Scheme Next, we consider a higher-order approximation scheme whereby A and B are approximated as linear functions in the interval $0 < \tau < \Delta t$:

$$A (t + \tau) = A (t) + \frac{\tau}{\Delta t} (A (t) - A (t - \Delta t)) = A (t) + \frac{\Delta a}{\Delta t} \tau,$$

$$B (t + \tau) = B (t) + \frac{\tau}{\Delta t} (B (t) - B (t - \Delta t)) = B (t) + \frac{\Delta b}{\Delta t} \tau,$$

where $\Delta a = A (t) - A (t - \Delta t)$ and $\Delta b = B (t) - B (t - \Delta t)$ are the first-order backward differences for A and B. Then,

$$\Delta A (t; \Delta t) = e^{-c\Delta t} \int_0^{\Delta t} e^{c\tau} A (t + \tau) d\tau$$

$$= \frac{A (t)}{c} \left(1 - e^{-c\Delta t} \right) + \frac{\Delta a}{\Delta t} e^{-c\Delta t} \int_0^{\Delta t} e^{c\tau} \tau d\tau$$

$$= \Delta A_1 (t; \Delta t) + \frac{c\Delta t - 1 + e^{-c\Delta t}}{c^2} \frac{\Delta a}{\Delta t}$$

$$= \Delta A_1 (t; \Delta t) + \Delta A_2 (t; \Delta t), \tag{9.175}$$

where $\Delta A_1 (t; \Delta t)$ has been defined in (9.169) and

$$\Delta A_2 (t; \Delta t) = \frac{c\Delta t - 1 + e^{-c\Delta t}}{c^2} \frac{\Delta a}{\Delta t}. \tag{9.176}$$

In addition $\Delta B\,(t;\,\Delta t)$ is given by

$$\Delta B\,(t;\,\Delta t) = e^{-c\Delta t} \int_0^{\Delta t} e^{c\tau} B\,(t+\tau)\,\theta\,(t+\tau)\,d\tau$$

$$= B\,(t)\,\Delta W_1 + \frac{\Delta b}{\Delta t} e^{-c\Delta t} \int_0^{\Delta t} \tau e^{c\tau}\theta\,(t+\tau)\,d\tau$$

$$= B\,(t)\,\Delta W_1 + \frac{\Delta b}{\Delta t}\,\Delta W_2, \tag{9.177}$$

where the Wiener process ΔW_1 has been defined in (9.171), and the Wiener process ΔW_2 is defined as

$$\Delta W_2\,(\Delta t) = e^{-c\Delta t} \int_0^{\Delta t} \tau e^{c\tau}\theta\,(t+\tau)\,d\tau. \tag{9.178}$$

It has zero mean

$$\langle \Delta W_2 \rangle = 0, \tag{9.179}$$

and its mean square is given by

$$\langle \Delta W_2 \Delta W_2 \rangle = e^{-2c\Delta t} \int_0^{\Delta t} dt' \int_0^{\Delta t} dt''\, t' e^{ct'} \left\langle \theta\,(t+t')\,\theta\,(t+t'') \right\rangle t'' e^{ct''}$$

$$= e^{-2c\Delta t} \int_0^{\Delta t} dt'\, t'^2 e^{2ct'}$$

$$= \frac{1}{4c^3} e^{-2c\Delta t} \left[e^{2c\Delta t} \left(2c^2 \Delta t^2 - 2c\Delta t + 1 \right) - 1 \right]$$

$$= \frac{1}{4c^3} \left[2c^2 \Delta t^2 - 2c\Delta t + 1 - e^{-2c\Delta t} \right]. \tag{9.180}$$

These result in an integration scheme called the second-order stochastic exponential time differencing, summarised as

$$\phi\,(t+\Delta t) = e^{-c\Delta t}\phi\,(t) + \Delta A_1\,(t;\,\Delta t) + \frac{c\Delta t - 1 + e^{-c\Delta t}}{c^2}\,\frac{\Delta a}{\Delta t}\,(t;\,\Delta t)$$

$$+ b\,(t)\,\Delta W_1\,(\Delta t) + \frac{\Delta b}{\Delta t}\,\Delta W_2\,(\Delta t). \tag{9.181}$$

In our DPD simulations, with mass as low as $m = 10^{-3}$, the first-order ETD scheme proved to be very accurate, and is to be preferred for its simplicity.

9.6.3 *Implementation of No-Slip Boundary Conditions*

Periodic boundary conditions are applied on the fluid boundaries of the computational domain. The solid wall is usually represented using frozen particles. Due to the soft repulsion between DPD particles, it is possible for fluid particles to penetrate the wall. Near-wall particles may not be sufficiently slowed down and slip with respect to the wall may then occur. To prevent this, a higher wall density and larger repulsive forces have been used to strengthen wall effects. These, however, result in large density distortions in the flow field near to the wall, similar to MD. Special treatments have been proposed to implement no-slip boundary condition in DPD simulation without using frozen wall particles. Revenga et al. [77] used effective forces to represent the effects of wall on fluid particles instead of using wall particles. For a planar wall, the effective forces can be obtained analytically. But these forces are not guaranteed to prevent fluid particles from crossing the wall. When particles cross the wall, a wall reflection is used to reflect particles back to the fluid. Willemsen et al. [91] added an extra layer of particles outside of the simulation domain. The position and velocity of particles in this layer are determined by the particles inside the simulation domain near the wall, such that the mean velocity of a pair of particles inside and outside the wall satisfies the given boundary conditions. The above mentioned methods may not work when the geometry of wall is complicated. We still use frozen particles to represent the wall. Near the wall we assume that there is a thin layer where the no-slip boundary condition holds. We enforce a random velocity distribution in this layer with zero mean corresponding to the given temperature. Similar to Revenga et al.'s [77] reflection law, we further require that particles in this layer always leave the wall. The velocity of particle i in the layer is

$$\mathbf{v}_i = \mathbf{v}_R + \mathbf{n} \left(\sqrt{(\mathbf{n} \cdot \mathbf{v}_R)^2} - \mathbf{n} \cdot \mathbf{v}_R \right), \tag{9.182}$$

where $\mathbf{v}_R$ is a random vector and $\mathbf{n}$ a unit vector normal to the wall and pointing to the fluid. The thickness of this layer and the strength of the repulsion between wall and fluid particles are chosen to minimise the velocity and density distortion. The layer thickness is chosen to be the minimum between 0.5% of the narrowest width of channels in the flow domain and 0.5 of length unit. A thin layer is necessary to prevent the frozen wall to cool down the fluid. This method is more flexible when dealing with a complex geometry.

A simpler reflection law is used in Duong et al. [19] to ensure no-slip boundary condition on a solid surface,

$$\mathbf{v}_{new} = 2\mathbf{V}_{wall} - \mathbf{v}_{old}, \tag{9.183}$$

where $\mathbf{V}_{wall}$ is the wall velocity vector. With this reflection law, no-slip boundary is guaranteed; however there is still a density fluctuation near the wall. Duong et al. [19] recommended a two-layer structure as illustrated in Fig. 9.5.

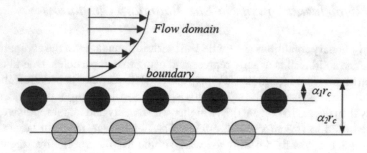

Fig. 9.5 Boundary DPD particles

In this scheme, the distances from the wall layers to the boundary are proportional to r_c, say $\alpha_1 r_c$ and $\alpha_2 r_c$, respectively, with $0 < \alpha_1 < \alpha_2 < 1$. Those layers of frozen particles also interact with free particles in the same way as the free particles interact between themselves. To prevent free particles from penetrating the wall, firstly we chose α_1 sufficient small, and secondly when particles enter in the wall layers then a bounce back is performed as:

$$\mathbf{r}_{new} = \mathbf{r}_{old} + 2d_r \mathbf{n}_w, \tag{9.184}$$

where d_r is the distance from the particle to the boundary and $\mathbf{n}_w$ is the normal vector on the wall directing into the simulation domain. The purpose of second layer is not only to strengthen the wall effect but also to make repulsive forces the wall particles exerting on free particles almost uniform. This layer plays a key role on maintaining a uniform density in the fluid near the wall. We typically use this approach in implementing no-slip boundary conditions in DPD.

9.6.4 Computation of Interparticle Forces

In simulation a large system the computational efficiency is an important consideration. The main effort in solving the equations of motion is to compute the interaction forces between the particles. Fortunately, there has been a considerable number of works done in MD, which are directly applicable to DPD simulations. In particular, the cell sub-division and linked-list approach as described in Rapaport [75] can be employed to reduce the computation time in the force calculation. The linked list associates a pointer with each data item and to provide a non-sequential path through the data. In the cell algorithm, the linked list associates particles with the cells in which they reside at any given instance. It needs a one dimension array to store the list and makes the search of neighbouring particles more efficient. To further save computational time, a neighbour-list method can also be employed. The neighbour-list remains valid over a number of time steps, typically 10–20 time steps. Hence the computation time in force calculation can be reduced further, with periodic refreshing the list.

9.6.5   Calculation of Stress Tensor

The flow domain is divided into grids and local data are collected in each bin. The flow properties are calculated by averaging over all sampled data in each bin. The stress tensor is calculated using the Irving–Kirkwood expression [40]

$$
\mathbf{S} = -\frac{1}{V} \left[\sum_i m \mathbf{V}_i \mathbf{V}_i + \frac{1}{2} \sum_i \sum_{j \neq i} \mathbf{r}_{ij} \mathbf{F}_{ij} \right] = -n \left(\langle m\mathbf{V}\mathbf{V} \rangle + \frac{1}{2} \langle \mathbf{r}\mathbf{F} \rangle \right), \quad (9.185)
$$

where n is the number density of particles, and $\mathbf{V}_i$ is the peculiar velocity of particle i (this should be compared to (9.49) and (9.50)). The first term on the right side of (9.185) denotes the contribution to the stress from the momentum (kinetic) transfer of DPD particles (c.f., (9.49)) and the second term from the interparticle forces. If particle i is an element of a complex particle (for example, a bead in a chain), the total force in the second term on the right of (9.185) should include constraint forces. The constitutive pressure can be determined from the trace of the stress tensor:

$$
p = -\frac{1}{3} tr \mathbf{S}. \quad (9.186)
$$

9.6.6   Complex-Structure Fluid

DPD particles are connected by various spring-force laws or constraints to simulate a complex-structure fluid. Various degrees of complexities in a complex-structure fluid may be incorporated in a DPD simulation. For example, a suspension may be modelled by constraining some DPD particles as rigid bodies in a DPD fluid (as a solvent) (i.e. frozen particle model, spring model), or by using two different species of DPD particles with different exclusion sizes: one of smaller size may represent solvent particles, and the other of larger size may represent suspended particles (i.e. single particle model). A polymer solution may be modelled in the same manner: a polymer chain is represented by connecting some DPD particles by some connector force law. Several kinds of spring-bead chains have been used in polymer rheology as coarse-grained models of macromolecules. For example, the Kramers' bead-rod chain, the FENE (Finitely Extendable Non-linear Elastic) dumbbell [6], worm-like chain [50]. These two types for force laws and the spring model are briefly reviewed below.

FENE Chains   The FENE chain is a simple coarse-grained model of a polymer molecule, which can capture most of the important nonlinear rheological properties of polymer solution. Its rheological properties are well known [88]. In the framework of DPD, the beads in a FENE chain are replaced by DPD particles. In addition to the three kinds of forces in (9.14), the connector (intermolecular) forces act on these

FENE particles and contribute to the right hand of (9.14). For the FENE chain, the force on bead i due to bead j is

$$\mathbf{F}_{ij}^S = -\frac{H\mathbf{r}_{ij}}{1 - \left(r_{ij}/r_m\right)^2}, \tag{9.187}$$

where H is the spring constant, $\mathbf{r}_{ij} = \mathbf{r}_i - \mathbf{r}_j$, $r_{ij} = |\mathbf{r}_{ij}|$, and r_m is the maximum length of one chain segment. The spring force increases unboundedly as r_{ij}/r_m approaches unity. This model can capture the finite extensibility of a polymer chain, and yields a shear-rate dependent viscosity along with a finite elongational viscosity (as opposed to an infinite elongational viscosity when the force law is linear for the Hookean dumbbell).

The time constant is important in characterizing molecular motion and can be formed from model parameters. Two constants with time dimension can be defined for the FENE spring [88],

$$\lambda_H = \frac{\zeta}{4H}, \tag{9.188}$$

where $\zeta = 6\pi\eta a_{eff}$ is the Stokes friction on a bead of size a_{eff}, and

$$\lambda_Q = \frac{\zeta r_m^2}{12 k_B T}. \tag{9.189}$$

The FENE parameter, b, is the ratio of these two time constants,

$$b = \frac{3\lambda_Q}{\lambda_H} = \frac{H r_m^2}{k_B T}. \tag{9.190}$$

Chain models (with more than two beads) have a spectrum of relaxation times [6]. There is no close-form expression for the relaxation time spectrum for FENE chains. However, a modification of FENE chain, called the FENE-PM chain, has the same spectrum as the Rouse chain (Hookean spring chain) [6]. A time constant can be defined for FENE-PM chain as [88]:

$$\lambda_{fene} = \frac{b}{b+3}\lambda_H \frac{N_b^2 - 1}{3}, \tag{9.191}$$

where N_b is the number of beads in the chain. The contour length is an appropriate parameter to represent the molecular size. If L_c denotes the length of one segment of a molecular chain, the contour length of the molecular chain is simply equal to $(N_b - 1) L_c$. We may use the equilibrium length of a segment as an estimation of L_c or just use the maximal length of a segment, r_m.

Fig. 9.6 A schematic diagram for the spring model: basic DPD particles (*circles*) connected to reference sites (*squares*) via linear springs. The reference sites are supposed to move as a rigid body

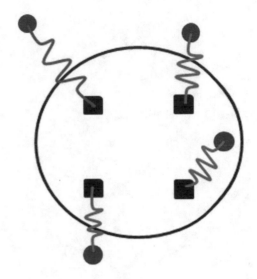

Wormlike Chains It was found that mechanical properties of DNA molecules in an aqueous solution can be realistically modelled by the wormlike chain model [13]. The spring force law of this model can be expressed as

$$\mathbf{F}_{ij}^S = -\frac{k_B T}{4\lambda_p^{eff}} \left[\left(1 - \frac{r_{ij}}{l}\right)^{-2} + \frac{4r_{ij}}{l} - 1 \right] \hat{\mathbf{r}}_{ij}, \tag{9.192}$$

where l is the maximum length of one chain segment, $\hat{\mathbf{r}}_{ij} = \mathbf{r}_{ij}/r_{ij}$, and λ_p^{eff} is the effective persistence length of the chain. If the total length of the chain is L and the number of bead in the chain is N_b, $l = L/(N_b - 1)$. Bustamante et al. [13] found the persistence length for their unlabelled DNA molecules, λ_p, is about $0.053\ \mu m$. When modelling the DNA molecules with wormlike spring chain, the beads increases the molecules' flexibility, since they do not transmit bending moment. This increase of the chain flexibility can effectively be compensated by slightly increasing the persistence length [49]. In their simulation, the length of the DNA molecule is 67.2 μm, Larson et al. [49] used $\lambda_p^{eff} = 0.061$ for a 40-bead DNA chain and $\lambda_p^{eff} = 0.07$ for a 80-bead DNA chain. Note that their chains are tethered, i.e., a bead in the tethered end is not required. Hence, the number of beads in their chains is equal to the number of chain segments.

Suspended Particles by Spring Model In the spring model ([55, 69]), a colloidal particle is modelled by using only a few basic DPD particles that are connected to the reference sites (on that colloidal particle) by linear springs of very large stiffness (Fig. 9.6). For example, a hard disc may be simply constructed using 3 or 4 basic DPD particles, whose associated reference sites are located at the vertices of an equilateral triangle or square, while a spherical particle may be represented using 6 or 8 basic DPD particles with their reference sites at the vertices of either an octahedron or a

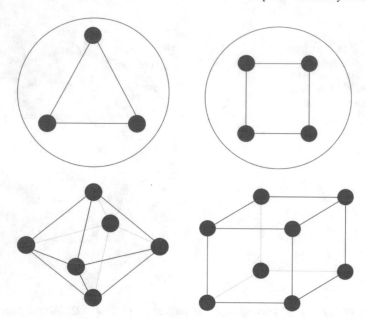

Fig. 9.7 A 2D circular disc can be modelled using 3 (equilateral *triangle*) and 4 (*square*) basic DPD particles (*top*), while a spherical particle can be represented using 6 (*octahedron*) and 8 (*cube*) basic DPD particles (*bottom*)

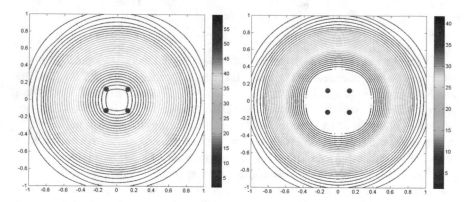

Fig. 9.8 2D circular disc, 4 basic DPD particles: contour plots of the superimposition of conservative/repulsive shape functions (*left*) and the radial component of the total conservative force vector (*right*)

cube (Fig. 9.7). The rationale behind the spring model is that the shape of a colloid is actually defined by the repulsion force field generated by constituent particles of that colloid (Fig. 9.8). The reference sites, collectively modelling a rigid body, move as a rigid body motion calculated through their Newton-Euler equations, using data from the previous time step,

$$M_c^k \frac{d\mathbf{V}_c^k}{dt} = \mathbf{F}^k(t - \Delta t), \tag{9.193}$$

$$\mathbf{I}^k \frac{d\boldsymbol{\omega}^k}{dt} = \mathbf{T}^k(t - \Delta t), \tag{9.194}$$

$$\mathbf{F}^k(t - \Delta t) = \sum_{i=1}^{p} \sum_{j=1, j \neq i}^{N} \left[\mathbf{F}_{ij,C}^k(t - \Delta t) + \mathbf{F}_{ij,D}^k(t - \Delta t) + \mathbf{F}_{ij,R}^k(t - \Delta t) \right], \tag{9.195}$$

$$\mathbf{T}^k(t - \Delta t) = \sum_{i=1}^{p} \left(\mathbf{r}_i^k(t - \Delta t) - \mathbf{R}_c^k(t - \Delta t) \right) \times$$

$$\sum_{j=1, j \neq i}^{N} \left[\mathbf{F}_{ij,C}^k(t - \Delta t) + \mathbf{F}_{ij,D}^k(t - \Delta t) + \mathbf{F}_{ij,R}^k(t - \Delta t) \right], \tag{9.196}$$

where $M_c^k, \mathbf{I}^k, \mathbf{R}_c^k, \mathbf{r}_i^k, \mathbf{V}_c^k, \boldsymbol{\omega}^k$ and p are the mass, moment of inertia tensor, centre of mass, position of particle i, centre-of-mass velocity, angular velocity, and the number of constituent particles of the kth colloidal particle, respectively.

The velocities of their associated DPD particles are found by solving the DPD equations at the current time level

$$m_i \frac{d\mathbf{v}_i^k(t)}{dt} = \sum_{j=1, j \neq i}^{N} \left[\mathbf{F}_{ij,C}^k(t) + \mathbf{F}_{ij,D}^k(t) + \mathbf{F}_{ij,R}^k(t) \right] + \mathbf{F}_{i,S}^k(t), \quad i = (1, 2, \ldots, p). \tag{9.197}$$

where $\mathbf{F}_{i,S}^k(t) = -H \left[\mathbf{r}_i^k(t) - \bar{\mathbf{r}}_i^k(t) \right]$ is the spring force with H being the stiffness of the spring and $\bar{\mathbf{r}}_i$ the position of the reference site i. Introducing springs into the model allows the constituent particles to fully participate in the system dynamics, leading to a better control of the system temperature. The spring stiffness is chosen large enough to ensure a good representation of rigid particles, but not so large because of stability consideration (the value of 3000 is chosen after some experimentations). Note that in the stiff limit ($H \to \infty$), the spring model exactly reduces to a frozen particle model.

Compared to the single DPD particle models, the spring model works well with larger time steps (only soft potentials are employed here), involves considerably less parameters (the dispersion phase is also based on basic DPD particles and conse-quently, the solvent-colloidal interaction can be decomposed into a set of solvent-solvent interactions), and can be extended to the case of suspended particles of non-spherical or non-circular shapes more straightforwardly. It was reported in [66] that the single particle model employs a time step in the range of 0.0002–0.0005. The time steps used in the spring model are 0.005–0.01 for 2D and 0.001–0.005 for 3D (i.e. about one order of magnitude higher). Compared to the frozen particle models, the spring model employs only a few DPD particles to represent a suspended particle and can maintain the specified temperature throughout the flow domain.

One distinguishing feature of the spring model is that the solvent particles and the constituent particles of the suspended particles use the same values of the DPD parameters including the Boltzmann temperature (contrasting to different sets of DPD parameters, solvent-solvent, solvent-colloidal, colloidal-colloidal, in one-single DPD particle models). Consequently, the resultant DPD system is based on identical basic particles, and the particle volume fraction can be simply defined as

$$\phi = \frac{N_c^0}{N_c^0 + N_s},$$ (9.198)

where N_c^0 and N_s are the numbers of the basic DPD particles used to represent the solvent and colloidal phases, respectively. Numerical results show that this definition of the volume fraction (9.198) (which is the number fraction) recovers Einstein's relation at low volume fraction irrespective of values of the input DPD parameters, and therefore is appropriate for DPD suspensions.

9.7   Flow Verifications and Some Typical Problems

Couette Flow In a Couette flow, in which the velocity is

$$\mathbf{u} = (u, 0, 0), \quad u = \dot{\gamma}y, \quad \dot{\gamma} = U/h,$$ (9.199)

the stress for a simple fluid is

$$[\mathbf{T}] = \begin{bmatrix} T_{11} & T_{12} & 0 \\ T_{12} & T_{22} & 0 \\ 0 & 0 & T_{33} \end{bmatrix},$$ (9.200)

with the shear stress is an odd function of the shear rate $\dot{\gamma}$, and the first (N_1) and the second (N_2) normal stress differences are even functions of the shear rate $\dot{\gamma}$,

$$\begin{aligned} T_{12} &= \eta(\dot{\gamma})\,\dot{\gamma}, \quad \eta(\dot{\gamma}) = \eta(-\dot{\gamma}), \\ N_1(\dot{\gamma}) &= T_{11} - T_{22} = N_1(-\dot{\gamma}), \\ N_2(\dot{\gamma}) &= T_{22} - T_{33} = N_2(-\dot{\gamma}), \end{aligned}$$ (9.201)

where the viscosity is an even function of the shear rate.

Our DPD flow domain consists of unit cells in the x, y and z directions, respectively, $L_x \times L_y \times L_z = 40 \times 10 \times 30$. Initially, a FCC arrangement of DPD particles was formed and was allowed to equilibrate before the simulation was started, with an initial density of 4 (4 particles per unit cell or volume); a total of 94400 DPD fluid particles and 12420 DPD wall particles were involved in the simulation. In our simulation, the wall boundary is constructed using three layers of frozen particles. In addition, we assume that there is a thin layer near the wall, in which a random

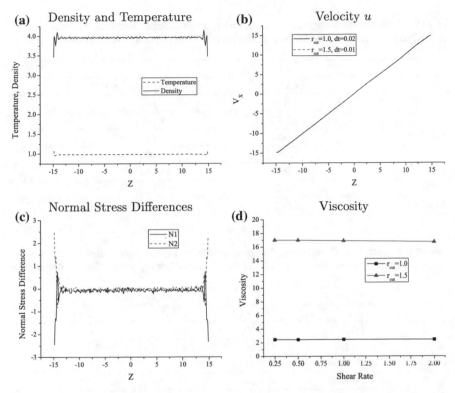

Fig. 9.9 Couette flow results **a** Density and temperature **b** Velocity u **c** Normal stress differences **d** Viscosity as function of shear rates

velocity distribution with zero mean corresponding to a given temperature is generated [19]. In order to prevent particles from penetrating the walls, we further require that the particles in this layer always leave the wall according to the reflection law reported in [77]. The central flow region across z direction was divided into 300 bins (for the averaging purpose), and the average was done in the bins over 104 time steps. The choice of the standard DPD parameters, $k_B T = 1, \gamma = 4.5, n = 4$, with a modified weighting function correspond to $s = \frac{1}{2}$ leads to the temperature, density, velocity, normal stress differences and viscosity profiles as shown in Fig. 9.9, with the expected fluctuations near the solid walls, using velocity-Verlet algorithm. The numerical results show a Newtonian behaviour with zero mean normal stress differences (except at the wall, where fluctuations occur), and constant viscosity with shear rate, in agreement with the expectation of a Newtonian behaviour of simple DPD fluid.

To explore the effects of mass, we consider four values of m, namely 1.0, 0.1, 0.01 and 0.001. This results in a fluid of $\eta = 2.4059$ for $m = 1$, $\eta = 2.2494$ for $m = 0.1$, $\eta = 2.2337$ for $m = 0.01$ and $\eta = 2.2322$ for $m = 0.001$.

Table 9.4 Couette flow: comparison of the mean equilibrium temperature of the ETD and velocity-Verlet algorithms for the case of $m = 0.1$. The velocity-Verlet algorithm fails to converge at $\Delta t > 0.009$

Δt	$\overline{k_B T}$	
	ETD	Velocity-Verlet
0.01	1.003	–
0.009	1.002	–
0.007	1.002	0.9538
0.005	1.002	0.9386
0.003	1.002	0.9607
0.001	1.002	0.9869
0.0009	1.002	0.9884

Table 9.5 Couette flow: Comparison of the mean equilibrium temperature of the ETD and velocity-Verlet algorithms for the case of $m = 0.01$. The velocity-Verlet algorithm fails to converge at $\Delta t > 0.0009$

Δt	ETD		Velocity-Verlet	
	$\overline{k_B T}$	Error (%)	$\overline{k_B T}$	Error (%)
0.007	0.5680	43.19	–	–
0.005	0.7633	23.66	–	–
0.003	0.9356	6.44	–	–
0.001	0.9835	1.64	–	–
0.0009	0.9863	1.36	–	–
0.0007	0.9916	0.83	0.9187	8.12
0.0005	0.9958	0.41	0.9322	6.77
0.0003	0.9987	0.12	0.9576	4.23
0.0001	0.9992	0.07	0.9840	1.60

Tables 9.4, 9.5 and 9.6 show the behaviour of the temperature against the time step for $m = 0.1$, $m = 0.01$ and $m = 0.001$, respectively, using the first-order ETD scheme (9.174). Results by the velocity-Verlet algorithm are also included. It can be seen that the first-order ETD algorithm works effectively for relatively-large time steps. Furthermore, for a given small time step, the ETD algorithm is much more accurate than the velocity-Verlet algorithm. In the case of $m = 0.1$, the ETD algorithm produces the equilibrium temperature that is accurate up to 3 significant digits. In the case of $m = 0.01$ and $m = 0.001$, equipartition is consistently improved as the time step reduces. Velocity-Verlet algorithm fails to converge except at small time steps, and the associated errors are much larger than those produced by the ETD algorithm. Velocity, temperature and number density results obtained with the first-order ETD scheme are presented in Fig. 9.9. We obtain a linear velocity profile in the x direction, and uniform temperature and density using a time step $\Delta t = 0.02$

Table 9.6 Couette flow: comparison of the mean equilibrium temperature of the ETD and velocity-Verlet algorithms for the case of $m = 0.001$. The velocity-Verlet algorithm fails to converge at $\Delta t > 0.00007$

| Δt | ETD | | Velocity-Verlet | |
	$\overline{k_B T}$	Error (%)	$\overline{k_B T}$	Error (%)
0.0002	0.9501	4.98	–	–
0.0001	0.9805	1.94	–	–
0.00009	0.9837	1.62	–	–
0.00008	0.9865	1.34	–	–
0.00007	0.9891	1.08	–	–
0.00006	0.9916	0.83	0.9127	8.72
0.00005	0.9934	0.65	0.9235	7.64
0.00004	0.9951	0.48	0.9365	6.34
0.00003	0.9958	0.41	0.9502	4.97

for $m = 1.0$, $\Delta t = 0.005$ for $m = 0.1$, $\Delta t = 0.001$ for $m = 0.01$ and $\Delta t = 0.0002$ for $m = 0.001$.

Poiseuille Flow Poiseuille flow in a channel of $L_x \times L_y \times L_z = 40 \times 10 \times 30$ is also assessed at four different values of m, 1.0, 0.1, 0.01 and 0.001. Other parameters used are kept the same as before, $r_c = 1.0$, $n = 4$, $a_{ij} = 18.75$, $\sigma = 3.0$, $s = 1/2$ and $k_B T = 1.0$, resulting in a fluid of $\eta = 2.4059$ for $m = 1$, $\eta = 2.2494$ for $m = 0.1$, $\eta = 2.2337$ for $m = 0.01$ and $\eta = 2.2322$ for $m = 0.001$. Pressure gradient is simulated by applying a body force $\mathbf{F} = (0.1, 0, 0)^T$ to each particle. Analytical parabolic velocity profile for a Newtonian fluid is

$$v_x = \frac{n F_x}{2\eta} \left(\frac{L_z}{2} - z \right) \left(\frac{L_z}{2} + z \right). \tag{9.202}$$

Results for the velocity, temperature and number density are presented in Fig. 9.10. We obtain a parabolic velocity profile in the x direction, as expected. Percentage errors (relative to the exact value obtained by (9.202) using the approximate viscosity values quoted) for the maximum value of v_x at the centreline are 8.71% for $m = 1$ and 10.57% for $m = 0.001$. Note also that the Sc number at $m = 0.001$ is estimated to be of $O\left(6.8 \times 10^3\right)$, which is much more than a water-like liquid ($Sc \sim 2 \times 10^3$).

Convergent-Divergent Channel Flow Next we consider a more complex flow: a simple DPD fluid flows through a periodic channel with abrupt contraction and diffusion shown in Fig. 9.11. The DPD parameters used in the velocity-Verlet simulation are, $r_c = 1.5$, $n = 4$, $a_{ij} = 18.75$, $\sigma = 3.0$, $s = 1/2$, $k_B T = 1.0$ and $m = 1$. The total length of one period is 90 and the length of the contraction segment is 40. The width of the contraction segment is 20 and the contraction ratio is 9:4. When a particle passes through this channel, it would be experienced acceleration and deceleration, and the dynamic behaviour of the system becomes important. A system with

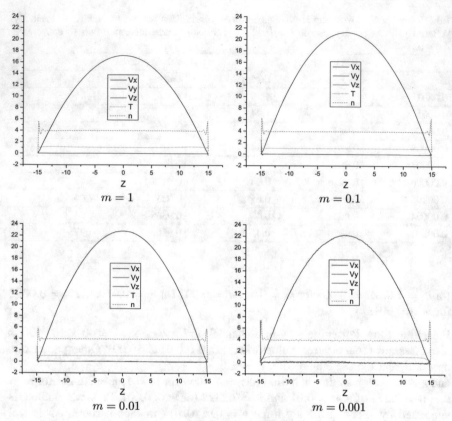

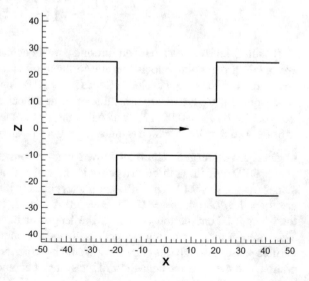

Fig. 9.10 Poiseuille flow: Profiles of velocity, temperature and number density for several values of the mass

Fig. 9.11 The geometry of a periodic abrupt expansion and diffusion channel

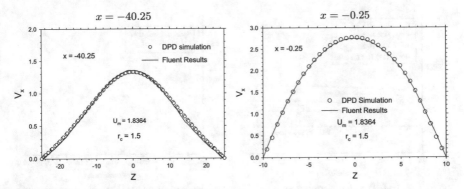

Fig. 9.12 Velocity profiles at two different stations as compared to Fluent's predictions, U_m is the mean velocity

the total number of 87886 particles was simulated, in which there were 8688 wall particles and 79198 fluid particles. To check the results of DPD simulation, we used a commercial software, Fluent, to simulate this flow. The same physical parameters for the DPD fluid, flow geometry and periodic boundary conditions are used in the Fluent simulation.

From the results in Fig. 9.12, it is clear that the DPD simulation captures the flow kinematics well, especially the dynamic behaviour of a particle as it moves through an unsteady Lagrangian flow field.

Hookean Dumbbell Model in DPD Simulation A dilute polymer solution could be modelled by a dilute suspension of Hookean dumbbells, in which the connector force between particles i and j on a dumbbell is given by

$$\mathbf{F}_{ij}^S = -H\mathbf{r}_{ij}, \tag{9.203}$$

where H is a spring constant. In Sect. 7, it was shown that this microstructure leads to the Oldroyd-B fluid in which the total stress tensor is

$$\mathbf{T} = \mathbf{S}^{(s)} + \mathbf{S}^{(p)}, \quad \mathbf{L} = (\nabla \mathbf{u})^T$$

$$\mathbf{S}^{(p)} + \lambda \left\{ \frac{d}{dt}\mathbf{S}^{(p)} - \mathbf{L}\mathbf{S}^{(p)} - \mathbf{S}^{(p)}\mathbf{L}^T \right\} = G\mathbf{I}, \quad G = \nu k_B T, \tag{9.204}$$

where ν is the number density of dumbbells, and $\lambda = \zeta/4H$ is the relaxation time, with ζ the frictional factor of a dumbbell's bead. ζ may be estimated as $6\pi\eta_s a_{eff}$, where a_{eff} is the effective size of a dumbbell bead (DPD particle).

In a simple shear flow, the solution viscosity is a constant, its 1st normal stress difference is quadratic in the shear rate, and its second normal stress difference is zero:

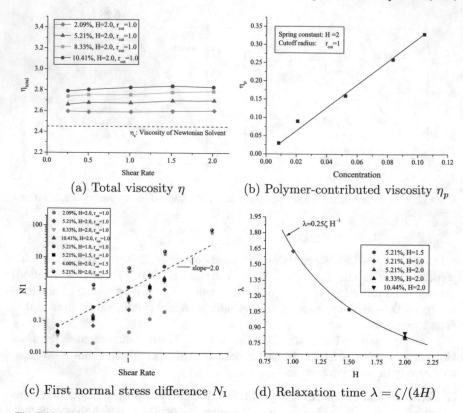

(a) Total viscosity η

(b) Polymer-contributed viscosity η_p

(c) First normal stress difference N_1

(d) Relaxation time $\lambda = \zeta/(4H)$

Fig. 9.13 DPD model for a dilute polymer solution

$$\eta = \frac{S_{12}}{\dot\gamma} = \eta_s + \eta_p, \quad N_1 = S_{11}^{(p)} - S_{22}^{(p)} = 2\eta_p\lambda\dot\gamma^2, \quad N_2 = S_{22}^{(p)} - S_{33}^{(p)} = 0,$$
(9.205)

where the polymer-contributed viscosity is $\eta_p = G\lambda$, which is proportional to the number density of the dumbbell. Note that λ may estimated from the normal stress to shear stress ratio,

$$\lambda = \frac{N_1}{2\dot\gamma S_{12}^{(p)}}.$$
(9.206)

In the velocity-Verlet DPD simulation, a total of 94400 DPD fluid plus 12420 wall DPD particles was used in a (40, 20, 30) cells domain, with the standard DPD parameters, $r_c = 1, n = 4, a_{ij} = 18.75, \sigma = 3.0, s = 1/2, k_B T = 1.0$ and $m = 1$. In the simulation, a dumbbell is modelled by connecting two DPD particles with a linear spring of constant stiffness H. From the solvent and the total viscosity (Fig. 9.13a), the polymer-contributed viscosity can be calculated, and the results are plotted in Fig. 9.13b. This quantity is seen proportional to the number density of the dumbbells as expected. Furthermore, from the first normal stress difference, Fig. 9.13c, which is quadratic in the shear rate, the relaxation time can be derived according to (9.206) and is plotted in Fig. 9.13c against the spring stiffness H. Plotted in the same figure is the predicted relaxation time from kinetic theory, $\zeta/(4H)$; a good agreement is clearly seen.

Particulate Suspensions In DPD, monodispersed suspensions are modelled through two sets of particles: one for the solvent phase (basic DPD particles freely movable) and the other for the dispersion phase (e.g. constrained basic DPD particles for the spring model). The kinetic theory [56], confirmed by numerical results, shows that there is an exclusion zone associated with a basic DPD particle. The size of this exclusion zone may be sensibly defined as the particle size. This particle size can be assessed by means of the radial distribution function. Note that the cutoff radius does not necessarily represent the size of the DPD particle - a larger cutoff radius may result in a smaller size of the particle. If the solvent particle size is significant (compared to colloidal size), the maximal packing fraction of the colloidal particles will be reduced. Consequently, the relative viscosity is expected to diverge earlier as the volume fraction increases approaching the maximal packing fraction, and the shear thinning behaviour becomes stronger. Controlling the solvent particle size to have a correct representation of the solvent phase (in the sense that the colloidal/solvent size ratio is very large) is a vital issue in the DPD modelling of colloidal suspensions.

As analysed in Sect. 9.3.3 (i.e. roles of interaction forces), there are many possible combinations of n and r_c that reduce a_{eff} of a basic DPD particle. One can employ $r_c = 1$ with a larger value of n (fine coarse-graining level with a standard cutoff radius). Or one can employ, for example $n = 3$ with a larger value of $r_c > 1$ (upper coarse graining level with a larger cutoff radius).

In what follows, we investigate numerically the effects of the number density, cutoff radius and thermodynamic temperature on the rheological properties of monodispersed suspensions. The problem domain is chosen as $L_x \times L_y = 20 \times 20$ and the input parameters employed are $\sigma = 3$, $s = 1/2$, $n = 3 - 9$, $r_c = 1 - 2$ and $k_B T = 0.25 - 1$. To represent the suspended particles, we utilise the spring model using 4–6 basic particles per colloid with their spring stiffness $H = 3000$. Their rheological properties are predicted by conducting the simulation in a simple shear flow. The relative viscosity (suspension/solvent) is calculated in an average sense from ten simulations - each simulation consists of 300,000 time steps. For "zero-shear-rate" viscosity, we compute it at a shear rate of 0.1 for the volume fraction $\phi \leq 0.1$ and 0.01 for $\phi > 0.1$.

Simulation of suspensions with large number densities

A wide range of the number density n are employed, while other DPD parameters are taken as standard values (e.g., $k_B T = 1$, $\sigma = 3$, $r_c = 1$ and 4 basic DPD particles per colloid). Results concerning the zero-shear-rate relative viscosity against the volume fraction are shown in Figs. 9.14 and 9.15.

Figure 9.14 reveals that the relative viscosity curves collapse onto a single curve at large values of the number density. It implies that the size effect of the particles

Fig. 9.14 Suspension:
Effects of the number
density

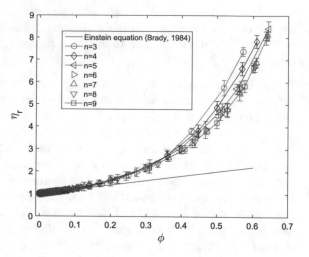

Fig. 9.15 Suspension:
Relative viscosities by DPD,
SPH and empirical model

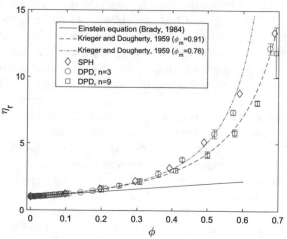

size ratio (colloidal/solvent) becomes negligible at fine coarse graining levels, i.e.
at $n \geq 6$. Theoretical estimate for the relative viscosity in the dilute regime [11] is
included.

Figure 9.15 shows a comparison of the present relative viscosities and those predicted by the empirical model of Krieger and Dougherty [45], defined as

$$\eta_r = \left(1 - \frac{\phi}{\phi_m}\right)^{-\phi_m[\eta]}, \tag{9.207}$$

where ϕ_m is the maximal packing fraction and $[\eta]$ the intrinsic viscosity. In 2D, the
maximal volume fraction is 0.91 for hexagonal close packing and the intrinsic viscosity is 2 for rigid cylinders. It can be seen that a fine coarse graining level $n = 9$

Fig. 9.16 Suspension:
Effects of the cutoff radius in
the upper coarse-graining
level $n = 3$

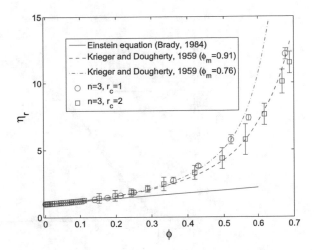

results in viscosities that are located along the empirical curve of $\phi_m = 0.91$ over
the whole range of volume fraction employed. In contrast, results at a coarse level
($n = 3$) fail to follow the correlation; it is in close agreement with the correlation
using $\phi_m = 0.76$ up to a semidilute regime, but under-predicts the correlation in
the concentrated regime. SPH results [5], where short range lubrication forces are
included explicitly, are also shown in the figure. Thus it is seen that the DPD method
at a fine coarse graining level follows the established correlation well.

Simulation of suspensions with large cutoff radii

Different values of the cutoff radius r_c are employed with their effects being seen
clearer for the case of using smaller number densities (e.g., an upper coarse graining
at $n = 3$). As the cutoff radius r_c increases, the solvent particle size is significantly
reduced (Sect. 9.3.3). The relative viscosity - volume fraction relation for $n = 3$ is
shown in Fig. 9.16, where a colloidal particle is model using 6 basic DPD particles.
By increasing r_c, a coarse level $n = 3$ is able to produce results following Krieger-
Dougherty correlation.

Simulation of suspensions with low thermodynamic temperature

Unlike the number density n and cutoff radius r_c, reducing the temperature $k_B T$
does not affect the solvent particle size significantly (Sect. 9.3.3). We now examine
the effects of $k_B T$ on the size of a colloidal particle modelled by 4 basic DPD particles and the results obtained in a fine coarse graining level ($n = 9$) are shown in
Fig. 9.17. From these results, it is expected that changing $k_B T$ has little effect on the
size ratio of the colloidal to solvent particle, and this is confirmed in Fig. 9.18 for the
relative viscosity versus volume fraction.

Remarks. In DPD, the fluid (solvent) and colloidal particles are replaced by a set
of DPD particles and therefore their relative sizes (as measured by their exclusion

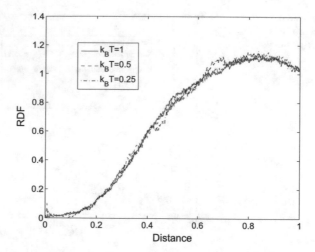

Fig. 9.17 Suspension: the colloidal particle size over a range of $k_B T$

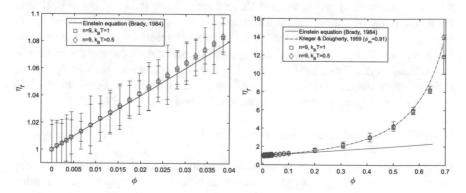

Fig. 9.18 Suspension: Effects of the thermodynamic temperature in the fine coarse graining level $n = 9$

zones) can affect the maximal packing fraction of the colloidal particles. To better mimic the physical system, the DPD system should be designed to have as small a solvent particle as possible in order to make the colloidal/solvent size ratio as large as possible (e.g., a few orders of magnitude). The size of DPD particles is found to be decided not only by the conservative force but also by the dissipative and random forces. By keeping the compressibility of the system unchanged, it is shown that the solvent phase can be modelled correctly (in the sense just mentioned above) at both low (large number density) and high (low number density) coarse-graining levels. In the former, one can simply employ standard values of the other input DPD parameters, while in the latter, a larger value of the cutoff radius is required. It is found that the solvent particle size is a decreasing function of the cutoff radius and varying the temperature is not an effective way of controlling the solvent particle size. When the requirement of large colloidal/solvent particles size is met, the DPD

. results for the reduced viscosity are basically identical for any values of the input DPD parameters.

There are other complex flows that have been successfully simulated by DPD, or some variants of DPD method, including DNA, droplet suspension, gas/vapour suspension, porous media flow, red blood cell modelling. We shall not review of the recent development in DPD in this compact book, but refer the reader to the review by Pivkin et al. [73].

9.8 Matlab Program

The following DPD code, written in MATLAB, is for the numerical simulation of viscometric flows in two dimensions. Its main purpose is to illustrate how the DPD equations are solved in practice. For an easy reading, only plain MATLAB functions (i.e. no toolboxes/user-defined functions) are used. The program consists of 4 parts: input, initialisation, computation and output in a single file. Tasks for computing the positions and velocities of the DPD particles and computing the velocities and stresses of the flow are highlighted. An efficient cell linked list approach is utilised to compute the interaction forces between the particles - the most time consuming part in a simulation. One can choose Couette flow or Poiseuille flow through the input values.

```
%
% Dissipative Particle Dynamics
% 2D viscometric flows and Periodic boundary conditions
% Couette flow: specify the imposed velocity (e.g. Vx=1)
%               and set the body force zero (Fex=0)
% Poiseuille flow: specify the body force (e.g. Fex=1)
%               and set the imposed velocity zero (Vx=0)
% No flow: set Vx=0 and Fex=0
close all; clear;

%%%%%%%%%%%%%%%%%%%%% INPUT %%%%%%%%%%%%%%%%%%%%%%%%%
% Flow domain and Imposed conditions (Vx: velocity and Fex: body force)
Lx=10; Ly=Lx;
Vx=0; Fex=1;

% Input DPD parameters and Time step
rCut=1; sigma=3; kBT=1;
numDensityX=1; numDensityY=3; numDen=numDensityX*numDensityY;
aij=57.23*kBT/numDen/rCut^3;
deltaT=0.01;

% Step numbers for thermal equilibrium and for flow simulation
stepEquil=500; stepSample=3000; stepLimit=stepEquil+stepSample;

%%%%%%%%%%%%%%%%%%%%% INITILISATION %%%%%%%%%%%%%%%%%%%%
% Number of DPD particles and Number of bins
nFreeAtom=Lx*Ly*numDen;
nxBin=4*Lx; nyBin=4*Ly; nBin=nxBin*nyBin;
% Initial configuration
```

```
r=zeros(2,nFreeAtom);
X=linspace(-Lx/2+1/2/numDensityX,Lx/2-1/2/numDensityX,numDensityX*Lx);
Y=linspace(-Ly/2+1/2/numDensityY,Ly/2-1/2/numDensityY,numDensityY*Ly);
[xm,ym]=meshgrid(X,Y);
r(:,:)=[xm(:),ym(:)]';
% Initial velocities
rv=zeros(2,nFreeAtom);
velMag=sqrt(2*(1-1/nFreeAtom)*kBT);
phim=2*pi*rand(numDensityX*Lx,numDensityY*Ly);
vxm=cos(phim); vym=sin(phim);
rv(:,:)=velMag*[vxm(:),vym(:)]'; vSum=sum(rv,2);
rv(1,:)=rv(1,:)-vSum(1)/nFreeAtom;
rv(2,:)=rv(2,:)-vSum(2)/nFreeAtom;
figure; plot(r(1,:),r(2,:),'mo'); hold on;
plot([-Lx/2,Lx/2,Lx/2,-Lx/2,-Lx/2],[-Ly/2,-Ly/2,Ly/2,Ly/2,-Ly/2],'b-');
quiver(r(1,:),r(2,:),rv(1,:),rv(2,:),'b'); hold off;
title('Initial configuration and Initial particles velocities');
% Initial accelerations and forces
ra=zeros(2,nFreeAtom);
rforce=zeros(3,nFreeAtom);
% Friction strength
gamma=sigma^2/(2*kBT);

areaBin=(Lx/nxBin)*(Ly/nyBin);
velGrid=zeros(nBin,4); % n, u^2, ux, uy
strsGrid=zeros(nBin,4); % Sxx, Syy, Sxy, p
voff=[0,1,1,0,-1,2;...
      0,0,1,1,1,1];
region=[Lx;Ly];
cells=floor((1/rCut)*region); nCell=cells(1)*cells(2);
%%%%%%%%%%% COMPUTATION %%%%%%%%%%%%%%%%%%%%%%%%%%
stepCount=0; timeNow=0;
while stepCount < stepLimit
    stepCount=stepCount+1; disp(stepCount);
    %%%%%%%%%%%%%%%%%%%%%%%%%%%%%%%%%%%%%%%%%%%%%%
    % Compute the particles' positions and velocities
    %%%%%%%%%%%%%%%%%%%%%%%%%%%%%%%%%%%%%%%%%%%%%%
    temp = deltaT*ra;
    % Particles' positions
    r = r + deltaT*(rv+0.5*temp);
    % Imposition of velocity conditions for Couette flow
    bdySlide=rem(2*Vx*(stepCount*deltaT)+Lx/2,Lx)-Lx/2;
    for i=1:nFreeAtom
        if (abs(r(1,i))>Lx || abs(r(2,i))>Ly)
            error('Particle moves too far, reduce time step')
        end
        if r(2,i) > 0.5*Ly
            r(2,i)=r(2,i)-Ly;
            r(1,i)=r(1,i)-bdySlide;
            rv(1,i) = rv(1,i)-2*Vx;
        elseif r(2,i) < -0.5*Ly
            r(2,i)=r(2,i)+Ly;
            r(1,i)=r(1,i)+bdySlide;
            rv(1,i) = rv(1,i)+2*Vx;
        end

        if r(1,i) > 0.5*Lx
            r(1,i)=r(1,i)-Lx;
        elseif r(1,i) < -0.5*Lx
```

```
            r(1,i)=r(1,i)+Lx;
        end
    end
    rvm = rv; rv = rv + 0.65*temp; rvm = rvm + 0.5*temp;
    % Interaction force calculations
    cellShiftX = fix(cells(1)*(1.-bdySlide/region(1))) - cells(1);
    ra=zeros(2,nFreeAtom); rforce=zeros(3,nFreeAtom);
    invWid = cells./region;
    cellList = -ones(nFreeAtom+nCell,1);
    rs = [r(1,:)+0.5*region(1); r(2,:)+0.5*region(2)];
    cc = ceil([rs(1,:)*invWid(1);rs(2,:)*invWid(2)]);
    c = (cc(2,:)-1)*cells(1)+cc(1,:) + nFreeAtom;
    for i=1:nFreeAtom
        cellList(i) = cellList(c(i));
        cellList(c(i)) = i;
    end
    for m1y = 1:cells(2)
        for m1x = 1:cells(1)
            m1v = [m1x;m1y];
            m1 = (m1v(2)-1)*cells(1)+m1v(1) + nFreeAtom;
            %------------------------------------------------------------
            offsetHi=size(voff,2);
            if m1y < cells(2)
                offsetHi=offsetHi-1;
            end
            for offset=1:offsetHi
                m2v = m1v+voff(:,offset);
                shift = zeros(2,1);
                velshift = zeros(2,1);
                if m1v(2)==cells(2) && voff(2,offset)==1
                    m2v(1)=m2v(1)+cellShiftX;
                    shift(1)=bdySlide;
                    if m2v(1) > cells(1)
                        m2v(1) = m2v(1)-cells(1);
                        shift(1) = shift(1)+region(1);
                    elseif m2v(1) < 1
                        m2v(1) = m2v(1)+cells(1);
                        shift(1) = shift(1)-region(1);
                    end
                else
                    if m2v(1) > cells(1)
                        m2v(1)=1; shift(1)=region(1);
                    elseif m2v(1) < 1
                        m2v(1)=cells(1); shift(1)=-region(1);
                    end
                end
                if m2v(2) > cells(2)
                    m2v(2)=1; shift(2)=region(2);
                    velshift(1)=2*Vx;
                elseif m2v(2) < 1
                    m2v(2)=cells(2); shift(2)=-region(2);
                    velshift(1)=-2*Vx;
                end
                %------------------------------------------------------------
                m2 = (m2v(2)-1)*cells(1)+m2v(1) + nFreeAtom;
                j1=cellList(m1);
                while j1>=1
                    j2=cellList(m2);
                    while j2>=1
```

```
                    if m1~=m2 || j2<j1
                        dr = r(:,j1)-(r(:,j2)+shift);
                        rr=dr'*dr;
                        if rr < rCut^2
                            rij = sqrt(rr); dr = dr/rij;
                            rdvij=dot(dr,rv(:,j1)-(rv(:,j2)+velshift));
                            weight  = 1 - rij/rCut;
                            weightc = weight;
                            weightd = sqrt(weight);
                            weightr = sqrt(weightd);
                            fcVal =  aij*weightc;
                            fdVal = -gamma*weightd*rdvij;
                            frVal=sigma*weightr*randn*(1/sqrt(deltaT));
                            ftot = fcVal+fdVal+frVal;
                            ra(:,j1) = ra(:,j1) + ftot*dr;
                            ra(:,j2) = ra(:,j2) - ftot*dr;

                            fstl = (fcVal + fdVal)*rij;
                            rforce(:,j1) = rforce(:,j1) + ...
                                fstl*[dr(1)*dr(1);dr(2)*dr(2);dr(1)*dr(2)];

                            rforce(:,j2) = rforce(:,j2) + ...
                                fstl*[dr(1)*dr(1);dr(2)*dr(2);dr(1)*dr(2)];
                        end
                    end
                    j2=cellList(j2);
                end
                j1=cellList(j1);
            end
        end
    end
end
% Imposition of body forces for Poiseuille flow
j=0;
for i=1:nFreeAtom
    if r(2,i)>0; j=j+1; end
end
for i=1:nFreeAtom
    if r(2,i)>0
        ra(1,i) = ra(1,i) + Fex*(0.5*nFreeAtom/j);
    else
        ra(1,i) = ra(1,i) - Fex*(0.5*nFreeAtom/(nFreeAtom-j));
    end
end
% Particles' velocities
rv = rvm + 0.5*deltaT*ra;
%%%%%%%%%%%%%%%%%%%%%%%%%%%%%%%%%%%%%%%%%%%%%%%
% Compute the flow velocities and stresses
%%%%%%%%%%%%%%%%%%%%%%%%%%%%%%%%%%%%%%%%%%%%%%%
if stepCount > stepEquil
    timeNow=timeNow+deltaT;
    for i=1:nFreeAtom
        c=floor((r(2,i)+Ly/2)*nyBin/Ly)*nxBin + ...
            floor((r(1,i)+Lx/2)*nxBin/Lx) + 1;
        velGrid(c,:)=velGrid(c,:)+[1,rv(:,i)'*rv(:,i),rv(:,i)'];
        strsGrid(c,1:3)=strsGrid(c,1:3)+ ...
        [rv(1,i)*rv(1,i),rv(2,i)*rv(2,i),rv(1,i)*rv(2,i)]+0.5*rforce(:,i)';
    end
    if (stepCount-stepEquil) == stepSample
```

```
                id = find(velGrid(:,1)>0);
                for j=2:4
                    velGrid(id,j)=velGrid(id,j)./velGrid(id,1);
                end
                velGrid(:,2)=velGrid(:,2)-(velGrid(:,3).^2+velGrid(:,4).^2);

                for j=1:3
                    strsGrid(id,j)=strsGrid(id,j)./velGrid(id,1);
                end
                strsGrid(:,1) = strsGrid(:,1) - velGrid(:,3).^2;
                strsGrid(:,2) = strsGrid(:,2) - velGrid(:,4).^2;
                strsGrid(:,3) = strsGrid(:,3) - velGrid(:,3).*velGrid(:,4);

                velGrid(:,1)=velGrid(:,1)/(stepSample*areaBin);
                strsGrid(:,1) = -strsGrid(:,1).*velGrid(:,1);
                strsGrid(:,2) = -strsGrid(:,2).*velGrid(:,1);
                strsGrid(:,3) = -strsGrid(:,3).*velGrid(:,1);
                strsGrid(:,4) = -(strsGrid(:,1) + strsGrid(:,2))/2;
        end
    end
end
%%%%%%%%%%%%%%%%% OUTPUT %%%%%%%%%%%%%%%%%%%%%
% Plot the particles' positions and velocities at the final time
figure; plot(r(1,:),r(2,:),'mo'); hold on
plot([-Lx/2,Lx/2,Lx/2,-Lx/2,-Lx/2],[-Ly/2,-Ly/2,Ly/2,Ly/2,-Ly/2],'b-');
quiver(r(1,:),r(2,:),rv(1,:),rv(2,:)); hold off;
title('Configuration and Particles velocities at the final time')
axis off

% Plot the flow fields: velocity and stress
Y=-Ly/2+Ly/(2*nyBin):Ly/nyBin:Ly/2-Ly/(2*nyBin);
R=zeros(nyBin,4);
for k=1:nyBin
    R(k,:)=sum(velGrid(nxBin*(k-1)+1:nxBin*k,:),1)/nxBin;
end
figure;
plot(Y,R(:,1),'b-',Y,0.5*R(:,2),'m-.',Y,R(:,3),'k--o',Y,R(:,4),'g--');
legend('n','T','u_x','u_y'); xlabel('y'); title('Flow field calculations');

R=zeros(nyBin,4);
for k=1:nyBin
    R(k,:)=sum(strsGrid(nxBin*(k-1)+1:nxBin*k,:),1)/nxBin;
end
figure; plot(Y,R(:,3),'b-o',Y,R(:,1)-R(:,2),'r-s');
legend('Shear stress','Normal stress difference'); xlabel('y');
title('Flow field calculations');
```

9.9 Epilogue

This compact book outlines the main basic developments in viscoelasticity, from continuum principles to microstructure modelling. We have not mentioned the reputation concept of Doi and Edwards [18], the modelling effort in fibre suspensions

[27], in biological materials [3], in electro-rheological fluids, as well as transition phenomena.

The constant theme emphasised throughout is that relevant evolution equations for the microstructure should be derived from well-established physics. This, together with relevant statistical mechanics linking the microstructure evolution to a macroscaled stress induced by the microstructure, should provide a useful constitutive equation for the fluid. Having a relevant constitutive relation is only half of the story; one needs to be able to make predictions with it, and that usually means a numerical implementation, a vast open area that we only touch on briefly with the Dissipative Particle Dynamics.

It is hoped that the readers find the book useful in their research works.

Problems

Problem 9.1 In Problems 9.1–9.4, we deal with the 1-D system (9.1). Re-define the state variable to reduce (9.1) to (9.4) in the inertial time scale, and show that its formal solution is

$$v(t) = \int_0^t e^{m^{-1}\gamma w_D(t'-t)} m^{-1} \sigma w_R \theta(t') \, dt'.$$

Then calculate the mean square velocity and show that

$$
\begin{aligned}
\langle v(t) v(t) \rangle &= \frac{\sigma^2}{m^2} \int_0^t \int_0^t e^{m^{-1}\gamma w_D(t'-t)} w_R^2 \langle \theta(t') \theta(t'') \rangle e^{m^{-1}\gamma w_D(t''-t)} dt' dt'' \\
&= \frac{\sigma^2}{m^2} \int_0^t e^{m^{-1}\gamma w_D(t'-t)} w_R e^{m^{-1}\gamma w_D(t'-t)} w_R dt', \\
&= \frac{1}{2} m^{-1} \sigma^2 w_R^2 \gamma^{-1} w_D^{-1}.
\end{aligned}
\tag{9.208}
$$

This leads directly to (9.6), assuming (9.3).

Problem 9.2 Show that the drift velocity and the diffusivity of the process (9.12) are given by

$$
\begin{aligned}
\frac{\langle \Delta r \rangle}{\Delta t} &= \gamma^{-1} w_D^{-1} F_c, \\
\frac{\langle \Delta r \Delta r \rangle}{2\Delta t} &= O(\Delta t) + k_B T \gamma^{-1} w_D^{-1},
\end{aligned}
\tag{9.209}
$$

leading to the Fokker–Planck equation (9.13).

Problem 9.3 Show that, from (9.4)

$$\frac{1}{2}m\frac{d}{dt}\left(\frac{d}{dt}\langle r^2\rangle\right) - m\langle v^2\rangle + \frac{1}{2}\gamma w_D \frac{d}{dt}\langle r^2\rangle = \sigma w_R \langle\theta(t)r\rangle. \qquad (9.210)$$

Define

$$e = d\langle r^2\rangle/dt,$$

and show that

$$\dot{e} + m^{-1}\gamma w_D e = 2k_B T m^{-1}, \quad e(0) = 0.$$

Show that this has the solution, for the assumed initial condition,

$$e = \frac{d}{dt}\langle r^2\rangle = 2k_B T\gamma^{-1}w_D^{-1}\left[1 - e^{-m^{-1}\gamma w_D t}\right].$$

Consequently, if $\Delta t \gg \tau_I = O\left(m^{-1}\gamma w_D\right)$, show that

$$\frac{\langle\Delta r\,\Delta r\rangle}{2\Delta t} = k_B T\gamma^{-1}w_D^{-1}.$$

Problem 9.4 Define the velocity correlation as

$$R(\tau) = \lim_{t\to\infty}\langle v(t+\tau)v(t)\rangle, \qquad (9.211)$$

where the limit refers to large time compared to the inertial time scale, but yet small compared to the relaxation time scale. From the solution (9.5), show that

$$R(\tau) = \lim_{t\to\infty}\int_0^{t+\tau} dt'\int_0^t e^{m^{-1}\gamma w_D(t'-t-\tau)}\left(m^{-1}\sigma w_R\right)^2$$
$$\langle\theta(t')\theta(t'')\rangle e^{m^{-1}\gamma w_D(t''-t)}dt'' \qquad (9.212)$$
$$= e^{-m^{-1}\gamma w_D\tau}\lim_{t\to\infty}\langle v(t)v(t)\rangle = e^{-m^{-1}\gamma w_D\tau}R(0)$$
$$= k_B T m^{-1}e^{-m^{-1}\gamma w_D\tau}. \qquad (9.213)$$

That is, the velocity correlation decays after an inertial time scale, after which the velocity is independent to its previous state.

Next, the diffusivity can also be defined as

$$D = \lim_{t\to\infty}\frac{1}{2}\langle v(t)r(t) + r(t)v(t)\rangle.$$

Show that this leads to

$$D = \lim_{t \to \infty} \int_0^t \langle v(t)\, v(t+\tau) \rangle \, d\tau = \int_0^\infty R(\tau)\, d\tau$$

$$= k_B T \gamma^{-1} w_D^{-1},$$

consistent with previous results.

Problem 9.5 Show, with the aid of the Langevin equation (9.14), that the drift and the diffusion of the process are given by

$$\left\langle \frac{\Delta \mathbf{v}_i}{\Delta t} \right\rangle = -m^{-1} \sum_j \left(\mathbf{F}_{ij}^C + \mathbf{F}_{ij}^D \right) = -m^{-1} \sum_j \left(\mathbf{F}_{ij}^C - \gamma w_{ij}^D \mathbf{e}_{ij} \mathbf{e}_{ij} \cdot \mathbf{v}_{ij} \right), \quad (9.214)$$

and

$$\left\langle \frac{\Delta \mathbf{v}_\alpha \Delta \mathbf{v}_\beta}{2\Delta t} \right\rangle = \frac{\gamma k_B T}{m^2} \left[\delta_{\alpha\beta} \sum_k w_{\alpha k}^D \mathbf{e}_{\alpha k} \mathbf{e}_{\beta k} - w_{\alpha\beta}^D \mathbf{e}_{\alpha\beta} \mathbf{e}_{\alpha\beta} \right]. \quad (9.215)$$

Thus show that the Fokker–Planck equation is given by (9.32).

Problem 9.6 Show that the equilibrium distribution of the associate system to (9.14) is

$$f_{eq}(\chi, t) = \frac{1}{Z} \exp \left[-\frac{1}{k_B T} \left(\sum_i \frac{\mathbf{p}_i \cdot \mathbf{p}_i}{2m} + \frac{1}{2} \sum_{i,j} \varphi(r_{ij}) \right) \right] \quad (9.216)$$

$$= \frac{1}{Z} \exp \left[-\frac{\mathcal{H}}{k_B T} \right],$$

where Z is a normalizing constant, and

$$\mathcal{H} = \sum_i \frac{\mathbf{p}_i \cdot \mathbf{p}_i}{2m} + \frac{1}{2} \sum_{i,j} \varphi(r_{ij}) \quad (9.217)$$

is the Hamiltonian of the associate system to (9.14). Show that,

$$\sum_i \mathbf{v}_i \cdot \frac{\partial f_{eq}}{\partial \mathbf{r}_i} = \frac{1}{k_B T} \sum_{i,j} \mathbf{v}_i \cdot \mathbf{F}_{ij}^C f_{eq}, \quad \sum_{i,j} \mathbf{F}_{ij}^C \cdot \frac{\partial f_{eq}}{\partial \mathbf{p}_i} = -\frac{1}{k_B T} \sum_{i,j} \mathbf{F}_{ij}^C \cdot \mathbf{v}_i f_{eq},$$

$$\gamma \sum_{i,j} w_{ij}^D \mathbf{e}_{ij} \frac{\partial}{\partial \mathbf{p}_i} \cdot (\mathbf{e}_{ij} \cdot \mathbf{v}_{ij} f_{eq}) = \gamma \sum_{i,j} w_{ij}^D \left(-\frac{1}{k_B T} \mathbf{e}_{ij} \cdot \mathbf{v}_i \mathbf{e}_{ij} \cdot \mathbf{v}_{ij} + \frac{1}{m} \right) f_{eq},$$

$$\gamma k_B T \sum_{i,j} w_{ij}^D \mathbf{e}_{ij} \cdot \frac{\partial}{\partial \mathbf{p}_i} \cdot \left(\mathbf{e}_{ij} \cdot \left(\frac{\partial}{\partial \mathbf{p}_i} - \frac{\partial}{\partial \mathbf{p}_j} \right) f_{eq} \right)$$

$$= \gamma \sum_{i,j} w_{ij}^D \left(-\frac{1}{m} + \frac{1}{k_B T} \mathbf{e}_{ij} \cdot \mathbf{v}_i \mathbf{e}_{ij} \mathbf{v}_{ij} \right),$$

and conclude that f_{eq} is also a stationary solution (i.e. solution that is independent of time) of the Fokker–Planck equation (9.32).

Problem 9.7 Show the equivalence between (9.61), (9.63), (9.65) and (9.66), by expressing $f_2 \left(\mathbf{r} + \lambda \mathbf{R}, \mathbf{r} - (1 - \lambda)\mathbf{R}, \mathbf{v}', \mathbf{v}'', t \right)$ as a function of $(\mathbf{r} - \varepsilon \mathbf{R})$, where $\varepsilon = 1 - \lambda$, then integrating after taking a Taylor's series in ε.

Problem 9.8 Show that

$$\mathbf{q}_C \left(\mathbf{r}, t \right) = \frac{1}{4} \int d\mathbf{R} \int d\mathbf{v} \int d\mathbf{v}' \mathbf{F}^C \left(\mathbf{R} \right) \cdot \left(\mathbf{v} + \mathbf{v}' \right) \mathbf{R} \left\{ 1 - \frac{1}{2} \mathbf{R} \cdot \nabla + \cdots \right\}$$
$$\cdot f_2 \left(\mathbf{r} + \mathbf{R}, \mathbf{r}, \mathbf{v}, \mathbf{v}', t \right). \tag{9.218}$$

Problem 9.9 The stress contributed from the damping forces is, from (9.63),

$$\mathbf{S}_D \left(\mathbf{r}, t \right) = -\frac{1}{2} \int d\mathbf{R} \int d\mathbf{v} \int d\mathbf{v}' \gamma w^D \left(R \right) \hat{\mathbf{R}} \hat{\mathbf{R}} \cdot \left(\mathbf{v} - \mathbf{v}' \right) \mathbf{R} \left\{ 1 + O \left(R \right) \right\}$$
$$\cdot f_2 \left(\mathbf{r} + \mathbf{R}, \mathbf{r}, \mathbf{v}, \mathbf{v}', t \right).$$

Show that, for an homogeneously shear flow, $\mathbf{v} - \mathbf{v}' = \mathbf{L}\mathbf{R}$, where $\mathbf{L}$ is the velocity gradient, together with Groot and Warren's approximation, $f_2 = n^2 \left(1 + O \left(R \right) \right)$, the stress contributed by the damping forces is

$$S_{D,\alpha\beta} \left(\mathbf{r}, t \right) = -\frac{\gamma n^2}{2} \left\langle \hat{R}_\alpha \hat{R}_\beta \hat{R}_i \hat{R}_j L_{ij} \int R^2 w^D \left(R \right) 4\pi R^2 dR \right\rangle$$

$$= -\frac{2\pi \gamma n^2}{15} \left(\delta_{\alpha\beta}\delta_{ij} + \delta_{\alpha i}\delta_{\beta j} + \delta_{\alpha j}\delta_{\beta i} \right) L_{ij} \int R^4 w^D \left(R \right) dR.$$

For the standard weighting function (9.88) adopted in DPD, show that

$$S_{D,\alpha\beta} \left(\mathbf{r}, t \right) = -\frac{2\pi \gamma n^2}{15} \left(L_{\alpha\beta} + L_{\beta\alpha} + L_{ii}\delta_{\alpha\beta} \right) \int_0^{r_c} R^4 \left(1 - R/r_c \right)^2 dR$$

$$= -\frac{2\pi \gamma n^2 r_c^5}{1575} \left(L_{\alpha\beta} + L_{\beta\alpha} + L_{ii}\delta_{\alpha\beta} \right), \tag{9.219}$$

and consequently the damping-contributed viscosities are given by

$$\eta_D = \frac{2\pi \gamma n^2 r_c^5}{1575}, \quad \zeta_D = \frac{5}{3}\eta_D = \frac{2\pi \gamma n^2 r_c^5}{945}. \tag{9.220}$$

Problem 9.10 For the standard DPD weighting function (9.88), show that the viscosities and diffusivity are given as shown in (9.90).

Problem 9.11 For the modified DPD weighting function (9.125), derive the viscosities and diffusivity of the DPD fluid.

Problem 9.12 Write a Matlab routine to compute the radial distribution function (RDF) for DPD particles and integrate it into the DPD main program in Sect. 9.8. Study the effects of n, r_c and $k_B T$ on the particle's exclusion zone for the case of using modified weighting function.

Problem 9.13 Write a Matlab routine to compute the mean square displacements (MSDs) of DPD particles and integrate it into the DPD main program in Sect. 9.8. Study the effects of n, r_c and $k_B T$ on the self-diffusion coefficient of the DPD particles for a given noise amplitude $\sigma = 3$.

Solutions to Problems

Problems in Chap. 1

Problem 1.1

The components of vectors $\mathbf{u}$, $\mathbf{v}$, and $\mathbf{w}$ are given by u_i, v_i, w_i. By expanding all the summations, we have:

$$\mathbf{u} \cdot \mathbf{v} = u_1 v_1 + u_2 v_2 + u_3 v_3 = u_i v_i. \tag{S.221}$$

Similarly,

$$\begin{aligned}
\mathbf{u} \times \mathbf{v} &= u_i \mathbf{e}_i \times v_j \mathbf{e}_j \\
&= (u_1 v_2 - u_2 v_1)\, \mathbf{e}_3 + (u_3 v_1 - u_1 v_3)\, \mathbf{e}_2 + (u_2 v_3 - u_3 v_2)\, \mathbf{e}_1 \\
&= \varepsilon_{ijk} \mathbf{e}_i u_j v_k.
\end{aligned} \tag{S.222}$$

And, by using (S.221) and (S.222),

$$\begin{aligned}
(\mathbf{u} \times \mathbf{v}) \cdot \mathbf{w} &= \varepsilon_{ijk} \mathbf{e}_i u_j v_k \cdot w_l \mathbf{e}_l \\
&= \varepsilon_{ijk} u_j v_k w_i \\
&= \varepsilon_{ijk} u_i v_j w_k.
\end{aligned} \tag{S.223}$$

From the cyclic properties of $\varepsilon_{ijk} = \varepsilon_{jki}$, and use the last preceding result,

$$(\mathbf{u} \times \mathbf{v}) \cdot \mathbf{w} = \mathbf{u} \cdot (\mathbf{v} \times \mathbf{w}). \tag{S.224}$$

Next, consider the 4th order tensor $\varepsilon_{nij}\varepsilon_{nkm}$. Clearly, if $i = j$, or $k = m$, this tensor takes on a zero value. If $i = k$, the sum in n will return a zero value, unless $j = m$, and $\varepsilon_{nij} = \varepsilon_{nkm} = \pm 1$, and the product $\varepsilon_{nij}\varepsilon_{nkm} = +1$. Now, if $i = m$ and $j = k$ the product $\varepsilon_{nij}\varepsilon_{nkm}$ will return a zero value unless $\varepsilon_{nij} = \pm 1$, then

© Springer International Publishing AG 2017
N. Phan-Thien and N. Mai-Duy, *Understanding Viscoelasticity*,
Graduate Texts in Physics, DOI 10.1007/978-3-319-62000-8

$\varepsilon_{nkm} = \varepsilon_{nji} = \mp 1$, and the product $\varepsilon_{nij}\varepsilon_{nkm} = -1$. These are also the components of the tensor $\delta_{ik}\delta_{jm} - \delta_{im}\delta_{jk}$. Hence

$$\varepsilon_{nij}\varepsilon_{nkm} = \delta_{ik}\delta_{jm} - \delta_{im}\delta_{jk}. \tag{S.225}$$

Using (S.225) in the following:

$$\begin{aligned}
(\mathbf{u} \times \mathbf{v}) \times \mathbf{w} &= \varepsilon_{ijk}\mathbf{e}_i\varepsilon_{jmn}u_m v_n w_k = (\delta_{km}\delta_{in} - \delta_{kn}\delta_{im})\, \mathbf{e}_i u_m v_n w_k \\
&= \mathbf{e}_i u_k v_i w_k - \mathbf{e}_i u_i v_k w_k \\
(\mathbf{u} \times \mathbf{v}) \times \mathbf{w} &= (\mathbf{u} \cdot \mathbf{w})\, \mathbf{v} - (\mathbf{v} \cdot \mathbf{w})\, \mathbf{u}. \tag{S.226}
\end{aligned}$$

This could also be derived by expanding components of both sides.

Now, use the result (S.225) in the following

$$\begin{aligned}
(\mathbf{u} \times \mathbf{v})^2 &= \varepsilon_{ijk}u_j v_k \varepsilon_{imn}u_m v_n = \left(\delta_{jm}\delta_{kn} - \delta_{jn}\delta_{km}\right) u_j v_k u_m v_n \\
&= u_j u_j v_k v_k - u_j v_j v_k u_k \\
&= u^2 v^2 - (\mathbf{u} \cdot \mathbf{v})^2. \tag{S.227}
\end{aligned}$$

Problem 1.2

Let $\mathbf{A}$ be a matrix with entries A_{ij},

$$[\mathbf{A}] = \begin{bmatrix} A_{11} & A_{12} & A_{13} \\ A_{21} & A_{22} & A_{23} \\ A_{31} & A_{32} & A_{33} \end{bmatrix}.$$

By expanding

$$\begin{aligned}
\det[\mathbf{A}] &= A_{11}A_{22}A_{33} + A_{12}A_{23}A_{31} + A_{13}A_{21}A_{32} - A_{31}A_{22}A_{13} \\
&\quad - A_{32}A_{23}A_{11} - A_{33}A_{21}A_{12} \\
&= A_{11}(A_{22}A_{33} - A_{23}A_{32}) + A_{12}(A_{23}A_{31} - A_{21}A_{33}) \\
&\quad + A_{13}(A_{21}A_{32} - A_{31}A_{22}) \\
&= \varepsilon_{ijk}A_{1i}A_{2j}A_{3k} \\
&= A_{11}(A_{22}A_{33} - A_{23}A_{32}) + A_{21}(A_{32}A_{13} - A_{12}A_{33}) \\
&\quad + A_{31}(A_{12}A_{23} - A_{22}A_{13}) \\
&= \varepsilon_{ijk}A_{i1}A_{j2}A_{k3}.
\end{aligned}$$

First, note that

$$\varepsilon_{123}\det[\mathbf{A}] = \det[\mathbf{A}] = \varepsilon_{123}\varepsilon_{ijk}A_{1i}A_{2j}A_{3k}.$$

If we swap two rows, or two columns, determinant of the resulting matrix will be the negative of $\det[\mathbf{A}]$. For instance

$$\varepsilon_{ijk} A_{1i} A_{2k} A_{3j} = -\det[\mathbf{A}] = \varepsilon_{132} \det[\mathbf{A}] = \varepsilon_{132} \varepsilon_{ijk} A_{1i} A_{3j} A_{2k}.$$

By inspection,

$$\varepsilon_{lmn} \det[\mathbf{A}] = \varepsilon_{lmn} \varepsilon_{ijk} A_{1i} A_{2j} A_{3k} = \varepsilon_{ijk} A_{li} A_{mj} A_{nk}.$$

Furthermore, since $\varepsilon_{lmn} \varepsilon_{lmn} = 6$,

$$\varepsilon_{lmn} \varepsilon_{lmn} \det[\mathbf{A}] = 6 \det[\mathbf{A}] = \varepsilon_{lmn} \varepsilon_{ijk} A_{il} A_{jm} A_{kn},$$

$$\det[\mathbf{A}] = \frac{1}{6} \varepsilon_{ijk} \varepsilon_{lmn} A_{il} A_{jm} A_{kn}. \tag{S.228}$$

Problem 1.3

The following result has been shown in Problem 1.1:
  Given that two matrices of entries

$$[\mathbf{A}] = \begin{bmatrix} A_{11} & A_{12} & A_{13} \\ A_{21} & A_{22} & A_{23} \\ A_{31} & A_{32} & A_{33} \end{bmatrix}, \quad [\mathbf{B}] = \begin{bmatrix} B_{11} & B_{12} & B_{13} \\ B_{21} & B_{22} & B_{23} \\ B_{31} & B_{32} & B_{33} \end{bmatrix},$$

Then, be expanding the summation: $[\mathbf{C}] = [\mathbf{A}] \cdot [\mathbf{B}]$, $C_{ij} = A_{ik} B_{kj}$.
  $[\mathbf{D}] = [\mathbf{A}]^T [\mathbf{B}]$, $D_{ij} = A^T_{ik} B_{kj} = A_{ki} B_{kj}$.

Problem 1.4

Note that the components of $\mathbf{e}'_i$ in frame $\mathcal{F}$ are A_{ij}, i.e., $\mathbf{e}'_i = A_{ij} \mathbf{e}_j$. Thus

$$(\mathbf{e}'_1 \times \mathbf{e}'_2) \cdot \mathbf{e}'_3 = +1 = \varepsilon_{ijk} A_{1i} A_{2j} A_{3k} = \det[\mathbf{A}].$$

Problem 1.5

By expanding, it can be seen that

$$\varepsilon_{ijk} u_i v_j w_k = \det \begin{bmatrix} u_1 & u_2 & u_3 \\ v_1 & v_2 & v_3 \\ w_1 & w_2 & w_3 \end{bmatrix}.$$

Consider a two-tensor W_{ij} and vector defined as $u_i = \varepsilon_{ijk} W_{jk}$. By expanding again, it can be demostrated that, if $\mathbf{W}$ is symmetric, $W_{ij} = W_{ji}$, $\mathbf{u}$ is zero, and if $\mathbf{W}$ is anti-symmetric, $W_{ij} = -W_{ji}$,

$$[\mathbf{W}] = \begin{bmatrix} 0 & W_{12} & W_{13} \\ -W_{12} & 0 & W_{23} \\ -W_{13} & -W_{23} & 0 \end{bmatrix}, \quad \begin{bmatrix} u_1 \\ u_2 \\ u_3 \end{bmatrix} = \begin{bmatrix} \varepsilon_{1jk} W_{jk} \\ \varepsilon_{2jk} W_{jk} \\ \varepsilon_{3jk} W_{jk} \end{bmatrix} = \begin{bmatrix} 2W_{23} \\ 2W_{31} \\ 2W_{12} \end{bmatrix}.$$

the components of $\mathbf{u}$ are twice those of $\mathbf{W}$. This vector is said to be the axial vector of $\mathbf{W}$. If $\mathbf{W}$ represent the vorticity tensor,

$$[\mathbf{W}] = \frac{1}{2}\left(\nabla \mathbf{u}^T - \nabla \mathbf{u}\right) = \frac{1}{2} \begin{bmatrix} 0 & \dfrac{\partial u_1}{\partial x_2} - \dfrac{\partial u_2}{\partial x_1} & \dfrac{\partial u_1}{\partial x_3} - \dfrac{\partial u_3}{\partial x_1} \\ -\dfrac{\partial u_1}{\partial x_2} + \dfrac{\partial u_2}{\partial x_1} & 0 & \dfrac{\partial u_2}{\partial x_3} - \dfrac{\partial u_3}{\partial x_2} \\ -\dfrac{\partial u_1}{\partial x_3} + \dfrac{\partial u_3}{\partial x_1} & -\dfrac{\partial u_2}{\partial x_3} + \dfrac{\partial u_3}{\partial x_2} & 0 \end{bmatrix}$$

Then the axial vector of this vorticity tensor is

$$[\mathbf{w}] = \begin{bmatrix} \dfrac{\partial u_2}{\partial x_3} - \dfrac{\partial u_3}{\partial x_2} \\ \dfrac{\partial u_3}{\partial x_1} - \dfrac{\partial u_1}{\partial x_3} \\ \dfrac{\partial u_1}{\partial x_2} - \dfrac{\partial u_2}{\partial x_1} \end{bmatrix} = -\varepsilon_{ijk} \nabla_j u_k = [-\nabla \times \mathbf{u}].$$

Problem 1.6

If $\mathbf{D}$, $\mathbf{S}$, and $\mathbf{W}$ are two-tensors, $\mathbf{D}$ symmetric and $\mathbf{W}$ anti-symmetric,

$$\mathbf{D} : \mathbf{S} = D_{ij} S_{ji} = D_{ji} S_{ji} = D_{ji} S_{ij}^T = \mathbf{D} : \mathbf{S}^T \text{ (symmetric } \mathbf{D})$$
$$= \tfrac{1}{2} D_{ij} S_{ji} + \tfrac{1}{2} D_{ji} S_{ij}^T = \tfrac{1}{2} D_{ij} \left(S_{ji} + S_{ji}^T\right) = \mathbf{D} : \tfrac{1}{2}\left(\mathbf{S} + \mathbf{S}^T\right).$$

Furthermore,

$$\mathbf{W} : \mathbf{S} = W_{ij} S_{ji} = -W_{ji} S_{ji} = -W_{ji} S_{ij}^T = -\mathbf{W} : \mathbf{S}^T \text{ (anti-symmetric } \mathbf{W})$$
$$= \tfrac{1}{2}\left(W_{ij} S_{ji} - W_{ji} S_{ij}^T\right) = \tfrac{1}{2} W_{ij}\left(S_{ji} + S_{ji}^T\right) = \mathbf{W} : \tfrac{1}{2}\left(\mathbf{S} - \mathbf{S}^T\right),$$
$$\mathbf{D} : \mathbf{W} = \mathbf{W} : \tfrac{1}{2}\left(\mathbf{D} - \mathbf{D}^T\right) = \mathbf{D} : \tfrac{1}{2}\left(\mathbf{W} + \mathbf{W}^T\right) = 0.$$

In addition, if

$$\mathbf{T} : \mathbf{S} = 0 \ \forall \mathbf{S} \text{ then } \mathbf{T} = 0. \tag{S.229}$$

This is shown by choosing $\mathbf{S}$ to be unity at any particular entry ij and zero elsewhere. The corresponding ij component of $\mathbf{T}$ has to be zero. This implies $\mathbf{T} = 0$.

If

$$\mathbf{T} : \mathbf{S} = 0 \ \forall \text{ symmetric } \mathbf{S} \text{ then } \mathbf{S} : \frac{1}{2} (\mathbf{T} + \mathbf{T}^T) = 0.$$

This leads to $\mathbf{T} + \mathbf{T}^T = 0$, or $\mathbf{T}$ is anti-symmetric. In a similar manner,

if $\mathbf{T} : \mathbf{S} = 0 \ \forall$ anti-symmetric $\mathbf{S}$ then $\mathbf{T}$ is symmetric.

Problem 1.7

If $\mathbf{Q}$ is orthogonal,

$$\mathbf{Q}^{-1} = \mathbf{Q}^T$$
$$\Rightarrow \mathbf{H} + \mathbf{H}^T + \mathbf{H}\mathbf{H}^T = \mathbf{Q} + \mathbf{Q}^T - 2\mathbf{I} + \mathbf{Q}\mathbf{Q}^T - \mathbf{Q} - \mathbf{Q}^T + \mathbf{I} = 0$$
$$\mathbf{H}\mathbf{H}^T = 2\mathbf{I} - \mathbf{Q} - \mathbf{Q}^T = \mathbf{H}^T\mathbf{H}$$

Conversely, if

$$\mathbf{H} + \mathbf{H}^T + \mathbf{H}\mathbf{H}^T = 0, \quad \mathbf{H}\mathbf{H}^T = \mathbf{H}^T\mathbf{H},$$

then

$$\mathbf{Q}\mathbf{Q}^T - \mathbf{Q} - \mathbf{Q}^T + \mathbf{I} = \mathbf{Q}^T\mathbf{Q} - \mathbf{Q}^T - \mathbf{Q} + \mathbf{I} \Rightarrow \mathbf{Q}\mathbf{Q}^T = \mathbf{Q}^T\mathbf{Q}.$$
$$\mathbf{Q} + \mathbf{Q}^T - 2\mathbf{I} + \mathbf{Q}\mathbf{Q}^T - \mathbf{Q} - \mathbf{Q}^T + \mathbf{I} = 0 \Rightarrow \mathbf{Q}\mathbf{Q}^T = \mathbf{I} = \mathbf{Q}^T\mathbf{Q},$$

and $\mathbf{Q}$ is orthogonal.

Problem 1.8

To show that I, II, III are invariants, we note that

$$I' = S'_{ii} = A_{ij} A_{ik} S_{jk} = \delta_{jk} S_{jk} = S_{jj} = I.$$
$$II' = S'_{ij} S'_{ji} = A_{ik} A_{jm} S_{km} A_{jn} A_{ir} S_{nr} = \delta_{kr} \delta_{mn} S_{km} S_{nr}$$
$$= S_{km} S_{mk} = II.$$

$$III' = S'_{ij}S'_{jk}S'_{ki} = A_{i\alpha}A_{j\beta}S_{\alpha\beta}A_{j\gamma}A_{k\delta}S_{\gamma\delta}A_{k\varepsilon}A_{i\varphi}S_{\varepsilon\varphi}$$
$$= \delta_{\alpha\varphi}\delta_{\beta\gamma}\delta_{\delta\varepsilon}S_{\alpha\beta}S_{\gamma\delta}S_{\varepsilon\varphi} = S_{\alpha\beta}S_{\beta\delta}S_{\delta\alpha} = III.$$

Next,

$$\det[\mathbf{S} - \omega\mathbf{I}] = \det\begin{bmatrix} S_{11} - \omega & S_{12} & S_{13} \\ S_{21} & S_{22} - \omega & S_{23} \\ S_{31} & S_{32} & S_{33} - \omega \end{bmatrix}$$

$$= (S_{11} - \omega)\left(S_{22}S_{33} - \omega(S_{22} + S_{33}) + \omega^2\right) + S_{12}S_{23}S_{31}$$
$$+ S_{13}S_{21}S_{32} - S_{31}S_{22}S_{13} + \omega(S_{31}S_{13} + S_{32}S_{23} + S_{21}S_{12})$$
$$- S_{32}S_{23}S_{11} - S_{33}S_{21}S_{12}$$
$$= -\omega^3 + \omega^2(S_{11} + S_{22} + S_{33}) - \omega(S_{11}S_{22} + S_{11}S_{33} + S_{22}S_{33}$$
$$- S_{31}S_{13} - S_{32}S_{23} - S_{21}S_{12}) + S_{11}S_{22}S_{33} + S_{12}S_{23}S_{31}$$
$$+ S_{13}S_{21}S_{32} - S_{31}S_{22}S_{13} - S_{32}S_{23}S_{11} - S_{33}S_{21}S_{12}$$
$$= -\omega^3 + I_1\omega^2 - I_2\omega + I_3,$$

where

$$I_1 = I = tr\mathbf{S},$$
$$I_2 = \tfrac{1}{2}\left(I^2 - II\right) = S_{11}S_{22} + S_{11}S_{33} + S_{22}S_{33} - S_{12}S_{21} - S_{13}S_{31} - S_{23}S_{32}$$
$$I_3 = \tfrac{1}{6}\left(I^3 - 3I.II + 2III\right) = \det\mathbf{S}.$$

If $\mathbf{e}$ is an eigenvector of $\mathbf{S}$, with eigenvalue ω then

$$\mathbf{S}\mathbf{e} = \omega\mathbf{e}.$$

The condition for this to have non-trivial solutions is that

$$\det[\mathbf{S} - \omega\mathbf{I}] = 0 = -\omega^3 + I_1\omega^2 - I_2\omega + I_3. \tag{S.230}$$

This is said to be the characteristic equation for $\mathbf{S}$. According to the Cayley-Hamilton theorem, $\mathbf{S}$ satisfies its own characteristic equation.

Problem 1.9

Consider the 2×2 matrix

$$[\mathbf{C}] = \begin{bmatrix} 1 + \gamma^2 & \gamma \\ \gamma & 1 \end{bmatrix}.$$

Denote $\mathbf{U} = \mathbf{C}^{1/2}$, $\mathbf{U}$ satisfies its own characteristic equation (in 2-D):

$$\mathbf{U}^2 - I_1\,(\mathbf{U})\,\mathbf{U} + \det\,[\mathbf{U}]\,\mathbf{I} = 0 \mathbf{U} = \frac{1}{I_1\,(\mathbf{U})}\,(\mathbf{C} + \det\,(\mathbf{U})\,\mathbf{I})\,.$$

Expressed in eigenspace,

$$\mathbf{U} = \begin{bmatrix} \lambda_1 & 0 \\ 0 & \lambda_2 \end{bmatrix}, \quad \mathbf{C} = \begin{bmatrix} \lambda_1^2 & 0 \\ 0 & \lambda_2^2 \end{bmatrix},$$

$I_1\,(\mathbf{U}) = \lambda_1 + \lambda_2, \quad I_1\,(\mathbf{C}) = \lambda_1^2 + \lambda_2^2, \quad I_2\,(\mathbf{U}) = \lambda_1\lambda_2, \quad I_2\,(\mathbf{C}) = \lambda_1^2\lambda_2^2,$

$I_2\,(\mathbf{U}) = \sqrt{I_2\,(\mathbf{C})}, \quad I_1\,(\mathbf{U}) = \sqrt{I_1\,(\mathbf{C}) + 2\sqrt{I_2\,(\mathbf{C})}}$

$I_1\,(\mathbf{C}) = 2 + \gamma^2, \quad I_2\,(\mathbf{C}) = \det\,[\mathbf{C}] = 1 + \gamma^2 - \gamma^2 = 1.$

Thus,

$$\mathbf{U} = \mathbf{C}^{1/2} = \frac{1}{\sqrt{4 + \gamma^2}} \begin{bmatrix} 2 + \gamma^2 & \gamma \\ \gamma & 2 \end{bmatrix}. \tag{S.231}$$

For the 3×3 matric case:

$$\mathbf{U}^3 - I_1\,(\mathbf{U})\,\mathbf{U}^2 + I_2\,(\mathbf{U})\,\mathbf{U} - I_3\,(\mathbf{U})\,\mathbf{I} = 0 \tag{S.232}$$

leading to

$$\mathbf{U}^4 + I_2\mathbf{U}^2 - I_3\mathbf{U} = I_1\mathbf{U}^3 = I_1^2\mathbf{U}^2 - I_1I_2\mathbf{U} + I_1I_3\mathbf{I}$$
$$\mathbf{U} = \frac{1}{I_3 - I_1I_2}\,\left(\mathbf{C}^2 + \left(I_2 - I_1^2\right)\mathbf{C} - I_1I_3\mathbf{I}\right), \tag{S.233}$$

where I_1, I_2, I_3 are the invariants of $\mathbf{U}$, and their dependence on $\mathbf{U}$ has been suppressed for brevity. In terms of the eigenvalues:

$$I_1 = \lambda_1 + \lambda_2 + \lambda_3, \quad I_2 = \lambda_1\lambda_2 + \lambda_1\lambda_3 + \lambda_2\lambda_3, \quad I_3 = \lambda_1\lambda_2\lambda_3.$$

These need to be expressed in terms of the invariants of $\mathbf{C}$. First

$$I_3\,(\mathbf{U}) = \sqrt{I_3\,(\mathbf{C})}. \tag{S.234}$$

Next, in terms of the eigenvalues of $\mathbf{U}$,

$$I_1^2\,(\mathbf{U}) = (\lambda_1 + \lambda_2 + \lambda_3)^2 = I_1\,(\mathbf{C}) + 2I_2\,(\mathbf{U})\,. \tag{S.235}$$

Next, from (S.232),

$$\lambda_1^4 + \lambda_2^4 + \lambda_3^4 = (I_3 - I_1 I_2)(\lambda_1 + \lambda_2 + \lambda_3) + (I_1^2 - I_2)(\lambda_1^2 + \lambda_2^2 + \lambda_3^2)$$
$$+ 3I_1 I_3$$
$$= (I_3 - I_1 I_2) I_1 + (I_1^2 - I_2)(I_1^2 - 2I_2) + 3I_1 I_3$$
$$= 4I_1 I_3 - 4I_1^2 I_2 + I_1^4 + 2I_2^2. \tag{S.236}$$

Left hand of (S.236) can be expressed as

$$I_1^2(\mathbf{C}) - 2I_2(\mathbf{C}) = (\lambda_1^2 + \lambda_2^2 + \lambda_3^2)^2 - 2(\lambda_1^2\lambda_2^2 + \lambda_1^2\lambda_3^2 + \lambda_2^2\lambda_3^2) = \lambda_1^4 + \lambda_2^4 + \lambda_3^4$$

Thus,

$$\frac{I_1^2(\mathbf{C}) - 2I_2(\mathbf{C})}{I_1^2(\mathbf{C})} = \frac{1}{I_1^2(\mathbf{C})}\left[4I_1 I_3 - 4I_1^2 I_2 + I_1^4 + 2I_2^2\right]$$

With some further algebra,

$$2\left[1 - \frac{I_1^2(\mathbf{C}) - 2I_2(\mathbf{C})}{I_1^2(\mathbf{C})}\right] + 8\sqrt{\frac{I_3(\mathbf{C})}{I_1^3(\mathbf{C})}}\frac{I_1(\mathbf{U})}{\sqrt{I_1(\mathbf{C})}} = 2$$
$$+ \frac{1}{I_1^2(\mathbf{C})}\left[8I_2(\mathbf{U}) I_1^2(\mathbf{U}) - 4I_2^2(\mathbf{U}) - 2I_1^4(\mathbf{U})\right]$$
$$= \frac{I_1^4(\mathbf{U})}{I_1^2(\mathbf{C})} - 2\frac{I_1^2(\mathbf{U})}{I_1(\mathbf{C})} + 1$$
$$= \left[\frac{I_1^2(\mathbf{U})}{I_1(\mathbf{C})} - 1\right]^2.$$

This is a quartic equation $(x^2 - 1)^2 = a + bx$ in x, with

$$x = \frac{I_1(\mathbf{U})}{\sqrt{I_1(\mathbf{C})}}, \quad a = 2\left[1 - \frac{I_1^2(\mathbf{C}) - 2I_2(\mathbf{C})}{I_1^2(\mathbf{C})}\right] = 2\left[1 - \frac{tr\mathbf{C}^2}{(tr\mathbf{C})^2}\right],$$
$$b = 8\sqrt{\frac{I_3(\mathbf{C})}{I_1^3(\mathbf{C})}} = 8\sqrt{\frac{\det \mathbf{C}}{(tr\mathbf{C})^3}}.$$

Once x is obtained, $tr(\mathbf{U})$ is determined in term of $tr(\mathbf{C})$, and from (S.235), the second invariant of $\mathbf{U}$ is determined in terms of $\mathbf{C}$.

Problem 1.10

The components of the strain rate tensor, $\mathbf{D} = \frac{1}{2}\left(\nabla\mathbf{u} + \nabla\mathbf{u}^T\right)$, and the vorticity tensor, $\mathbf{W} = \frac{1}{2}\left(\nabla\mathbf{u}^T - \nabla\mathbf{u}\right)$ are

$$[\mathbf{L}] = [\nabla\mathbf{u}]^T = \begin{bmatrix} \dfrac{\partial u_1}{\partial x_1} & \dfrac{\partial u_1}{\partial x_2} & \dfrac{\partial u_1}{\partial x_3} \\[2mm] \dfrac{\partial u_2}{\partial x_1} & \dfrac{\partial u_2}{\partial x_2} & \dfrac{\partial u_2}{\partial x_3} \\[2mm] \dfrac{\partial u_3}{\partial x_1} & \dfrac{\partial u_3}{\partial x_2} & \dfrac{\partial u_3}{\partial x_3} \end{bmatrix},$$

$$[\mathbf{D}] = \frac{1}{2}\begin{bmatrix} 2\dfrac{\partial u_1}{\partial x_1} & \dfrac{\partial u_1}{\partial x_2} + \dfrac{\partial u_2}{\partial x_1} & \dfrac{\partial u_1}{\partial x_3} + \dfrac{\partial u_3}{\partial x_1} \\[2mm] \dfrac{\partial u_1}{\partial x_2} + \dfrac{\partial u_2}{\partial x_1} & 2\dfrac{\partial u_2}{\partial x_2} & \dfrac{\partial u_2}{\partial x_3} + \dfrac{\partial u_3}{\partial x_2} \\[2mm] \dfrac{\partial u_1}{\partial x_3} + \dfrac{\partial u_3}{\partial x_1} & \dfrac{\partial u_2}{\partial x_3} + \dfrac{\partial u_3}{\partial x_2} & 2\dfrac{\partial u_3}{\partial x_3} \end{bmatrix},$$

$$[\mathbf{W}] = \frac{1}{2}\begin{bmatrix} 0 & \dfrac{\partial u_1}{\partial x_2} - \dfrac{\partial u_2}{\partial x_1} & \dfrac{\partial u_1}{\partial x_3} - \dfrac{\partial u_3}{\partial x_1} \\[2mm] -\dfrac{\partial u_1}{\partial x_2} + \dfrac{\partial u_2}{\partial x_1} & 0 & \dfrac{\partial u_2}{\partial x_3} - \dfrac{\partial u_3}{\partial x_2} \\[2mm] -\dfrac{\partial u_1}{\partial x_3} + \dfrac{\partial u_3}{\partial x_1} & -\dfrac{\partial u_2}{\partial x_3} + \dfrac{\partial u_3}{\partial x_2} & 0 \end{bmatrix}.$$

Problem 1.11

First, we express $\mathbf{S}$ in cylindrical coordinates as

$$\begin{aligned} \mathbf{S} = & S_{rr}\mathbf{e}_r\mathbf{e}_r + S_{r\theta}\mathbf{e}_r\mathbf{e}_\theta + S_{rz}\mathbf{e}_r\mathbf{e}_z + S_{\theta r}\mathbf{e}_\theta\mathbf{e}_r \\ & +S_{\theta\theta}\mathbf{e}_\theta\mathbf{e}_\theta + S_{\theta z}\mathbf{e}_\theta\mathbf{e}_z + S_{zr}\mathbf{e}_z\mathbf{e}_r + S_{z\theta}\mathbf{e}_z\mathbf{e}_\theta + S_{zz}\mathbf{e}_z\mathbf{e}_z, \end{aligned}$$

and taking the gradient operation,

$$\nabla = \mathbf{e}_r\frac{\partial}{\partial r} + \mathbf{e}_\theta\frac{1}{r}\frac{\partial}{\partial\theta} + \mathbf{e}_z\frac{\partial}{\partial z},$$

keeping in mind that

$$\frac{\partial}{\partial r}\mathbf{e}_r = 0, \qquad \frac{\partial}{\partial r}\mathbf{e}_\theta = 0, \qquad \frac{\partial}{\partial r}\mathbf{e}_z = 0$$

$$\frac{\partial}{\partial \theta}\mathbf{e}_r = \mathbf{e}_\theta, \quad \frac{\partial}{\partial \theta}\mathbf{e}_\theta = -\mathbf{e}_r, \quad \frac{\partial}{\partial \theta}\mathbf{e}_z = 0$$

$$\frac{\partial}{\partial z}\mathbf{e}_r = 0, \quad \frac{\partial}{\partial z}\mathbf{e}_\theta = 0, \quad \frac{\partial}{\partial z}\mathbf{e}_z = 0$$

to result in

$$\nabla \cdot \mathbf{S} = \mathbf{e}_r \left(\frac{\partial S_{rr}}{\partial r} + \frac{S_{rr} - S_{\theta\theta}}{r} + \frac{1}{r}\frac{\partial S_{\theta r}}{\partial \theta} + \frac{\partial S_{zr}}{\partial z} \right)$$

$$+ \mathbf{e}_\theta \left(\frac{\partial S_{r\theta}}{\partial r} + \frac{2S_{r\theta}}{r} + \frac{1}{r}\frac{\partial S_{\theta\theta}}{\partial \theta} + \frac{\partial S_{z\theta}}{\partial z} + \frac{S_{\theta r} - S_{r\theta}}{r} \right)$$

$$+ \mathbf{e}_z \left(\frac{\partial S_{rz}}{\partial r} + \frac{S_{rz}}{r} + \frac{1}{r}\frac{\partial S_{\theta z}}{\partial \theta} + \frac{\partial S_{zz}}{\partial z} \right). \tag{S.237}$$

Problem 1.12

In cylindrical coordinates,

$$\mathbf{u} = u_r \mathbf{e}_r + u_\theta \mathbf{e}_\theta + u_z \mathbf{e}_z.$$

Thus

$$\nabla \mathbf{u} = \left(\mathbf{e}_r \frac{\partial}{\partial r} + \mathbf{e}_\theta \frac{1}{r}\frac{\partial}{\partial \theta} + \mathbf{e}_z \frac{\partial}{\partial z} \right) (u_r \mathbf{e}_r + u_\theta \mathbf{e}_\theta + u_z \mathbf{e}_z)$$

$$= \frac{\partial u_r}{\partial r}\mathbf{e}_r \mathbf{e}_r + \frac{\partial u_\theta}{\partial r}\mathbf{e}_r \mathbf{e}_\theta + \frac{\partial u_z}{\partial r}\mathbf{e}_r \mathbf{e}_z$$

$$+ \left(\frac{1}{r}\frac{\partial u_r}{\partial \theta} - \frac{u_\theta}{r} \right)\mathbf{e}_\theta \mathbf{e}_r + \left(\frac{1}{r}\frac{\partial u_\theta}{\partial \theta} + \frac{u_r}{r} \right)\mathbf{e}_\theta \mathbf{e}_\theta + \frac{1}{r}\frac{\partial u_z}{\partial \theta}\mathbf{e}_\theta \mathbf{e}_z$$

$$+ \frac{\partial u_r}{\partial z}\mathbf{e}_z \mathbf{e}_r + \frac{\partial u_\theta}{\partial z}\mathbf{e}_z \mathbf{e}_\theta + \frac{\partial u_z}{\partial z}\mathbf{e}_z \mathbf{e}_z,$$

and

$$\mathbf{u} \cdot \nabla \mathbf{u} = \mathbf{e}_r \left[u_r \frac{\partial u_r}{\partial r} + \frac{u_\theta}{r}\frac{\partial u_r}{\partial \theta} + u_z \frac{\partial u_r}{\partial z} - \frac{u_\theta u_\theta}{r} \right]$$

$$+ \mathbf{e}_\theta \left[u_r \frac{\partial u_\theta}{\partial r} + \frac{u_\theta}{r}\frac{\partial u_\theta}{\partial \theta} + u_z \frac{\partial u_\theta}{\partial z} + \frac{u_\theta u_r}{r} \right]$$

$$+ \mathbf{e}_z \left[u_r \frac{\partial u_z}{\partial r} + \frac{u_\theta}{r}\frac{\partial u_z}{\partial \theta} + u_z \frac{\partial u_z}{\partial z} \right]. \tag{S.238}$$

Additional problem: evaluate $\mathbf{u} \cdot \nabla \mathbf{S}$ in cylindrical coordinates. Now with

$$
\mathbf{S} = S_{rr}\mathbf{e}_r\mathbf{e}_r + S_{r\theta}\left(\mathbf{e}_r\mathbf{e}_\theta + \mathbf{e}_\theta\mathbf{e}_r\right) + S_{rz}\left(\mathbf{e}_r\mathbf{e}_z + \mathbf{e}_z\mathbf{e}_r\right)
$$
$$
+ S_{\theta\theta}\mathbf{e}_\theta\mathbf{e}_\theta + S_{\theta z}\left(\mathbf{e}_z\mathbf{e}_\theta + \mathbf{e}_\theta\mathbf{e}_z\right) + S_{zz}\mathbf{e}_z\mathbf{e}_z, \tag{S.239}
$$

$$
\nabla \mathbf{S} = \left(\mathbf{e}_r\frac{\partial}{\partial r} + \mathbf{e}_\theta\frac{\partial}{r\partial\theta} + \mathbf{e}_z\frac{\partial}{\partial z}\right)\mathbf{S}
$$
$$
= \mathbf{e}_r\left(\mathbf{e}_r\mathbf{e}_r\frac{\partial S_{rr}}{\partial r} + \left(\mathbf{e}_r\mathbf{e}_\theta + \mathbf{e}_\theta\mathbf{e}_r\right)\frac{\partial S_{r\theta}}{\partial r} + \left(\mathbf{e}_r\mathbf{e}_z + \mathbf{e}_z\mathbf{e}_r\right)\frac{\partial S_{rz}}{\partial r}\right.
$$
$$
\left. + \mathbf{e}_\theta\mathbf{e}_\theta\frac{\partial S_{\theta\theta}}{\partial r} + \left(\mathbf{e}_z\mathbf{e}_\theta + \mathbf{e}_\theta\mathbf{e}_z\right)\frac{\partial S_{\theta z}}{\partial r} + \mathbf{e}_z\mathbf{e}_z\frac{\partial S_{zz}}{\partial r}\right)
$$
$$
+ \mathbf{e}_\theta\left(\mathbf{e}_r\mathbf{e}_r\frac{\partial S_{rr}}{r\partial\theta} + \frac{S_{rr}}{r}\left(\mathbf{e}_\theta\mathbf{e}_r + \mathbf{e}_r\mathbf{e}_\theta\right) + \left(\mathbf{e}_r\mathbf{e}_\theta + \mathbf{e}_\theta\mathbf{e}_r\right)\frac{\partial S_{r\theta}}{r\partial\theta}\right.
$$
$$
+ 2\frac{S_{r\theta}}{r}\left(\mathbf{e}_\theta\mathbf{e}_\theta - \mathbf{e}_r\mathbf{e}_r\right) + \left(\mathbf{e}_r\mathbf{e}_z + \mathbf{e}_z\mathbf{e}_r\right)\frac{\partial S_{rz}}{r\partial\theta} + \frac{S_{rz}}{r}\left(\mathbf{e}_\theta\mathbf{e}_z + \mathbf{e}_z\mathbf{e}_\theta\right)
$$
$$
+ \mathbf{e}_\theta\mathbf{e}_\theta\frac{\partial S_{\theta\theta}}{r\partial\theta} - \frac{S_{\theta\theta}}{r}\left(\mathbf{e}_r\mathbf{e}_\theta + \mathbf{e}_\theta\mathbf{e}_r\right)
$$
$$
\left. + \left(\mathbf{e}_z\mathbf{e}_\theta + \mathbf{e}_\theta\mathbf{e}_z\right)\frac{\partial S_{\theta z}}{r\partial\theta} - \frac{S_{\theta z}}{r}\left(\mathbf{e}_z\mathbf{e}_r + \mathbf{e}_r\mathbf{e}_z\right) + \mathbf{e}_z\mathbf{e}_z\frac{\partial S_{zz}}{r\partial\theta}\right)
$$
$$
+ \mathbf{e}_z\left(\mathbf{e}_r\mathbf{e}_r\frac{\partial S_{rr}}{\partial z} + \left(\mathbf{e}_r\mathbf{e}_\theta + \mathbf{e}_\theta\mathbf{e}_r\right)\frac{\partial S_{r\theta}}{\partial z} + \left(\mathbf{e}_r\mathbf{e}_z + \mathbf{e}_z\mathbf{e}_r\right)\frac{\partial S_{rz}}{\partial z}\right.
$$
$$
\left. + \mathbf{e}_\theta\mathbf{e}_\theta\frac{\partial S_{\theta\theta}}{\partial z} + \left(\mathbf{e}_z\mathbf{e}_\theta + \mathbf{e}_\theta\mathbf{e}_z\right)\frac{\partial S_{\theta z}}{\partial z} + \mathbf{e}_z\mathbf{e}_z\frac{\partial S_{zz}}{\partial z}\right).
$$

Thus

$$
\mathbf{u} \cdot \nabla \mathbf{S} = u_r\left(\mathbf{e}_r\mathbf{e}_r\frac{\partial S_{rr}}{\partial r} + \left(\mathbf{e}_r\mathbf{e}_\theta + \mathbf{e}_\theta\mathbf{e}_r\right)\frac{\partial S_{r\theta}}{\partial r} + \left(\mathbf{e}_r\mathbf{e}_z + \mathbf{e}_z\mathbf{e}_r\right)\frac{\partial S_{rz}}{\partial r}\right.
$$
$$
\left. + \mathbf{e}_\theta\mathbf{e}_\theta\frac{\partial S_{\theta\theta}}{\partial r} + \left(\mathbf{e}_z\mathbf{e}_\theta + \mathbf{e}_\theta\mathbf{e}_z\right)\frac{\partial S_{\theta z}}{\partial r} + \mathbf{e}_z\mathbf{e}_z\frac{\partial S_{zz}}{\partial r}\right)
$$
$$
+ u_\theta\left(\mathbf{e}_r\mathbf{e}_r\frac{\partial S_{rr}}{r\partial\theta} + \frac{S_{rr}}{r}\left(\mathbf{e}_\theta\mathbf{e}_r + \mathbf{e}_r\mathbf{e}_\theta\right) + \left(\mathbf{e}_r\mathbf{e}_\theta + \mathbf{e}_\theta\mathbf{e}_r\right)\frac{\partial S_{r\theta}}{r\partial\theta}\right.
$$
$$
+ 2\frac{S_{r\theta}}{r}\left(\mathbf{e}_\theta\mathbf{e}_\theta - \mathbf{e}_r\mathbf{e}_r\right) + \left(\mathbf{e}_r\mathbf{e}_z + \mathbf{e}_z\mathbf{e}_r\right)\frac{\partial S_{rz}}{r\partial\theta} + \frac{S_{rz}}{r}\left(\mathbf{e}_\theta\mathbf{e}_z + \mathbf{e}_z\mathbf{e}_\theta\right)
$$
$$
+ \mathbf{e}_\theta\mathbf{e}_\theta\frac{\partial S_{\theta\theta}}{r\partial\theta} - \frac{S_{\theta\theta}}{r}\left(\mathbf{e}_r\mathbf{e}_\theta + \mathbf{e}_\theta\mathbf{e}_r\right)
$$
$$
\left. + \left(\mathbf{e}_z\mathbf{e}_\theta + \mathbf{e}_\theta\mathbf{e}_z\right)\frac{\partial S_{\theta z}}{r\partial\theta} - \frac{S_{\theta z}}{r}\left(\mathbf{e}_z\mathbf{e}_r + \mathbf{e}_r\mathbf{e}_z\right) + \mathbf{e}_z\mathbf{e}_z\frac{\partial S_{zz}}{r\partial\theta}\right)
$$

$$+ u_z \left(\mathbf{e}_r\mathbf{e}_r \frac{\partial S_{rr}}{\partial z} + (\mathbf{e}_r\mathbf{e}_\theta + \mathbf{e}_\theta\mathbf{e}_r) \frac{\partial S_{r\theta}}{\partial z} + (\mathbf{e}_r\mathbf{e}_z + \mathbf{e}_z\mathbf{e}_r) \frac{\partial S_{rz}}{\partial z} \right.$$

$$\left. + \mathbf{e}_\theta\mathbf{e}_\theta \frac{\partial S_{\theta\theta}}{\partial z} + (\mathbf{e}_z\mathbf{e}_\theta + \mathbf{e}_\theta\mathbf{e}_z) \frac{\partial S_{\theta z}}{\partial z} + \mathbf{e}_z\mathbf{e}_z \frac{\partial S_{zz}}{\partial z} \right).$$

$$\mathbf{u} \cdot \nabla \mathbf{S} = \mathbf{e}_r\mathbf{e}_r \left(u_r \frac{\partial S_{rr}}{\partial r} + u_\theta \frac{\partial S_{rr}}{r\partial\theta} + u_z \frac{\partial S_{rr}}{\partial z} - 2u_\theta \frac{S_{r\theta}}{r} \right)$$

$$+ (\mathbf{e}_r\mathbf{e}_\theta + \mathbf{e}_\theta\mathbf{e}_r) \left(u_r \frac{\partial S_{r\theta}}{\partial r} + u_\theta \frac{\partial S_{r\theta}}{r\partial\theta} + u_z \frac{\partial S_{r\theta}}{\partial z} + u_\theta \frac{S_{rr}}{r} - u_\theta \frac{S_{\theta\theta}}{r} \right)$$

$$+ (\mathbf{e}_r\mathbf{e}_z + \mathbf{e}_z\mathbf{e}_r) \left(u_r \frac{\partial S_{rz}}{\partial r} + u_\theta \frac{\partial S_{rz}}{r\partial\theta} + u_z \frac{\partial S_{rz}}{\partial z} - u_\theta \frac{S_{\theta z}}{r} \right)$$

$$+ \mathbf{e}_\theta\mathbf{e}_\theta \left(u_r \frac{\partial S_{\theta\theta}}{\partial r} + u_\theta \frac{\partial S_{\theta\theta}}{r\partial\theta} + u_z \frac{\partial S_{\theta\theta}}{\partial z} + 2u_\theta \frac{S_{r\theta}}{r} \right)$$

$$+ (\mathbf{e}_z\mathbf{e}_\theta + \mathbf{e}_\theta\mathbf{e}_z) \left(u_r \frac{\partial S_{\theta z}}{\partial r} + u_\theta \frac{\partial S_{\theta z}}{r\partial\theta} + u_z \frac{\partial S_{\theta z}}{\partial z} + u_\theta \frac{S_{rz}}{r} \right)$$

$$+ \mathbf{e}_z\mathbf{e}_z \left(u_r \frac{\partial S_{zz}}{\partial r} + u_\theta \frac{\partial S_{zz}}{r\partial\theta} + u_z \frac{\partial S_{zz}}{\partial z} \right).$$

Problem 1.13

The stress tensor in a material satisfies $\nabla \cdot \mathbf{S} = \mathbf{0}$. Thus

$$\nabla_j \left(x_k S_{ij} \right) = S_{ik}.$$

and

$$\langle S_{ik} \rangle = \frac{1}{V} \int_V S_{ik} dV = \frac{1}{V} \int_V \nabla_j \left(x_k S_{ij} \right) dV = \frac{1}{V} \int_S x_k S_{ij} n_j dA$$
$$= \frac{1}{2V} \int_S \left(x_k t_i + x_i t_k \right) dA$$

after a symmetrisation.

Problem 1.14

We can regard

$$\langle \mathbf{nn} \rangle = \frac{1}{S} \int_S \mathbf{nn} dS$$

and

$$\langle \mathbf{nnnn} \rangle = \frac{1}{S} \int_S \mathbf{nnnn} \, dS$$

as averages of various moments of the normal unit vector, uniformly distributed in space.

We further note that $< \mathbf{nn} >$ and $< \mathbf{nnnn} >$ are isotropic tensors (unchanged with coordinate rotation) and thus their general forms are given by

$$\langle \mathbf{nn} \rangle = \alpha \mathbf{I} = \frac{1}{S} \int_S \mathbf{nn} \, dS,$$

$$\langle n_i n_j n_k n_l \rangle = \beta \left(\delta_{ij} \delta_{kl} + \delta_{ik} \delta_{jl} + \delta_{il} \delta_{jk} \right) = \frac{1}{S} \int_S n_i n_j n_k n_l \, dS.$$

The scalars α and β are found by contracting indices and performing some simple integrations:

$$3\alpha = \frac{1}{S} \int_S dS = 1 \; \alpha = \frac{1}{3}.$$

$$15\beta = \frac{1}{S} \int_S dS = 1 \; \beta = \frac{1}{15}.$$

Another approach (here a is the radius of the sphere): for $\langle \mathbf{nn} \rangle$:

$$\frac{1}{S} \int n_i n_j \, dS = \frac{1}{aS} \int_S x_i n_j \, dS$$

$$= \frac{1}{aS} \int_V \frac{\partial}{\partial x_j} (x_i) \, dV$$

$$= \frac{1}{aS} \delta_{ij} \int_V dV = \frac{V}{aS} \delta_{ij}$$

$$= \frac{1}{3} \delta_{ij}$$

And for $\langle n_i n_j n_k n_l \rangle$:

$$\frac{1}{S} \int n_i n_j n_k n_l \, dS = \frac{1}{a^3 S} \int_S x_i x_j x_k n_l \, dS$$

$$= \frac{1}{a^3 S} \int_V \frac{\partial}{\partial x_l} (x_i x_j x_k) \, dV$$

$$= \frac{1}{a^3 S} \int_V \left(\delta_{il} x_j x_k + \delta_{jl} x_i x_k + \delta_{kl} x_i x_j \right) dV$$

Also, from

$$\frac{\partial}{\partial x_k}\left(x_i x_j x_k\right) = \delta_{ik} x_j x_k + \delta_{jk} x_i x_k + \delta_{kk} x_i x_j = 5 x_i x_j$$

Thus

$$\begin{aligned}
\frac{1}{a^3 S}\int_V \delta_{il} x_j x_k dV &= \frac{1}{5a^3 S}\int_V \delta_{il}\frac{\partial}{\partial x_m}\left(x_j x_k x_m\right) dV \\
&= \frac{1}{5a^3 S}\delta_{il}\int_S x_j x_k x_m n_m dS \\
&= \frac{a^3}{5a^3 S}\delta_{il}\int_S n_j n_k dS \\
&= \frac{1}{15}\delta_{il}\delta_{jk}
\end{aligned}$$

Similarly

$$\frac{1}{a^3 S}\int_V \delta_{jl} x_i x_k dV = \frac{1}{15}\delta_{jl}\delta_{ik}, \quad \frac{1}{a^3 S}\int_V \delta_{kl} x_i x_j dV = \frac{1}{15}\delta_{kl}\delta_{ij}$$

and

$$\frac{1}{S}\int n_i n_j n_k n_l dS = \frac{1}{15}\left(\delta_{il}\delta_{jk} + \delta_{jl}\delta_{ik} + \delta_{kl}\delta_{ij}\right).$$

Problems in Chap. 3

Problem 3.1

From $\mathbf{F}\mathbf{F}^{-1} = \mathbf{I}$,

$$\mathbf{F}\frac{d}{dt}\mathbf{F}^{-1} + \left(\frac{d}{dt}\mathbf{F}\right)\mathbf{F}^{-1} = \mathbf{0} \quad \mathbf{F}\frac{d}{dt}\mathbf{F}^{-1} = -\left(\frac{d}{dt}\mathbf{F}\right)\mathbf{F}^{-1} = -\mathbf{L}\mathbf{F}\mathbf{F}^{-1}$$

yielding

$$\frac{d}{dt}\mathbf{F}^{-1} = -\mathbf{F}^{-1}\mathbf{L}, \quad \mathbf{F}\left(0\right) = \mathbf{I} = \mathbf{F}^{-1}\left(0\right).$$

Problem 3.2

For a simple shear flow, where the velocity field takes the form

$$u = \dot{\gamma}y, \quad v = 0, \quad w = 0,$$

the velocity gradient is

$$[\mathbf{L}] = [\nabla \mathbf{u}]^T = \begin{bmatrix} 0 & \dot{\gamma} & 0 \\ 0 & 0 & 0 \\ 0 & 0 & 0 \end{bmatrix}.$$

The deformation gradient obeys

$$\frac{d\mathbf{F}}{dt} = \mathbf{LF}, \quad \mathbf{F}(0) = \mathbf{I},$$

which has the solution for a constant $\mathbf{L}$,

$$\mathbf{F}(t) = \exp(t\mathbf{L}) = \mathbf{I} + \sum_{n=1} \frac{1}{n!} (t\mathbf{L})^n = \mathbf{I} + t\mathbf{L} \text{ for } \mathbf{L}^n = 0, n \geq 2$$

The particle X is carried to $\mathbf{x}$ at time t according to

$$\mathbf{x}(t) = \mathbf{F}(t)\,\mathbf{X} \text{ or } \mathbf{X} = \mathbf{F}^{-1}(t)\,\mathbf{x}$$

At time τ if the particle is at point ζ, then

$$\zeta(\tau) = \mathbf{F}(\tau)\,\mathbf{X} = \mathbf{F}(\tau)\,\mathbf{F}^{-1}(t)\,\mathbf{x} = \exp((\tau - t)\mathbf{L})\,\mathbf{x}$$
$$= \mathbf{x} + (\tau - t)\,\mathbf{Lx}. \tag{S.240}$$

This shows the linearly stretching nature of a fluid filament.

Problem 3.3

For an elongational flow, the velocity field is

$$u = ax, \quad v = by, \quad w = cz, \quad a + b + c = 0 \text{ (for incompressibility).} \quad \text{(S.241)}$$

the velocity gradient is

$$[\mathbf{L}] = \begin{bmatrix} a & 0 & 0 \\ 0 & b & 0 \\ 0 & 0 & c \end{bmatrix}, \quad [e^{t\mathbf{L}}] = \begin{bmatrix} e^{at} & 0 & 0 \\ 0 & e^{bt} & 0 \\ 0 & 0 & e^{ct} \end{bmatrix}.$$

The path lines are given by

$$[\boldsymbol{\xi}(\tau)] = \begin{bmatrix} \xi \\ \psi \\ \zeta \end{bmatrix} = \begin{bmatrix} e^{a(\tau-t)} & 0 & 0 \\ 0 & e^{b(\tau-t)} & 0 \\ 0 & 0 & e^{c(\tau-t)} \end{bmatrix} \begin{bmatrix} x \\ y \\ z \end{bmatrix}, \qquad \text{(S.242)}$$

and thus exponential flow can stretch the fluid element exponentially fast.

Problem 3.4

Consider a super-imposed oscillatory shear flow:

$$u = \dot{\gamma}_m y, \quad v = 0, \quad w = \omega \gamma_a y \cos \omega t. \qquad \text{(S.243)}$$

The path lines are defined by

$$\frac{d}{dt}(x, y, z) = (\dot{\gamma}_m y, 0, \omega \gamma_a y \cos \omega t), \quad x(0) = X, \ y(0) = Y, \ z(0) = Z$$

This has the solution

$$x(t) = X + t\dot{\gamma}_m Y, \ y(t) = Y, \ z(t) = Z + \gamma_a Y \sin \omega t$$

The path line, $(\xi(\tau), \psi(\tau), \zeta(\tau))$,

$$\begin{aligned}
(\xi(\tau), \psi(\tau), \zeta(\tau)) &= (X + \tau \dot{\gamma}_m Y, Y, Z + \gamma_a Y \sin \omega \tau) \\
&= (x(t) + (\tau - t)\dot{\gamma}_m y, y, z(t) + \gamma_a y (\cos \omega \tau - \cos \omega t)).
\end{aligned}$$

Problem 3.5

We want to calculate Rivlin–Ericksen tensor for an elongation flow (S.241) where the velocity gradient is

$$[\mathbf{L}] = \operatorname{diag}(a, b, c).$$

The first Rivlin–Ericksen tensor is

$$\mathbf{A}_1 = \mathbf{L} + \mathbf{L}^T = 2\mathbf{L}.$$

The subsequent two Rivlin–Ericksen tensors are, because of $\mathbf{L}$ being diagonal,

$$\mathbf{A}_2 = \mathbf{A}_1\mathbf{L} + \mathbf{L}^T\mathbf{A}_1 = \mathbf{A}_1^2, \quad \mathbf{A}_3 = \mathbf{A}_2\mathbf{L} + \mathbf{L}^T\mathbf{A}_2 = \mathbf{A}_1^3.$$

By induction,

$$\mathbf{A}_n = \mathbf{A}_1^n = \operatorname{diag}\left((2a)^n, (2b)^n, (2c)^n\right).$$

Problem 3.6

For the velocity field of (S.243) takes the form

$$u = \dot{\gamma}_m y, \quad v = 0, \quad w = \omega\gamma_a y \cos\omega t.$$

The velocity gradient tensor is

$$[\mathbf{L}] = \begin{bmatrix} 0 & \dot{\gamma}_m & 0 \\ 0 & 0 & 0 \\ 0 & \omega\gamma_a\cos\omega t & 0 \end{bmatrix}.$$

The first Rivlin–Ericksen tensor is

$$[\mathbf{A}_1] = [\mathbf{L} + \mathbf{L}^T] = \begin{bmatrix} 0 & \dot{\gamma}_m & 0 \\ \dot{\gamma}_m & 0 & \omega\gamma_a\cos\omega t \\ 0 & \omega\gamma_a\cos\omega t & 0 \end{bmatrix}.$$

The second Rivlin–Ericksen tensor is

$$[\mathbf{A}_2] = \left[\tfrac{d}{dt}\mathbf{A}_1 + \mathbf{A}_1\mathbf{L} + \mathbf{L}^T\mathbf{A}_1\right] = \begin{bmatrix} 0 & 0 & 0 \\ 0 & 0 & -\omega^2\gamma_a\sin\omega t \\ 0 & -\omega^2\gamma_a\sin\omega t & 0 \end{bmatrix}$$

$$+ \begin{bmatrix} 0 & \dot{\gamma}_m & 0 \\ \dot{\gamma}_m & 0 & \omega\gamma_a\cos\omega t \\ 0 & \omega\gamma_a\cos\omega t & 0 \end{bmatrix}\begin{bmatrix} 0 & \dot{\gamma}_m & 0 \\ 0 & 0 & 0 \\ 0 & \omega\gamma_a\cos\omega t & 0 \end{bmatrix}$$

$$+ \begin{bmatrix} 0 & 0 & 0 \\ \dot{\gamma}_m & 0 & \omega\gamma_a\cos\omega t \\ 0 & 0 & 0 \end{bmatrix}\begin{bmatrix} 0 & \dot{\gamma}_m & 0 \\ \dot{\gamma}_m & 0 & \omega\gamma_a\cos\omega t \\ 0 & \omega\gamma_a\cos\omega t & 0 \end{bmatrix}$$

$$= \begin{bmatrix} 0 & 0 & 0 \\ 0 & 2\left(\dot{\gamma}_m^2 + \omega^2\gamma_a^2\cos^2\omega t\right) & -\omega^2\gamma_a\sin\omega t \\ 0 & -\omega^2\gamma_a\sin\omega t & 0 \end{bmatrix}.$$

The third Rivlin–Ericksen tensor is

$$[A_3] = \left[\frac{d}{dt}A_2 + A_2L + L^T A_2\right] = \begin{bmatrix} 0 & 0 & 0 \\ 0 & -2\omega^3\gamma_a^2\sin 2\omega t & -\omega^3\gamma_a\cos\omega t \\ 0 & -\omega^3\gamma_a\cos\omega t & 0 \end{bmatrix}$$

$$+ \begin{bmatrix} 0 & 0 & 0 \\ 0 & 2\left(\dot\gamma_m^2 + \omega^2\gamma_a^2\cos^2\omega t\right) & -\omega^2\gamma_a\sin\omega t \\ 0 & -\omega^2\gamma_a\sin\omega t & 0 \end{bmatrix} \begin{bmatrix} 0 & \dot\gamma_m & 0 \\ 0 & 0 & 0 \\ 0 & \omega\gamma_a\cos\omega t & 0 \end{bmatrix}$$

$$+ \begin{bmatrix} 0 & 0 & 0 \\ \dot\gamma_m & 0 & \omega\gamma_a\cos\omega t \\ 0 & 0 & 0 \end{bmatrix} \begin{bmatrix} 0 & 0 & 0 \\ 0 & 2\left(\dot\gamma_m^2 + \omega^2\gamma_a^2\cos^2\omega t\right) & -\omega^2\gamma_a\sin\omega t \\ 0 & -\omega^2\gamma_a\sin\omega t & 0 \end{bmatrix}$$

$$= \begin{bmatrix} 0 & 0 & 0 \\ 0 & -3\omega^3\gamma_a^2\sin 2\omega t & -\omega^3\gamma_a\cos\omega t \\ 0 & -\omega^3\gamma_a\cos\omega t & 0 \end{bmatrix}.$$

The 4th Rivlin–Ericksen tensor:

$$[A_4] = \left[\frac{d}{dt}A_3 + A_3L + L^T A_3\right] = \begin{bmatrix} 0 & 0 & 0 \\ 0 & -6\omega^4\gamma_a^2\cos 2\omega t & \omega^4\gamma_a\sin\omega t \\ 0 & \omega^4\gamma_a\sin\omega t & 0 \end{bmatrix}$$

$$+ \begin{bmatrix} 0 & 0 & 0 \\ 0 & -3\omega^3\gamma_a^2\sin 2\omega t & -\omega^3\gamma_a\cos\omega t \\ 0 & -\omega^3\gamma_a\cos\omega t & 0 \end{bmatrix} \begin{bmatrix} 0 & \dot\gamma_m & 0 \\ 0 & 0 & 0 \\ 0 & \omega\gamma_a\cos\omega t & 0 \end{bmatrix}$$

$$+ \begin{bmatrix} 0 & 0 & 0 \\ \dot\gamma_m & 0 & \omega\gamma_a\cos\omega t \\ 0 & 0 & 0 \end{bmatrix} \begin{bmatrix} 0 & 0 & 0 \\ 0 & -3\omega^3\gamma_a^2\sin 2\omega t & -\omega^3\gamma_a\cos\omega t \\ 0 & -\omega^3\gamma_a\cos\omega t & 0 \end{bmatrix}$$

$$= \begin{bmatrix} 0 & 0 & 0 \\ 0 & 2\omega^4\gamma_a^2\left(3 - 7\cos^2\omega t\right) & \omega^4\gamma_a\sin\omega t \\ 0 & \omega^4\gamma_a\sin\omega t & 0 \end{bmatrix}.$$

The 5th Rivlin–Ericksen tensor:

$$[A_5] = \left[\frac{d}{dt}A_4 + A_4L + L^T A_4\right] = \begin{bmatrix} 0 & 0 & 0 \\ 0 & 14\omega^5\gamma_a^2\sin 2\omega t & \omega^5\gamma_a\cos\omega t \\ 0 & \omega^5\gamma_a\cos\omega t & 0 \end{bmatrix}$$

$$+ \begin{bmatrix} 0 & 0 & 0 \\ 0 & 2\omega^4\gamma_a^2\left(3 - 7\cos^2\omega t\right) & \omega^4\gamma_a\sin\omega t \\ 0 & \omega^4\gamma_a\sin\omega t & 0 \end{bmatrix} \begin{bmatrix} 0 & \dot\gamma_m & 0 \\ 0 & 0 & 0 \\ 0 & \omega\gamma_a\cos\omega t & 0 \end{bmatrix}$$

$$+ \begin{bmatrix} 0 & 0 & 0 \\ \dot\gamma_m & 0 & \omega\gamma_a\cos\omega t \\ 0 & 0 & 0 \end{bmatrix} \begin{bmatrix} 0 & 0 & 0 \\ 0 & 2\omega^4\gamma_a^2\left(3 - 7\cos^2\omega t\right) & \omega^4\gamma_a\sin\omega t \\ 0 & \omega^4\gamma_a\sin\omega t & 0 \end{bmatrix}$$

$$= \begin{bmatrix} 0 & 0 & 0 \\ 0 & 15\omega^5\gamma_a^2\sin 2\omega t & \omega^5\gamma_a\cos\omega t \\ 0 & \omega^5\gamma_a\cos\omega t & 0 \end{bmatrix}.$$

The 6th Rivlin–Ericksen tensor is

$$[\mathbf{A}_6] = \left[\frac{d}{dt}\mathbf{A}_5 + \mathbf{A}_5\mathbf{L} + \mathbf{L}^T\mathbf{A}_5\right] = \begin{bmatrix} 0 & 0 & 0 \\ 0 & 30\omega^6\gamma_a^2\cos 2\omega t & -\omega^6\gamma_a\sin\omega t \\ 0 & -\omega^6\gamma_a\sin\omega t & 0 \end{bmatrix}$$

$$+ \begin{bmatrix} 0 & 0 & 0 \\ 0 & 15\omega^5\gamma_a^2\sin 2\omega t & \omega^5\gamma_a\cos\omega t \\ 0 & \omega^5\gamma_a\cos\omega t & 0 \end{bmatrix} \begin{bmatrix} 0 & \dot\gamma_m & 0 \\ 0 & 0 & 0 \\ 0 & \omega\gamma_a\cos\omega t & 0 \end{bmatrix}$$

$$+ \begin{bmatrix} 0 & 0 & 0 \\ \dot\gamma_m & 0 & \omega\gamma_a\cos\omega t \\ 0 & 0 & 0 \end{bmatrix} \begin{bmatrix} 0 & 0 & 0 \\ 0 & 15\omega^5\gamma_a^2\sin 2\omega t & \omega^5\gamma_a\cos\omega t \\ 0 & \omega^5\gamma_a\cos\omega t & 0 \end{bmatrix}$$

$$= \begin{bmatrix} 0 & 0 & 0 \\ 0 & 30\omega^6\gamma_a^2\cos 2\omega t & -\omega^6\gamma_a\sin\omega t \\ 0 & -\omega^6\gamma_a\sin\omega t & 0 \end{bmatrix} + \begin{bmatrix} 0 & 0 & 0 \\ 0 & 2\omega^6\gamma_a^2\cos^2\omega t & 0 \\ 0 & 0 & 0 \end{bmatrix}$$

$$= \begin{bmatrix} 0 & 0 & 0 \\ 0 & -2\omega^6\gamma_a^2\left(15 - 31\cos^2\omega t\right) & \omega^6\gamma_a\sin\omega t \\ 0 & \omega^6\gamma_a\sin\omega t & 0 \end{bmatrix}.$$

By induction, the general form for Rivlin–Ericksen tensors is ($n \geq 1$)

$$[\mathbf{A}_{2n+1}] = \begin{bmatrix} 0 & 0 & 0 \\ 0 & (-1)^n\,\alpha\omega^{2n+1}\gamma_a^2\sin 2\omega t & (-1)^n\,\omega^{2n+1}\gamma_a\cos\omega t \\ 0 & (-1)^n\,\omega^{2n+1}\gamma_a\cos\omega t & 0 \end{bmatrix}$$

$$[\mathbf{A}_{2n+2}] = \begin{bmatrix} 0 & 0 & 0 \\ 0 & 2(-1)^{n+1}\omega^{2n+2}\gamma_a^2\left(\alpha - \beta\cos^2\omega t\right) & (-1)^{n+1}\omega^{2n+2}\gamma_a\sin\omega t \\ 0 & (-1)^{n+1}\omega^{2n+2}\gamma_a\sin\omega t & 0 \end{bmatrix}$$

$$\alpha = 3.5\ldots(2n+1), \quad \beta = 2\alpha + 1.$$

Problem 3.7

Cylindrical Coordinates. Conservation of mass: $\frac{\partial\rho}{\partial t} + \nabla\cdot(\rho\mathbf{u}) = 0$

$$\frac{\partial\rho}{\partial t} + \frac{1}{r}\frac{\partial}{\partial r}(r\rho u_r) + \frac{1}{r}\frac{\partial}{\partial\theta}(\rho u_\theta) + \frac{\partial}{\partial z}(\rho u_z) = 0. \tag{S.244}$$

Conservation of linear momentum $\rho\mathbf{a} = \rho\mathbf{b} + (\nabla\cdot\mathbf{S})^T$ where

$$\mathbf{a} = \left(\frac{\partial}{\partial t}\mathbf{u} + \mathbf{u}\cdot\nabla\mathbf{u}\right)$$

is the fluid acceleration, $\mathbf{b}$ is the body force and $\mathbf{S}$ is the (symmetric) stress tensor:

$$a_r = \frac{\partial u_r}{\partial t} + u_r \frac{\partial u_r}{\partial r} + \frac{u_\theta}{r} \frac{\partial u_r}{\partial \theta} - \frac{u_\theta^2}{r} + u_z \frac{\partial u_r}{\partial z},$$

$$a_\theta = \frac{\partial u_\theta}{\partial t} + u_r \frac{\partial u_\theta}{\partial r} + \frac{u_\theta u_r}{r} + \frac{u_\theta}{r} \frac{\partial u_\theta}{\partial \theta} + u_z \frac{\partial u_\theta}{\partial z}, \qquad (\text{S}.245)$$

$$a_z = \frac{\partial u_z}{\partial t} + u_r \frac{\partial u_z}{\partial r} + \frac{u_\theta}{r} \frac{\partial u_z}{\partial \theta} + u_z \frac{\partial u_z}{\partial z},$$

$$\rho a_r = \rho b_r + \frac{1}{r} \frac{\partial}{\partial r} (r S_{rr}) - \frac{S_{\theta\theta}}{r} + \frac{1}{r} \frac{\partial}{\partial \theta} S_{r\theta} + \frac{\partial}{\partial z} S_{rz},$$

$$\rho a_\theta = \rho b_\theta + \frac{1}{r^2} \frac{\partial}{\partial r} (r^2 S_{\theta r}) + \frac{1}{r} \frac{\partial}{\partial \theta} S_{\theta\theta} + \frac{\partial}{\partial z} S_{\theta z}, \qquad (\text{S}.246)$$

$$\rho a_z = \rho b_z + \frac{1}{r} \frac{\partial}{\partial r} (r S_{zr}) + \frac{1}{r} \frac{\partial}{\partial \theta} S_{z\theta} + \frac{\partial}{\partial z} S_{zz}.$$

Spherical Coordinates. Conservation of mass: $\frac{\partial \rho}{\partial t} + \nabla \cdot (\rho \mathbf{u}) = 0$

$$\frac{\partial \rho}{\partial t} + \frac{1}{r^2} \frac{\partial}{\partial r} (r^2 \rho u_r) + \frac{1}{r \sin \theta} \frac{\partial}{\partial \theta} (\rho u_\theta \sin \theta) + \frac{1}{r \sin \theta} \frac{\partial}{\partial \phi} (\rho u_\phi) = 0. \quad (\text{S}.247)$$

Conservation of linear momentum $\rho \mathbf{a} = \rho \mathbf{b} + (\nabla \cdot \mathbf{S})^T$ where

$$\mathbf{a} = \left(\frac{\partial}{\partial t} \mathbf{u} + \mathbf{u} \cdot \nabla \mathbf{u} \right)$$

is the fluid acceleration, $\mathbf{b}$ is the body force and $\mathbf{S}$ is the (symmetric) stress tensor:

$$a_r = \frac{\partial u_r}{\partial t} + u_r \frac{\partial u_r}{\partial r} + \frac{u_\theta}{r} \frac{\partial u_r}{\partial \theta} - \frac{u_\theta^2}{r} + u_\phi \left(\frac{1}{r \sin \theta} \frac{\partial u_r}{\partial \phi} - \frac{u_\phi}{r} \right),$$

$$a_\theta = \frac{\partial u_\theta}{\partial t} + u_r \frac{\partial u_\theta}{\partial r} + \frac{u_\theta u_r}{r} + \frac{u_\theta}{r} \frac{\partial u_\theta}{\partial \theta}$$

$$+ u_\phi \left(\frac{1}{r \sin \theta} \frac{\partial u_\theta}{\partial \phi} - \frac{u_\phi}{r} \cot \theta \right), \qquad (\text{S}.248)$$

$$a_\phi = \frac{\partial u_\phi}{\partial t} + u_r \frac{\partial u_\phi}{\partial r} + \frac{u_\theta}{r} \frac{\partial u_\phi}{\partial \theta} + u_\phi \left(\frac{1}{r \sin \theta} \frac{\partial u_\phi}{\partial \phi} + \frac{u_r}{r} + \frac{u_\theta}{r} \cot \theta \right),$$

$$\rho a_r = \rho b_r + \frac{1}{r^2} \frac{\partial}{\partial r} \left(r^2 S_{rr} \right) - \frac{S_{\theta\theta} + S_{\phi\phi}}{r} + \frac{1}{r \sin \theta} \frac{\partial}{\partial \theta} \left(S_{r\theta} \sin \theta \right)$$

$$+ \frac{1}{r \sin \theta} \frac{\partial}{\partial \phi} S_{r\phi},$$

$$\rho a_\theta = \rho b_\theta + \frac{1}{r^3} \frac{\partial}{\partial r} \left(r^3 S_{\theta r} \right) + \frac{1}{r \sin \theta} \frac{\partial}{\partial \theta} \left(S_{\theta\theta} \sin \theta \right) + \frac{1}{r \sin \theta} \frac{\partial}{\partial \phi} S_{\theta\phi}$$

$$- \frac{S_{\phi\phi}}{r} \cot \theta, \tag{S.249}$$

$$\rho a_\phi = \rho b_\phi + \frac{1}{r^3} \frac{\partial}{\partial r} \left(r^3 S_{\phi r} \right) + \frac{1}{r \sin \theta} \frac{\partial}{\partial \theta} \left(S_{\phi\theta} \sin \theta \right) + \frac{1}{r \sin \theta} \frac{\partial}{\partial \phi} S_{\phi\phi}$$

$$+ \frac{S_{\theta\phi}}{r} \cot \theta.$$

Problems in Chap. 4

Problem 4.1

In a simple shear deformation of a linearly elastic body (4.7), in which the displacement field takes the form

$$v_1 = \gamma y, \quad v_2 = 0, \quad v_3 = 0, \tag{S.250}$$

where γ is the amount of shear, the infinitesimal strain tensor is

$$[\varepsilon] = \frac{1}{2} \begin{bmatrix} 0 & \gamma & 0 \\ \gamma & 0 & 0 \\ 0 & 0 & 0 \end{bmatrix}, \tag{S.251}$$

corresponding to the stress tensor

$$[\mathbf{T}] = \mu \begin{bmatrix} 0 & \gamma & 0 \\ \gamma & 0 & 0 \\ 0 & 0 & 0 \end{bmatrix}. \tag{S.252}$$

The only non-trivial stress component is the shear stress, $T_{12} = \mu\gamma$. Thus μ is called the shear modulus.

Problem 4.2

In a uni-axial extension of a linearly elastic material (4.7), in which the displacement field is given by

$$v_1 = \varepsilon x, \quad v_2 = -\nu \varepsilon y, \quad v_3 = -\nu \varepsilon z, \tag{S.253}$$

where ε is the elongational strain, and ν is the amount of lateral contraction due to the axial elongation, called Poisson's ratio. The infinitesimal strain tensor is

$$[\varepsilon] = diag\,[\varepsilon, -\nu\varepsilon, -\nu\varepsilon], \tag{S.254}$$

leading to the stress tensor

$$[\mathbf{T}] = \lambda\,(1 - 2\nu)\,\varepsilon\mathbf{I} + 2\mu\,diag\,[\varepsilon, -\nu\varepsilon, -\nu\varepsilon], \tag{S.255}$$

or

$$
\begin{aligned}
T_{xx} &= [\lambda\,(1 - 2\nu) + 2\mu]\,\varepsilon, \\
T_{yy} &= [\lambda\,(1 - 2\nu) - 2\mu\nu]\,\varepsilon, \\
T_{zz} &= [\lambda\,(1 - 2\nu) - 2\mu\nu]\,\varepsilon.
\end{aligned}
\tag{S.256}
$$

All the other components of $\mathbf{T}$ are zero. If the lateral stresses are zero, then

$$\nu = \frac{\lambda}{2\,(\lambda + \mu)}. \tag{S.257}$$

In addition, if we set

$$\lambda\,(1 - 2\nu) + 2\mu = 2\mu\,(1 + \nu) = E, \tag{S.258}$$

then

$$T_{xx} = E\varepsilon. \tag{S.259}$$

E is called the Young's modulus of the material.

Problem 4.3

Consider a deformation characterised by

$$\mathbf{F} = \frac{\partial \mathbf{x}}{\partial \mathbf{X}},$$

$\mathbf{F}$ takes an element $d\mathbf{X}$ into $d\mathbf{x} = \mathbf{F}d\mathbf{X}$. Thus, if $d\mathbf{X} = dX\mathbf{P}$, where $\mathbf{P}$ is a unit vector,

$$
\begin{aligned}
dx^2 &= F_{ik}dX_k F_{il}dX_l = F_{ik}F_{il}P_k P_l dX^2 \\
&= C_{kl}P_k P_l dX^2.
\end{aligned} \tag{S.260}
$$

If $\mathbf{P}$ is distributed randomly in space, then on the average, $\langle P_k P_l \rangle = \delta_{kl}/3$ and one has

$$
\langle dx^2 \rangle = \frac{1}{3}dX^2 tr\ \mathbf{C}, \tag{S.261}
$$

and a measure of the amount of stretch (Weissenberg number) could be defined as

$$
Wi = \frac{1}{3}tr\ \mathbf{C}. \tag{S.262}
$$

Problem 4.4

Let $\mathbf{f}$ be a vector-valued, isotropic polynomial of a symmetric tensor $\mathbf{S}$ and a vector $\mathbf{v}$. By definition, it satisfies, $\forall$ orthogonal $\mathbf{Q}$,

$$
\mathbf{Qf}\,(\mathbf{S}, \mathbf{v}) = \mathbf{f}\left(\mathbf{QSQ}^T, \mathbf{Qv}\right). \tag{S.263}
$$

Now, define a scalar function $g = g\,(\mathbf{u}, \mathbf{S}, \mathbf{v})$ as

$$
g\,(\mathbf{u}, \mathbf{S}, \mathbf{v}) = \mathbf{u} \cdot \mathbf{f}\,(\mathbf{S}, \mathbf{v}). \tag{S.264}
$$

Then,

$$
g\left(\mathbf{Qu}, \mathbf{QSQ}^T, \mathbf{Qv}\right) = \mathbf{Qu} \cdot \mathbf{f}\left(\mathbf{QSQ}^T, \mathbf{Qv}\right), \tag{S.265}
$$

and from the properties of $\mathbf{f}$, (S.263),

$$
\begin{aligned}
g\left(\mathbf{Qu}, \mathbf{QSQ}^T, \mathbf{Qv}\right) &= \mathbf{Qu} \cdot \mathbf{Qf}\,(\mathbf{S}, \mathbf{v}) \\
&= \mathbf{u} \cdot \mathbf{f}\,(\mathbf{S}, \mathbf{v}) = g\,(\mathbf{u}, \mathbf{S}, \mathbf{v}) \tag{S.266}
\end{aligned}
$$

Thus $g\,(\mathbf{u}, \mathbf{S}, \mathbf{v})$ is an isotropic scalar function in all of its arguments. If $\mathbf{f}$ is a polynomial in its arguments then g is also a polynomial in its arguments - its integrity basis has been listed in (4.36):

$$
\begin{aligned}
&tr\ \mathbf{S},\ tr\ \mathbf{S}^2,\ tr\ \mathbf{S}^3,\ \mathbf{u} \cdot \mathbf{u},\ \mathbf{u} \cdot \mathbf{v},\ \mathbf{v} \cdot \mathbf{v},\ \mathbf{u} \cdot \mathbf{Su},\ \mathbf{u} \cdot \mathbf{S}^2\mathbf{u}, \\
&\mathbf{v} \cdot \mathbf{Sv},\ \mathbf{v} \cdot \mathbf{S}^2\mathbf{v},\ \mathbf{u} \cdot \mathbf{Sv},\ \mathbf{u} \cdot \mathbf{S}^2\mathbf{v}. \tag{S.267}
\end{aligned}
$$

But g is linear in $\mathbf{u}$ and cannot depend on any non-linear manner on $\mathbf{u}$. Consequently,

$$g = \mathbf{u} \cdot \mathbf{f} \left(\text{tr } \mathbf{S}, \ \text{tr } \mathbf{S}^2, \ \text{tr } \mathbf{S}^3, \ \mathbf{v} \cdot \mathbf{v}, \ \mathbf{v} \cdot \mathbf{Sv}, \ \mathbf{v} \cdot \mathbf{S}^2\mathbf{v} \right) \tag{S.268}$$
$$= \mathbf{u} \cdot [f_0 \mathbf{1} + f_1 \mathbf{S} + f_2 \mathbf{S}^2]\mathbf{v}.$$

This choice is dictated by the fact the term multiplied with $\mathbf{u}$ is a vector. This leads to the form

$$\mathbf{f}(\mathbf{S}, \mathbf{v}) = [f_0 \mathbf{1} + f_1 \mathbf{S} + f_2 \mathbf{S}^2]\mathbf{v}, \tag{S.269}$$

where the scalar valued coefficients are polynomials in the six invariants involving only $\mathbf{S}$ and $\mathbf{v}$ in the list (S.268).

Problem 4.5

Consider a simple shear deformation of a rubber-like material (4.55), where

$$x = X + \gamma Y, \quad y = Y, \quad z = Z. \tag{S.270}$$

The deformation gradient is

$$[\mathbf{F}] = \begin{bmatrix} 1 & \gamma & 0 \\ 0 & 1 & 0 \\ 0 & 0 & 1 \end{bmatrix},$$

leading to the Finger tensor

$$[\mathbf{B}] = \left[\mathbf{F}\mathbf{F}^T \right] = \begin{bmatrix} 1 & \gamma & 0 \\ 0 & 1 & 0 \\ 0 & 0 & 1 \end{bmatrix} \begin{bmatrix} 1 & 0 & 0 \\ \gamma & 1 & 0 \\ 0 & 0 & 1 \end{bmatrix}$$
$$= \begin{bmatrix} 1+\gamma^2 & \gamma & 0 \\ \gamma & 1 & 0 \\ 0 & 0 & 1 \end{bmatrix} \tag{S.271}$$

$$[\mathbf{B}^{-1}] = \begin{bmatrix} 1 & -\gamma & 0 \\ -\gamma & 1+\gamma^2 & 0 \\ 0 & 0 & 1 \end{bmatrix}. \tag{S.272}$$

Consequently, the stress tensor is given by

$$[\mathbf{T}] = -P\,[\mathbf{I}] + \begin{bmatrix} \beta_1\left(1+\gamma^2\right) - \beta_2 & (\beta_1+\beta_2)\,\gamma & 0 \\ (\beta_1+\beta_2)\,\gamma & \beta_1 - \beta_2\left(1+\gamma^2\right) & 0 \\ 0 & 0 & \beta_1 - \beta_2 \end{bmatrix}, \tag{S.273}$$

where P is the hydrostatic pressure. The shear stress and the normal stress differences are

$$S = (\beta_1 + \beta_2)\,\gamma, \quad N_1 = (\beta_1 + \beta_2)\,\gamma^2, \quad N_2 = -\beta_2\gamma^2. \tag{S.274}$$

Deduce that the linear shear modulus of elasticity is

$$G = \lim_{\gamma \to 0} (\beta_1 + \beta_2). \tag{S.275}$$

The ratio

$$\frac{N_1}{S} = \gamma \tag{S.276}$$

is independent of the material properties. Such a relation is called *universal*.

Problem 4.6

In a uniaxial elongational deformation of a rubber-like material (4.55), where (in cylindrical coordinates)

$$R = \lambda^{1/2}r, \quad \Theta = \theta, \quad Z = \lambda^{-1}z, \tag{S.277}$$

show that the inverse deformation gradient is

$$[\mathbf{F}^{-1}] = \begin{bmatrix} \dfrac{\partial R}{\partial r} & \dfrac{1}{r}\dfrac{\partial R}{\partial \theta} & \dfrac{\partial R}{\partial z} \\[2mm] 0 & \dfrac{R}{r} & 0 \\[2mm] \dfrac{\partial Z}{\partial r} & \dfrac{1}{r}\dfrac{\partial Z}{\partial \theta} & \dfrac{\partial Z}{\partial z} \end{bmatrix} = \begin{bmatrix} \lambda^{1/2} & 0 & 0 \\ 0 & \lambda^{1/2} & 0 \\ 0 & 0 & \lambda^{-1} \end{bmatrix}. \tag{S.278}$$

Consequently, the strains are

$$[\mathbf{B}^{-1}] = [\mathbf{F}^{-T}\mathbf{F}^{-1}] = \begin{bmatrix} \lambda & 0 & 0 \\ 0 & \lambda & 0 \\ 0 & 0 & \lambda^{-2} \end{bmatrix}, \quad [\mathbf{B}] = \begin{bmatrix} \lambda^{-1} & 0 & 0 \\ 0 & \lambda^{-1} & 0 \\ 0 & 0 & \lambda^{2} \end{bmatrix}. \tag{S.279}$$

Thus the total stress tensor for a rubber-like material (4.55) is

$$[\mathbf{T}] = -P\,[\mathbf{I}] + \begin{bmatrix} \beta_1\lambda^{-1} - \beta_2\lambda & 0 & 0 \\ 0 & \beta_1\lambda^{-1} - \beta_2\lambda & 0 \\ 0 & 0 & \beta_1\lambda^{2} - \beta_2\lambda^{-2} \end{bmatrix}. \tag{S.280}$$

Under the condition that the lateral tractions are zero, i.e., $T_{rr} = 0$, the pressure can be found

$$P = \beta_1 \lambda^{-1} - \beta_2 \lambda, \tag{S.281}$$

and thus the tensile stress is

$$T_{zz} = -P + \beta_1 \lambda^2 - \beta_2 \lambda^{-2} = \beta_1 \lambda^2 - \beta_2 \lambda^{-2} - \beta_1 \lambda^{-1} + \beta_2 \lambda$$
$$= \left(\lambda^2 - \lambda^{-1}\right)\left(\beta_1 + \beta_2 \lambda^{-1}\right). \tag{S.282}$$

This tensile stress is the force per unit area in the deformed configuration. As $r = \lambda^{1/2} R$, the corresponding force per unit area in the undeformed configuration is

$$T_{ZZ} = T_{zz} \lambda^{-1} = \left(\lambda - \lambda^{-2}\right)\left(\beta_1 + \beta_2 \lambda^{-1}\right). \tag{S.283}$$

Problem 4.7

Note that

$$\mathbf{C}_t(\tau) = \mathbf{F}_t(\tau)^T \mathbf{F}_t(\tau), \tag{S.284}$$

but (use subscript notation for clarity)

$$\mathbf{F}_t(\tau) = \frac{\partial \mathbf{x}(\tau)}{\partial \mathbf{x}(t)} = \frac{\partial \mathbf{x}(\tau)}{\partial \mathbf{X}} \frac{\partial \mathbf{X}}{\partial \mathbf{x}(t)} = \mathbf{F}(\tau) \mathbf{F}(t)^{-1}$$

and therefore

$$\mathbf{C}_t(\tau) = \mathbf{F}(t)^{-T} \mathbf{F}(\tau)^T \mathbf{F}(\tau) \mathbf{F}(t)^{-1} = \mathbf{F}(t)^{-T} \mathbf{C}(\tau) \mathbf{F}(t)^{-1}$$

leading to the desired result

$$\mathbf{C}(\tau) = \mathbf{F}(t)^T \mathbf{C}_t(\tau) \mathbf{F}(t). \tag{S.285}$$

Problem 4.8

Consider a simple shear flow

$$u = \dot{\gamma} y, \quad v = 0, \quad w = 0. \tag{S.286}$$

The first and second Rivlin–Ericksen tensors are

$$[\mathbf{A}_1] = \begin{bmatrix} 0 & \dot{\gamma} & 0 \\ \dot{\gamma} & 0 & 0 \\ 0 & 0 & 0 \end{bmatrix}, \quad [\mathbf{A}_1^2] = \begin{bmatrix} \dot{\gamma}^2 & 0 & 0 \\ 0 & \dot{\gamma}^2 & 0 \\ 0 & 0 & 0 \end{bmatrix}$$

$$[\mathbf{A}_2] = \left[\mathbf{A}_1 \mathbf{L} + \mathbf{L}^T \mathbf{A}_1 \right] = \begin{bmatrix} 0 & 0 & 0 \\ 0 & 2\dot{\gamma}^2 & 0 \\ 0 & 0 & 0 \end{bmatrix}.$$

Thus, the stress tensor in the second-order model is given by

$$[\mathbf{S}] = \eta_0 \begin{bmatrix} 0 & \dot{\gamma} & 0 \\ \dot{\gamma} & 0 & 0 \\ 0 & 0 & 0 \end{bmatrix} + (\nu_1 + \nu_2) \begin{bmatrix} \dot{\gamma}^2 & 0 & 0 \\ 0 & \dot{\gamma}^2 & 0 \\ 0 & 0 & 0 \end{bmatrix} - \frac{\nu_1}{2} \begin{bmatrix} 0 & 0 & 0 \\ 0 & 2\dot{\gamma}^2 & 0 \\ 0 & 0 & 0 \end{bmatrix}, \quad \text{(S.287)}$$

and the three viscometric functions are

$$S = \eta_0 \dot{\gamma}, \quad N_1 = S_{11} - S_{22} = \nu_1 \dot{\gamma}^2, \quad N_2 = S_{22} - S_{33} = \nu_2 \dot{\gamma}^2. \quad \text{(S.288)}$$

Problem 4.9

In an elongational flow

$$u = \dot{\varepsilon}x, \quad v = -\frac{\dot{\varepsilon}}{2}y, \quad w = -\frac{\dot{\varepsilon}}{2}z, \quad \text{(S.289)}$$

the first and second Rivlin–Ericksen tensors are

$$[\mathbf{A}_1] = \begin{bmatrix} 2\dot{\varepsilon} & 0 & 0 \\ 0 & -\dot{\varepsilon} & 0 \\ 0 & 0 & -\dot{\varepsilon} \end{bmatrix}, \quad [\mathbf{A}_1^2] = \begin{bmatrix} 4\dot{\varepsilon}^2 & 0 & 0 \\ 0 & \dot{\varepsilon}^2 & 0 \\ 0 & 0 & \dot{\varepsilon}^2 \end{bmatrix},$$

$$[\mathbf{A}_2] = \left[\mathbf{A}_1 \mathbf{L} + \mathbf{L}^T \mathbf{A}_1 \right] = \begin{bmatrix} 4\dot{\varepsilon}^2 & 0 & 0 \\ 0 & \dot{\varepsilon}^2 & 0 \\ 0 & 0 & \dot{\varepsilon}^2 \end{bmatrix}.$$

Thus the stress is given by

$$[S] = \eta_0 \begin{bmatrix} 2\dot{\varepsilon} & 0 & 0 \\ 0 & -\dot{\varepsilon} & 0 \\ 0 & 0 & -\dot{\varepsilon} \end{bmatrix} + (\nu_1 + \nu_2) \begin{bmatrix} 4\dot{\varepsilon}^2 & 0 & 0 \\ 0 & \dot{\varepsilon}^2 & 0 \\ 0 & 0 & \dot{\varepsilon}^2 \end{bmatrix}$$
$$- \frac{\nu_1}{2} \begin{bmatrix} 4\dot{\varepsilon}^2 & 0 & 0 \\ 0 & \dot{\varepsilon}^2 & 0 \\ 0 & 0 & \dot{\varepsilon}^2 \end{bmatrix}. \tag{S.290}$$

Consequently the elongational viscosity is given by

$$\eta_E = \frac{S_{xx} - S_{yy}}{\dot{\varepsilon}} = 3\eta_0 + 3\left(\frac{\nu_1}{2} + \nu_2\right)\dot{\varepsilon}. \tag{S.291}$$

Problem 4.10

For potential flows, the velocity field is a gradient of a potential:

$$\mathbf{u} = \nabla\phi, \tag{S.292}$$

from which, incompressibility demands

$$\nabla \cdot \mathbf{u} = \nabla^2 \phi = 0. \tag{S.293}$$

Now, since

$$L_{ij} = \frac{\partial u_i}{\partial x_j} = \phi_{,ij}, \quad A_{1ij} = \frac{\partial u_i}{\partial x_j} + \frac{\partial u_j}{\partial x_i} = 2\phi_{,ij} \tag{S.294}$$

where the comma denotes a spatial derivative, it follows that

$$\nabla \cdot \mathbf{A}_1 = 0. \tag{S.295}$$

Next,

$$\left(\nabla \cdot \mathbf{A}_1^2\right)_i = \left(A_{1ik} A_{1kj}\right)_{,j} = 4\left(\phi_{,ik}\phi_{,kj}\right)_{,j} = 4\phi_{,kj}\phi_{,ikj}$$
$$= \frac{1}{2}\left(4\phi_{,kj}\phi_{,kj}\right)_{,i}$$
$$\nabla \cdot \mathbf{A}_1^2 = \frac{1}{2}\left(\nabla(tr\,\mathbf{A}_1^2)\right). \tag{S.296}$$

Furthermore,

$$A_{2ij} = u_k A_{1ij,k} + A_{1ik} L_{kj} + L_{ki} A_{1kj}$$
$$= 2\phi_{,k}\phi_{,ijk} + 4\phi_{,ik}\phi_{,kj}$$

$$A_{2ij,j} = 2\phi_{,kj}\phi_{,ijk} + 4\phi_{,kj}\phi_{,ikj} = 6\phi_{,kj}\phi_{,ijk}$$
$$= 3\left(\phi_{,kj}\phi_{,kj}\right)_i$$
$$= \frac{3}{4}\left(4\phi_{,kj}\phi_{,kj}\right)_i$$
$$\nabla \cdot \mathbf{A}_2 = \frac{3}{4}\nabla(\text{tr } \mathbf{A}_1^2). \tag{S.297}$$

In a second-order fluid model, the stress is given by

$$\mathbf{T} = -P\mathbf{I} + \eta_0 \mathbf{A}_1 + (\nu_1 + \nu_2)\,\mathbf{A}_1^2 - \frac{\nu_1}{2}\mathbf{A}_2,$$

and conservation of momentum yields

$$\nabla P = \nabla \cdot \left[\eta_0 \mathbf{A}_1 + (\nu_1 + \nu_2)\,\mathbf{A}_1^2 - \frac{\nu_1}{2}\mathbf{A}_2\right] - \rho\mathbf{a}, \tag{S.298}$$

where $\mathbf{a}$ is the acceleration field. If the same flow, of the same kinematics occurs in a Newtonian fluid of viscosity η_0, then we must have

$$\nabla P_N = \eta_0 \nabla \cdot \mathbf{A}_1 - \rho\mathbf{a}, \tag{S.299}$$

where P_N is the Newtonian pressure field. Thus, for this to occur, one must have an "extra" pressure field

$$\nabla P_E = \nabla \cdot \left[(\nu_1 + \nu_2)\,\mathbf{A}_1^2 - \frac{\nu_1}{2}\mathbf{A}_2\right]. \tag{S.300}$$

From the results (S.296) and (S.297), this is possible, and thus in potential flows, the velocity fields for a Newtonian and a second-order fluid are identical, with the extra pressure given by

$$\nabla P_E = \nabla \cdot \left[(\nu_1 + \nu_2)\,\mathbf{A}_1^2 - \frac{\nu_1}{2}\mathbf{A}_2\right]$$
$$= \left[\frac{1}{2}(\nu_1 + \nu_2) - \frac{3\nu_1}{8}\right]\nabla\left(tr\mathbf{A}_1^2\right),$$

or that

$$P_E = \frac{1}{8} (\nu_1 + 4\nu_2) \, tr \mathbf{A}_1^2, \tag{S.301}$$

to within a constant which can be absorbed in P_N.

Problem 4.11

For steady two-dimensional incompressible flows, a stream function $\psi = \psi(x, y)$ can be defined such that the velocity components u and v can be expressed as

$$u = \frac{\partial \psi}{\partial y}, \quad v = -\frac{\partial \psi}{\partial x}. \tag{S.302}$$

If the fluid is incompressible Newtonian, then for a flow at zero Reynolds number, the Newtonian stresses are

$$[\mathbf{S}_N] = \eta \begin{bmatrix} 2\psi_{,xy} & \psi_{,yy} - \psi_{,xx} \\ \psi_{,yy} - \psi_{,xx} & -2\psi_{,xy} \end{bmatrix}$$

and the balance of momentum requires

$$\begin{aligned}
\frac{\partial P}{\partial x} &= 2\eta \frac{\partial^2 u}{\partial x^2} + \eta \left(\frac{\partial^2 u}{\partial y^2} + \frac{\partial^2 v}{\partial x \partial y} \right) \\
&= \eta \left[2\psi_{,yxx} + \psi_{,yyy} - \psi_{,xxy} \right], \\
\frac{\partial P}{\partial y} &= \eta \left(\frac{\partial^2 u}{\partial x \partial y} + \frac{\partial^2 v}{\partial x^2} \right) + 2\eta \frac{\partial^2 v}{\partial y^2} \\
&= \eta \left[\psi_{,yyx} - \psi_{,xxx} - 2\psi_{,xyy} \right].
\end{aligned}$$

For compatibility, the stream function must satisfy a bi-harmonic equation:

$$\begin{aligned}
\psi_{,xxy} y + \psi_{,yyyy} &= -\psi_{,xxxx} - \psi_{,xxyy} \\
\nabla^2 \nabla^2 \psi &= \Delta^2 \psi = 0.
\end{aligned} \tag{S.303}$$

Now,

$$[\mathbf{L}] = \begin{bmatrix} \psi_{,xy} & \psi_{,yy} \\ -\psi_{,xx} & -\psi_{xy} \end{bmatrix}$$

$$[\mathbf{A}_1] = \begin{bmatrix} 2\psi_{,xy} & \psi_{,yy} - \psi_{,xx} \\ \psi_{,yy} - \psi_{,xx} & -2\psi_{,xy} \end{bmatrix}, \tag{S.304}$$

$$[\mathbf{A}_1^2] = \left(4\psi_{,xy}^2 + \psi_{,yy}^2 - 2\psi_{,xx}\psi_{,yy} + \psi_{,xx}^2\right)\begin{bmatrix} 1 & 0 \\ 0 & 1 \end{bmatrix}, \qquad \text{(S.305)}$$

$$[\mathbf{A}_2] = [\mathbf{u} \cdot \nabla \mathbf{A}_1] + [\mathbf{A}_1 \mathbf{L}] + [\mathbf{L}^T \mathbf{A}_1] \qquad \text{(S.306)}$$

$$= \left(\psi_y \frac{\partial}{\partial x} - \psi_x \frac{\partial}{\partial y}\right)\begin{bmatrix} 2\psi_{,xy} & \psi_{,yy} - \psi_{,xx} \\ \psi_{,yy} - \psi_{,xx} & -2\psi_{,xy} \end{bmatrix}$$

$$+ \begin{bmatrix} 2\psi_{,xy}^2 - \psi_{,xx}\left(\psi_{,yy} - \psi_{,xx}\right) & 2\psi_{,xy}\psi_{,yy} - \psi_{,xy}\left(\psi_{,yy} - \psi_{,xx}\right) \\ 2\psi_{,xy}\psi_{,xx} + \psi_{,xy}\left(\psi_{,yy} - \psi_{,xx}\right) & 2\psi_{,xy}^2 + \psi_{,yy}\left(\psi_{,yy} - \psi_{,xx}\right) \end{bmatrix}$$

$$+ \begin{bmatrix} 2\psi_{,xy}^2 - \psi_{,xx}\left(\psi_{,yy} - \psi_{,xx}\right) & 2\psi_{,xy}\psi_{,xx} + \psi_{,xy}\left(\psi_{,yy} - \psi_{,xx}\right) \\ 2\psi_{,xy}\psi_{,yy} - \psi_{,xy}\left(\psi_{,yy} - \psi_{,xx}\right) & 2\psi_{,xy}^2 + \psi_{,yy}\left(\psi_{,yy} - \psi_{,xx}\right) \end{bmatrix},$$

from which the second-order fluid stresses (see 4.71) can be determined as

$$S_{xx} = 2\eta\psi_{,xy} + (\nu_1 + \nu_2)\left(4\psi_{,xy}^2 + \psi_{,yy}^2 - 2\psi_{,xx}\psi_{,yy} + \psi_{,xx}^2\right) \qquad \text{(S.307)}$$

$$-\nu_1\left[\left(\psi_y \frac{\partial}{\partial x} - \psi_x \frac{\partial}{\partial y}\right)\psi_{,xy} + 2\psi_{,xy}^2 - \psi_{,xx}\left(\psi_{,yy} - \psi_{,xx}\right)\right]$$

$$S_{xy} = \eta\left(\psi_{,yy} - \psi_{,xx}\right) - \frac{\nu_1}{2}[\left(\psi_y \frac{\partial}{\partial x} - \psi_x \frac{\partial}{\partial y}\right)\left(\psi_{,yy} - \psi_{,xx}\right)$$

$$+ 2\psi_{,xy}\left(\psi_{,yy} + \psi_{,xx}\right)] \qquad \text{(S.308)}$$

$$S_{yy} = -2\eta\psi_{,xy} + (\nu_1 + \nu_2)\left(4\psi_{,xy}^2 + \psi_{,yy}^2 - 2\psi_{,xx}\psi_{,yy} + \psi_{,xx}^2\right)$$

$$- \nu_1\left[\left(\psi_x \frac{\partial}{\partial y} - \psi_y \frac{\partial}{\partial x}\right)\psi_{,xy} + 2\psi_{,xy}^2 + \psi_{,yy}\left(\psi_{,yy} - \psi_{,xx}\right)\right]. \qquad \text{(S.309)}$$

The balance of momentum requires

$$\frac{\partial P}{\partial x} = \frac{\partial S_{xx}}{\partial x} + \frac{\partial S_{xy}}{\partial y},$$

$$\frac{\partial P}{\partial y} = \frac{\partial S_{xy}}{\partial x} + \frac{\partial S_{yy}}{\partial y}.$$

Compatibility requires

$$\frac{\partial^2}{\partial x \partial y}\left(S_{xx} - S_{yy}\right) + \left(\frac{\partial^2}{\partial y^2} - \frac{\partial^2}{\partial x^2}\right)S_{xy} = 0. \qquad \text{(S.310)}$$

The second-order fluid stresses can be substituted in the preceding results to yield

$$0 = \frac{\partial^2}{\partial x \partial y} \left[4\eta \psi_{,xy} - \nu_1 \left(2 \left(\psi_y \frac{\partial}{\partial x} - \psi_x \frac{\partial}{\partial y} \right) \psi_{,xy} - (\psi_{,yy}^2 - \psi_{,xx}^2) \right) \right]$$
$$+ \left(\frac{\partial^2}{\partial y^2} - \frac{\partial^2}{\partial x^2} \right) [\eta (\psi_{,yy} - \psi_{,xx}) - \frac{\nu_1}{2}[(\psi_y \frac{\partial}{\partial x} - \psi_x \frac{\partial}{\partial y}) (\psi_{,yy} - \psi_{,xx})$$
$$+ 2\psi_{,xy} (\psi_{,yy} + \psi_{,xx})]].$$

This is simplified to

$$\eta \Delta^2 \psi - \frac{\nu_1}{2} \mathbf{u} \cdot \nabla (\Delta^2 \psi) = 0. \tag{S.311}$$

This shows that a Newtonian velocity field ($\Delta^2 \psi = 0$) is also a velocity field for the second-order fluid.

Problem 4.12

In a simple shear flow,

$$u = \dot{\gamma}(t) y, \quad v = 0, \quad w = 0,$$

the path lines $\mathbf{x}$ satisfy

$$\frac{d}{dt}(x, y, z) = (\dot{\gamma}(t) y, 0, 0), \quad \mathbf{x}(0) = \mathbf{X}.$$

That is,

$$x(t) = X + Y \int_0^t \dot{\gamma}(t') dt', \quad y = Y, \quad z = Z.$$

Thus the path lines $_{\mathbf{s}}(\tau) = (\xi, \psi, \zeta)$

$$\xi(\tau) = X + Y \int_0^\tau \dot{\gamma}(t') dt'$$
$$= x + y \int_t^\tau \dot{\gamma}(t') dt',$$
$$\psi = y,$$
$$\zeta = z,$$

or

$$\xi(\tau) = x + y\gamma(t, \tau), \quad \psi(\tau) = y, \quad \zeta(\tau) = z, \tag{S.312}$$

where

$$\gamma(t, \tau) = \int_t^\tau \dot{\gamma}(s)\, ds. \tag{S.313}$$

The relative deformation gradient:

$$\mathbf{F}_t(\tau) = (\nabla_{\mathbf{x}})^T = \begin{bmatrix} 1 & \gamma(t, \tau) & 0 \\ 0 & 1 & 0 \\ 0 & 0 & 1 \end{bmatrix}$$

$$\mathbf{C}_t(\tau) = \mathbf{F}_t(\tau)^T \mathbf{F}_t(\tau) = \begin{bmatrix} 1 & \gamma(t, \tau) & 0 \\ \gamma(t, \tau) & 1 + \gamma^2(t, \tau) & 0 \\ 0 & 0 & 1 \end{bmatrix}.$$

$$\mathbf{S}(t) = \int_0^\infty \mu(s)\,(\mathbf{C}_t(t - s) - \mathbf{I})\, ds$$

$$= \int_0^\infty \mu(s) \begin{bmatrix} 0 & \gamma(t, t - s) & 0 \\ \gamma(t, t - s) & \gamma^2(t, t - s) & 0 \\ 0 & 0 & 0 \end{bmatrix}.$$

That the shear stress and the normal stress differences are given by

$$S_{12}(t) = \int_0^\infty \mu(s)\,\gamma(t, t - s)\, ds, \tag{S.314}$$

$$N_1(t) = -\int_0^\infty \mu(s)\,[\gamma(t, t - s)]^2\, ds = -N_2. \tag{S.315}$$

For a constant shear rate $\dot{\gamma}$,

$$\gamma(t, \tau) = \dot{\gamma}(\tau - t),$$

and with the memory function $\mu(s) = -\dfrac{G}{\lambda} e^{-s/\lambda}$,

$$S_{12}(t) = \int_0^\infty \mu(s)\,\gamma(t, t - s)\, ds$$

$$= -\int_0^\infty \mu(s)\dot{\gamma}s\, ds$$

$$= \frac{G}{\lambda}\dot{\gamma}\int_0^\infty e^{-s/\lambda} s\, ds$$

$$= G\dot{\gamma}. \tag{S.316}$$

$$N_1(t) = \frac{G}{\lambda}\dot{\gamma}^2 \int_0^\infty e^{-s/\lambda}s^2 ds$$

$$= 2G(\lambda\dot{\gamma})^2. \tag{S.317}$$

For a sinusoidal shear rate $\dot{\gamma} = \dot{\gamma}_0 \cos(\omega t)$

$$\gamma(t,\tau) = \int_t^\tau \dot{\gamma}_0 \cos\omega t \, dt = \frac{\dot{\gamma}_0}{\omega}(\sin\omega\tau - \sin\omega t).$$

$$S_{12}(t) = \int_0^\infty \mu(s)\gamma(t, t-s) \, ds$$

$$= -\frac{G\dot{\gamma}_0}{\lambda\omega}\int_0^\infty e^{-s/\lambda}(\sin\omega(t-s) - \sin\omega t) \, ds$$

$$= \frac{G\dot{\gamma}_0}{\lambda\omega}\int_0^\infty e^{-s/\lambda}(\sin\omega t - \sin\omega t\cos\omega s + \cos\omega t\sin\omega s) \, ds$$

$$= \frac{G\dot{\gamma}_0}{\omega}\left[\sin\omega t - \frac{\sin\omega t}{1+\lambda^2\omega^2} + \frac{\omega\lambda\cos\omega t}{1+\lambda^2\omega^2}\right]$$

$$= G\dot{\gamma}_0\lambda\frac{\lambda\omega\sin\omega t + \cos\omega t}{1+\lambda^2\omega^2}. \tag{S.318}$$

$$N_1 = \frac{G\dot{\gamma}_0^2}{\lambda\omega^2}\int_0^\infty e^{-s/\lambda}(\sin\omega t - \sin\omega t\cos\omega s + \cos\omega t\sin\omega s)^2 ds$$

$$= \frac{G\dot{\gamma}_0^2}{\lambda\omega^2}\int_0^\infty e^{-s/\lambda}[\sin^2\omega t - 2\sin^2\omega t\cos\omega s + \sin 2\omega t\sin\omega s$$

$$+ \frac{1}{2}\sin 2\omega t\sin 2\omega s + \sin^2\omega t\cos^2\omega s + \cos^2\omega t\sin^2\omega s] ds$$

$$= \frac{G\dot{\gamma}_0^2}{\omega^2}\left[\sin^2\omega t - \frac{2\sin^2\omega t}{1+\lambda^2\omega^2} + \frac{\omega\lambda\sin 2\omega t}{1+\lambda^2\omega^2}\right.$$

$$\left. + \frac{\omega\lambda\sin 2\omega t}{1+4\lambda^2\omega^2} + \frac{\sin^2\omega t\left(1+2\lambda^2\omega^2\right)}{1+4\lambda^2\omega^2} + \frac{2\lambda^2\omega^2\cos^2\omega t}{1+4\lambda^2\omega^2}\right]. \tag{S.319}$$

Problems in Chap. 5

Problem 5.1

Assume the relaxation modulus function (5.13)

$$G(t) = \sum_{j=1}^N G_j e^{-t/\lambda_j},$$

the relation (5.11) is

$$\mathbf{S}(t) = 2 \int_{-\infty}^{t} \sum_{j=1}^{N} G_j e^{-(t-t')/\lambda_j} \mathbf{D}(t') \, dt'$$

$$= \sum_{j=1}^{N} \mathbf{S}^{(j)}$$

$$\mathbf{S}^{(j)} = 2 \int_{-\infty}^{t} G_j e^{-(t-t')/\lambda_j} \mathbf{D}(t') \, dt'.$$

Each component $\mathbf{S}^{(j)}$ satisfies, by direct differentiation

$$\dot{\mathbf{S}}^{(j)} = 2 G_j \mathbf{D}(t) - 2 \frac{G_j}{\lambda_j} \int_{-\infty}^{t} e^{-(t-t')/\lambda_j} \mathbf{D}(t') \, dt',$$

or that

$$\mathbf{S}^{(j)} + \lambda_j \dot{\mathbf{S}}^{(j)} = 2\eta_j \mathbf{D}, \quad \eta_j = G_j \lambda_j.$$

This relation is called the linear Maxwell equation.

Problem 5.2

In an oscillatory flow where the shear rate and the shear strain are

$$\dot{\gamma} = \dot{\gamma}_0 \cos \omega t, \quad \gamma = \gamma_0 \sin \omega t, \quad \dot{\gamma}_0 = \omega \gamma_0 \tag{S.320}$$

the only non-zero component of the stress is

$$
\begin{aligned}
S_{12} &= \int_{-\infty}^{t} G(t - t') \, \dot{\gamma}_0 \cos \omega t' \, dt', \\
&= \int_{0}^{\infty} \dot{\gamma}_0 G(s) \cos \omega (t - s) \, ds, \\
&= \int_{0}^{\infty} \dot{\gamma}_0 G(s) \left[\cos \omega t \cos \omega s + \sin \omega t \sin \omega s \right] ds, \\
&= G'(\omega) \gamma_0 \sin \omega t + \eta'(\omega) \dot{\gamma}_0 \cos \omega t, \tag{S.321}
\end{aligned}
$$

where the coefficients in the strain is the *storage modulus*,

$$G'(\omega) = \int_{0}^{\infty} \omega G(s) \sin \omega s \, ds, \tag{S.322}$$

and in the strain rate, the *dynamic viscosity*,

$$\eta'(\omega) = \int_0^\infty G(s) \cos \omega s \, ds.$$

The shear stress can be written as

$$S_{12} = G'(\omega) \gamma_0 \sin \omega t + G''(\omega) \gamma_0 \cos \omega t, \qquad (S.323)$$

where the *loss modulus* G'' is defined as

$$G''(\omega) = \omega \eta'(\omega). \qquad (S.324)$$

Re-write the shear stress as

$$\begin{aligned} S_{12} &= S_{12}^0 \sin(\omega t + \phi) \\ &= S_{12}^0 (\sin \omega t \cos \phi + \sin \phi \cos \omega t). \end{aligned}$$

Identify this with the previous result for S_{12} :

$$S_{12}^0 \cos \phi = \gamma_0 G'(\omega), \quad S_{12}^0 \sin \phi = \gamma_0 G''(\omega),$$

or that

$$S_{12}^0 = \gamma_0 \sqrt{G'^2(\omega) + G''^2(\omega)}, \quad \tan \phi = \frac{G''(\omega)}{G'(\omega)}, \qquad (S.325)$$

where $\tan \phi$ is called the loss tangent. Sometimes it is more convenient to work with complex numbers, and the complex modulus G^* and the complex viscosity η^* are thus defined as

$$G^*(\omega) = G'(\omega) + i G''(\omega), \quad \eta^*(\omega) = \eta'(\omega) - i\eta''(\omega). \qquad (S.326)$$

One can define the complex shear strain as

$$\gamma^* = \gamma_0 e^{-i\omega t}, \qquad (S.327)$$

then the shear stress is

$$\begin{aligned} S_{12} &= \mathrm{Re}\,(G^* \gamma^*) = \gamma_0 \, \mathrm{Re}\left[(G'(\omega) + i G''(\omega))(\sin \omega t - i \cos \omega t)\right] \\ &= \gamma_0 \left(G'(\omega) \sin \omega t + G''(\omega) \cos \omega t\right). \end{aligned} \qquad (S.328)$$

Problem 5.3

Recall the definition of the spectrum $H(\lambda)$:

$$G'(\omega) = \int_{-\infty}^{\infty} \frac{\omega^2 \lambda^2}{1 + \omega^2 \lambda^2} H(\lambda)\, d\ln\lambda, \qquad (S.329)$$

or, equivalently,

$$\eta'(\omega) = \int_{0}^{\infty} \frac{H(\lambda)}{1 + \omega^2 \lambda^2} d\lambda. \qquad (S.330)$$

For the spectrum of the form

$$H(\lambda) = \cos^2(n\lambda),$$

$$\eta'(\omega) = \int_{0}^{\infty} \frac{H(\omega)}{1 + \lambda^2 \omega^2} d\omega$$

$$= \frac{\pi}{4\omega}\left(1 - e^{-2n/\omega}\right). \qquad (S.331)$$

At large n, the data η' is smooth ($e^{-2n/\omega}$ goes to zero), but the spectrum is highly oscillatory. Thus, the inverse problem of finding $H(\lambda)$, given the data η' in the chosen form is ill-conditioned – that is, a small variation in the data (in the exponentially small term) may lead to a large variation in the solution.

Problem 5.4

For the Maxwell discrete relaxation spectrum (5.13),

$$G(t) = \sum_{j=1}^{N} G_j e^{-t/\lambda_j}$$

the storage modulus and the dynamic viscosity are given by (5.17):

$$G'(\omega) = \int_{0}^{\infty} \omega G(s) \sin \omega s\, ds$$

$$= \sum_{j=1}^{N} \int_{0}^{\infty} \omega G_j e^{-s/\lambda_j} \sin \omega s\, ds$$

$$= \sum_{j=1}^{N} \frac{\omega^2 G_j \lambda_j^2}{1 + \omega^2 \lambda_j^2}, \tag{S.332}$$

$$\eta'(\omega) = \int_0^\infty G(s) \cos \omega s \, ds$$

$$= \sum_{j=1}^{N} \int_0^\infty G_j e^{-s/\lambda_j} \cos \omega s \, ds$$

$$= \sum_{j=1}^{N} \frac{G_j \lambda_j}{1 + \omega^2 \lambda_j^2}. \tag{S.333}$$

In particular, with one relaxation mode $\lambda = \lambda_1$,

$$G'(\omega) = \frac{\omega^2 G_1 \lambda_1^2}{1 + \omega^2 \lambda_1^2}, \quad \eta'(\omega) = \frac{G_1 \lambda_1}{1 + \omega^2 \lambda_1^2}, \quad \tan \phi = \frac{\omega \eta'(\omega)}{G'(\omega)} = \frac{1}{\omega \lambda_1} \tag{S.334}$$

As as $\omega \lambda_1 = 0$, $\eta'(\omega) = G_1 \lambda_1$, $G'(\omega) = 0$ (Newtonian fluid-like response), and when $\omega \lambda_1 \to \infty$, $\eta' \to 0$, $G' \to G_1$ corresponds to a solid-like response.

Problem 5.5

Suppose we have a Maxwell material with one relaxation time,

$$G(t) = \frac{\eta_0}{\lambda} e^{-t/\lambda}.$$

and Ω_i = constant. For the circular Couette flow problem, the torque on the inner cylinder is $\Gamma = M(R_o - R_i)$, where

$$M(t) = \frac{4\pi R_o^2}{1 - R_o^2/R_i^2} \int_0^t G(t - t') \Omega_i(t') \, dt'$$

$$= \frac{4\pi R_o^2 \eta_0 \Omega_i}{(1 - R_o^2/R_i^2) \lambda} \int_0^t e^{-(t-t')/\lambda} dt'$$

$$= \frac{4\pi R_o^2 \eta_0 \Omega_i}{(1 - R_o^2/R_i^2)} (1 - e^{-t/\lambda}).$$

With the Newtonian result (corresponds to $\lambda \to 0$),

$$M_N = \frac{4\pi R_o^2 \eta_0 \Omega_i}{(1 - R_o^2/R_i^2)},$$

we thus have

$$\frac{M(t)}{M_N} = 1 - e^{-t/\lambda}. \tag{S.335}$$

Problem 5.6

Working in Laplace transform domain (s is the Laplace transform variable, and the overbar denotes the Laplace transform function), and denote the displacement across the spring G_1 as x_1, across the Kelvin-Voigt element (G_2, η_2) as $x_2 - x_1$, and across the dashpot η_1 as $x - x_2$, the relationship between force and various displacements so defined are

- across the spring G_1:

$$\frac{\bar{F}}{G_1} = \bar{x}_1,$$

- across the dashpot of η_1:

$$\frac{\bar{F}}{s\eta_1} = \bar{x} - \bar{x}_2$$

- across the Kelvin–Voigt–Meyer element (G_2, η_2):

$$\frac{\bar{F}}{G_2 + s\eta_2} = \bar{x}_2 - \bar{x}_1.$$

These may be summed up to yield

$$\frac{\bar{F}}{G_1} + \frac{\bar{F}}{G_2 + s\eta_2} + \frac{\bar{F}}{s\eta_1} = \bar{x}$$

$$[s\eta_1 (G_2 + s\eta_2) + s\eta_1 G_1 + G_1 (G_2 + s\eta_2)] \bar{F} = s\eta_1 G_1 (G_2 + s\eta_2) \bar{x}$$

leading to

$$\left[1 + a_1 s + a_2 s^2\right] \bar{F} = (b_1 + b_2 s) s\bar{x},$$

where

$$a_1 = \frac{\eta_1}{G_2} \left(1 + \frac{G_2}{G_1} + \frac{\eta_2}{\eta_1}\right), \quad a_2 = \frac{\eta_1 \eta_2}{G_1 G_2}, \quad b_1 = \eta_1, \quad b_2 = \frac{\eta_1 \eta_2}{G_2}.$$

In time domain, this is equivalent to

$$F + a_1\dot{F} + a_2\ddot{F} = b_1\dot{x} + b_2\ddot{x},$$

which corresponds to the following stress-strain relation for the four-element model

$$S_{ij} + a_1\dot{S}_{ij} + a_2\ddot{S}_{ij} = b_1\dot{\gamma}_{ij} + b_2\ddot{\gamma}_{ij}. \tag{S.336}$$

Problems in Chap. 6

Problem 6.1

The velocity field for (6.3) is

$$\mathbf{u} = \dot{\gamma}\,(\mathbf{b} \cdot \mathbf{x})\,\mathbf{a}, \tag{S.337}$$

where $\mathbf{a}$, $\mathbf{b}$, $\mathbf{c}$ are a set of orthonormal vectors. The velocity gradient is

$$\begin{aligned}
L_{ij} &= \frac{\partial u_i}{\partial x_j} = \frac{\partial}{\partial x_j}\,(\dot{\gamma}b_k x_k a_i) \\
&= \dot{\gamma}b_j a_i,
\end{aligned}$$

leading to the stated result

$$\mathbf{L} = \dot{\gamma}\mathbf{ab}. \tag{S.338}$$

Since

$$tr\,\mathbf{L} = 0,$$

all the flows represented by (6.3) are isochoric (volume-conserving). Now

$$\begin{aligned}
4\mathbf{D}^2 &= \dot{\gamma}^2\,(\mathbf{ab} + \mathbf{ba}) \cdot (\mathbf{ab} + \mathbf{ba}) \\
&= \dot{\gamma}^2\,(\mathbf{aa} + \mathbf{bb}),
\end{aligned}$$

the shear rate is

$$\sqrt{2tr\,\mathbf{D}^2} = \sqrt{\dot{\gamma}^2} = |\dot{\gamma}|. \tag{S.339}$$

Problem 6.2

For the helicoidal flow (6.14), the velocity field is

$$\mathbf{u} = (r\mathbf{e}_\theta + c\mathbf{e}_z)\,\omega\,(r, z - c\theta), \tag{S.340}$$

where c is a constant. The velocity gradient is

$$
\begin{aligned}
(\nabla\mathbf{u})^T &= \left[\nabla\,(r\omega\mathbf{e}_\theta + c\omega\mathbf{e}_z)\right]^T \\
&= \left[\nabla\omega\,(r\mathbf{e}_\theta + c\mathbf{e}_z) + \omega\nabla\,(r\mathbf{e}_\theta + c\mathbf{e}_z)\right]^T \\
&= \left[\nabla\omega\,(r\mathbf{e}_\theta + c\mathbf{e}_z) + \omega\,(\mathbf{e}_r\mathbf{e}_\theta - \mathbf{e}_\theta\mathbf{e}_r)\right]^T \\
&= (r\mathbf{e}_\theta + c\mathbf{e}_z)\,\nabla\omega + \omega\,(\mathbf{e}_\theta\mathbf{e}_r - \mathbf{e}_r\mathbf{e}_\theta) \\
&= \dot{\gamma}\mathbf{ab} = \mathbf{L} \tag{S.341}
\end{aligned}
$$

We first note that

$$
\begin{aligned}
2\mathbf{D} &= (r\mathbf{e}_\theta + c\mathbf{e}_z)\,\nabla\omega + \nabla\omega\,(r\mathbf{e}_\theta + c\mathbf{e}_z) \\
&\quad + \omega\,(\mathbf{e}_\theta\mathbf{e}_r - \mathbf{e}_r\mathbf{e}_\theta) + \omega\,(-\mathbf{e}_\theta\mathbf{e}_r + \mathbf{e}_r\mathbf{e}_\theta) \\
&= (r\mathbf{e}_\theta + c\mathbf{e}_z)\,\nabla\omega + \nabla\omega\,(r\mathbf{e}_\theta + c\mathbf{e}_z). \tag{S.342}
\end{aligned}
$$

Thus

$$
\begin{aligned}
4\mathbf{D}^2 &= (r\mathbf{e}_\theta + c\mathbf{e}_z)\,\nabla\omega \cdot (r\mathbf{e}_\theta + c\mathbf{e}_z)\,\nabla\omega \\
&\quad + 2\nabla\omega\,(r\mathbf{e}_\theta + c\mathbf{e}_z) \cdot (r\mathbf{e}_\theta + c\mathbf{e}_z)\,\nabla\omega \\
&\quad + \nabla\omega\,(r\mathbf{e}_\theta + c\mathbf{e}_z) \cdot \nabla\omega\,(r\mathbf{e}_\theta + c\mathbf{e}_z) \\
&= (r\mathbf{e}_\theta + c\mathbf{e}_z)\left(\frac{\partial\omega}{\partial\theta} + c\frac{\partial\omega}{\partial z}\right)\nabla\omega \\
&\quad + 2\left(r^2 + c^2\right)\nabla\omega\nabla\omega \\
&\quad + \nabla\omega\left(\frac{\partial\omega}{\partial\theta} + c\frac{\partial\omega}{\partial z}\right)(r\mathbf{e}_\theta + c\mathbf{e}_z).
\end{aligned}
$$

Since $\omega = \omega\,(r, z - c\theta)$ is a function of r and $z - c\theta = Z$,

$$\frac{\partial\omega}{\partial\theta} + c\frac{\partial\omega}{\partial z} = -c\frac{\partial\omega}{\partial Z} + c\frac{\partial\omega}{\partial Z} = 0,$$

and the shear rate squared given by

$$\dot{\gamma}^2 = \left(r^2 + c^2\right)\nabla\omega \cdot \nabla\omega. \tag{S.343}$$

The vectors $\mathbf{a}$ and $\mathbf{b}$ are identified as

$$\dot{\gamma}\mathbf{a} = \mathbf{L}\mathbf{L}^T, \quad \dot{\gamma}\mathbf{b} = \mathbf{L}^T\mathbf{L}, \tag{S.344}$$

neither is constant.

Problem 6.3

This flow is one of the fan flows - the flow is uniaxial and by inspection of the boundary conditions we find that the velocity field is

$$\mathbf{u} = U\frac{\theta}{\theta_0}\mathbf{e}_z. \tag{S.345}$$

This velocity field may be written as

$$\mathbf{u} = \dot{\gamma}\,(\mathbf{b}\cdot\mathbf{x})\,\mathbf{a}, \tag{S.346}$$

where $\mathbf{a} = \mathbf{e}_z$. To identify $\mathbf{b}$ we note

$$\mathbf{L} = \nabla\mathbf{u}^T = \left(\mathbf{e}_\theta\frac{1}{r}\frac{\partial}{\partial\theta}\mathbf{u}\right)^T$$

$$= \mathbf{e}_z\mathbf{e}_\theta\frac{U}{r\theta_0} = \dot{\gamma}\mathbf{a}\mathbf{b}.$$

Clearly

$$\dot{\gamma} = \frac{U}{r\theta_0}, \quad \mathbf{b} = \mathbf{e}_\theta.$$

The stress is given by (see equation preceding 6.20)

$$\mathbf{T} = -P\mathbf{I} + \dot{\gamma}\eta\,(\mathbf{a}\mathbf{b} + \mathbf{b}\mathbf{a}) + (N_1 + N_2)\,(\mathbf{a}\mathbf{a} + \mathbf{b}\mathbf{b}) - N_1\mathbf{b}\mathbf{b},$$

or

$$\mathbf{T} = -P\mathbf{I} + \eta\dot{\gamma}\,(\mathbf{e}_z\mathbf{e}_\theta + \mathbf{e}_\theta\mathbf{e}_z) + (N_1 + N_2)\,\mathbf{e}_z\mathbf{e}_z + N_2\mathbf{e}_\theta\mathbf{e}_\theta. \tag{S.347}$$

In full, the non-trivial components for the stress are

$$T_{rr} = -P, \quad T_{\theta\theta} = -P + N_2, \quad T_{zz} = -P + N_1 + N_2, \quad T_{z\theta} = T_{\theta z} = \eta\dot{\gamma}.$$

To find the pressure field, we note that the velocity field is uni-directional, with zero inertia force $\mathbf{u} \cdot \nabla \mathbf{u} = \mathbf{0}$. Thus, the non-trivial component of the conservation of momentum equation is

$$0 = \frac{\partial T_{rr}}{\partial r} + \frac{T_{rr} - T_{\theta\theta}}{r} + \frac{1}{r}\frac{\partial T_{\theta r}}{\partial \theta} + \frac{\partial T_{zr}}{\partial z}$$

$$\frac{\partial P}{\partial r} = -\frac{N_2}{r} = -\frac{\dot{\gamma}^2 \nu_2\,(\dot{\gamma})}{r}$$

Noting that

$$\dot{\gamma} = \frac{U}{r\theta_0} = \frac{r_0\dot{\gamma}_0}{r}, \quad \dot{\gamma}_0 = \frac{U}{r_0\theta_0}$$

and $d\dot{\gamma}/\dot{\gamma} = -dr/r$, which can be used in integrating for the pressure:

$$P = P\,(r_0) - \int_{r_0}^{r} \frac{\dot{\gamma}^2 \nu_2\,(\dot{\gamma})}{r} dr$$

$$= P\,(r_0) + \int_{\dot{\gamma}_0}^{\dot{\gamma}} \dot{\gamma}\nu_2\,(\dot{\gamma})\,.$$

Alternatively,

$$P = P\,(r_0) + I_2\,(\dot{\gamma}) - I_2\,(\dot{\gamma}_0)\,, \quad I_2\,(\dot{\gamma}) = \int_0^{\dot{\gamma}} \dot{\gamma}\nu_2 d\dot{\gamma}. \tag{S.348}$$

Suppose we have a pressure measurement $P = P\,(\dot{\gamma})$ at different shear rates. Then according to the preceding

$$\frac{dP}{d\dot{\gamma}} = \frac{dI_2}{d\dot{\gamma}} = \dot{\gamma}\nu_2\,(\dot{\gamma})\,.$$

This may be used to deduce ν_2, or $N_2 = \dot{\gamma}^2\nu_2\,(\dot{\gamma})\,.$

Problem 6.4

In the flow between two parallel, coaxial disks, or the *torsional flow*, the flow is a sub-class of the helicoidal flow (6.14), and by the boundary conditions on the plates, the velocity may be shown that (another way would be to show that the kinematics below satisfy conservation of mass and momentum)

$$\mathbf{u} = \Omega r \frac{z}{h}\mathbf{e}_\theta = \dot{\gamma} z \mathbf{e}_\theta, \quad \dot{\gamma} = \Omega\frac{r}{h}. \tag{S.349}$$

This velocity is already in the form (6.3)

$$\mathbf{u} = \dot{\gamma} \, (\mathbf{b} \cdot \mathbf{x}) \, \mathbf{a}, \tag{S.350}$$

where

$$\mathbf{a} = \mathbf{e}_\theta, \tag{S.351}$$

and

$$(\mathbf{b} \cdot \mathbf{x}) = z, \quad \mathbf{b} = \mathbf{e}_z. \tag{S.352}$$

The stress is

$$\mathbf{T} = -P\mathbf{I} + \mathbf{S} = -P\mathbf{I} + \eta\dot{\gamma} \, (\mathbf{ab} + \mathbf{ba}) + \dot{\gamma}^2 \, (\nu_1 + \nu_2) \, \mathbf{aa} + \dot{\gamma}^2 \nu_2 \mathbf{bb}, \tag{S.353}$$

with the non-trivial components

$$T_{rr} = -P, \quad T_{\theta\theta} = -P + N_1 + N_2, \quad T_{zz} = -P + N_2, \quad S_{z\theta} = S_{\theta z} = \eta\dot{\gamma}. \tag{S.354}$$

The conservation of momentum required, neglecting inertia terms,

$$\frac{\partial P}{\partial r} = \frac{\partial S_{rr}}{\partial r} + \frac{S_{rr} - S_{\theta\theta}}{r} = -\frac{N_1 + N_2}{r},$$

$$\frac{\partial P}{r\partial \theta} = 0,$$

$$\frac{\partial P}{\partial z} = 0.$$

Consequently, P is a function of r only. It is given by

$$P(r) = C - \int_0^r \frac{N_1(\zeta) + N_2(\zeta)}{\zeta} d\zeta \tag{S.355}$$

where C is a constant of integration. This is found by the assumption that at the free surface at $r = R$, the radial stress is zero:

$$T_{rr}(R) = -P(R) = 0. \tag{S.356}$$

This implies

$$C = \int_0^R \frac{N_1(\zeta) + N_2(\zeta)}{\zeta} d\zeta,$$

or that

$$
\begin{aligned}
P(r) &= \int_r^R \frac{N_1(\zeta) + N_2(\zeta)}{\zeta} d\zeta \\
&= \int_{\dot{\gamma}(r)}^{\dot{\gamma}_R} \frac{N_1(\dot{\gamma}) + N_2(\dot{\gamma})}{\dot{\gamma}} d\dot{\gamma} \\
&= \int_{\dot{\gamma}}^{\dot{\gamma}_R} \dot{\gamma}(\nu_1 + \nu_2) d\dot{\gamma},
\end{aligned}
\tag{S.357}
$$

where $\dot{\gamma}_R = \Omega R/h$ is the shear rate at the rim $r = R$.

The torque required to turn the top disk is $\dot{\gamma} = \Omega r/h = \dot{\gamma}_R r/R$

$$
\begin{aligned}
M &= \int_0^R 2\pi r^2 S_{z\theta} dr \\
&= 2\pi \int_0^R \dot{\gamma}\eta(\dot{\gamma}) r^2 dr \\
&= 2\pi \int_0^{\dot{\gamma}_R} \eta r^3 d\dot{\gamma} \\
&= 2\pi R^3 \dot{\gamma}_R^{-3} \int_0^{\dot{\gamma}_R} \dot{\gamma}^3 \eta(\dot{\gamma}) d\dot{\gamma}.
\end{aligned}
\tag{S.358}
$$

From the axial stress and the result for the pressure,

$$
\begin{aligned}
T_{zz} &= -P + N_2 \\
&= N_2 + \int_R^r \frac{N_1(\zeta) + N_2(\zeta)}{\zeta} d\zeta,
\end{aligned}
\tag{S.359}
$$

and therefore the normal force on the top disk is, by an integration by parts, keeping in mind that normal stresses are zero at the free surface $r = R$,

$$
\begin{aligned}
F &= -2\pi \int_0^R T_{zz} r \, dr \\
&= \pi \int_0^R T'_{zz} r^2 dr \\
&= \pi \int_0^R (-2r N_2 + r(N_1 + N_2)) \, dr \\
&= \pi \int_0^R r(N_1 - N_2) \, dr
\end{aligned}
\tag{S.360}
$$

Convert this into an integral with respect to the shear rate, the normal force is given by

$$F = \pi R^2 \dot{\gamma}_R^{-2} \int_0^{\dot{\gamma}_R} \dot{\gamma} \left(N_1 - N_2 \right) d\dot{\gamma}. \tag{S.361}$$

By normalizing the torque and the force as

$$m = \frac{M}{2\pi R^3}, \quad f = \frac{F}{\pi R^2}, \tag{S.362}$$

we have

$$m = \dot{\gamma}_R^{-3} \int_0^{\dot{\gamma}_R} \dot{\gamma}^3 \eta \left(\dot{\gamma} \right) d\dot{\gamma}.$$

Thus

$$\frac{dm}{d\dot{\gamma}_R} = -3\frac{m}{\dot{\gamma}_R} + \eta \left(\dot{\gamma}_R \right)$$

or

$$\eta \left(\dot{\gamma}_R \right) = \frac{m}{\dot{\gamma}_R} \left(3 + \frac{\dot{\gamma}_R}{m} \frac{dm}{d\dot{\gamma}_R} \right)$$

$$= \frac{m}{\dot{\gamma}_R} \left(3 + \frac{d \ln m}{d \ln \dot{\gamma}_R} \right), \tag{S.363}$$

as required to show.

In addition

$$f = \dot{\gamma}_R^{-2} \int_0^{\dot{\gamma}_R} \dot{\gamma} \left(N_1 - N_2 \right) d\dot{\gamma}.$$

And thus

$$\frac{df}{d\dot{\gamma}_R} = -2\frac{f}{\dot{\gamma}_R} + N_1 \left(\dot{\gamma}_R \right) - N_2 \left(\dot{\gamma}_R \right),$$

or

$$N_1 \left(\dot{\gamma}_R \right) - N_2 \left(\dot{\gamma}_R \right) = f \left(2 + \frac{d \ln f}{d \ln \dot{\gamma}_R} \right). \tag{S.364}$$

Relations (S.363) and (S.364) are the basis for the operation of the parallel-disk viscometer.

Problem 6.5

Pipe flow is a special case for helical flow, where the velocity field is (uniaxial flow)

$$\mathbf{u} = u\,(r)\,\mathbf{e}_z. \qquad\qquad (S.365)$$

This velocity is already in the viscometric form (6.3), and the non-trivial stresses are functions of r. The only non-trivial momentum equation (in the axial z direction) is

$$\frac{\partial P}{\partial z} = \frac{\partial S_{rz}}{\partial r} + \frac{S_{rz}}{r} = \frac{1}{r}\frac{d}{dr}\,(r\,S_{rz}), \qquad\qquad (S.366)$$

where

$$S_{rz} = \eta\frac{du}{dr}.$$

The right side of (S.366) is a function of r alone, thus

$$P = \left(\frac{\partial S_{rz}}{\partial r} + \frac{S_{rz}}{r}\right)z + P_0$$

$$= -\frac{\Delta P}{L}z + P_0, \qquad\qquad (S.367)$$

where $\Delta P/L$ is the pressure drop per unit length. This leads to

$$\frac{1}{r}\frac{d}{dr}\,(r\,S_{rz}) = -\frac{\Delta P}{L}$$

$$S_{rz} = -\frac{\Delta P}{2L}r. \qquad\qquad (S.368)$$

noting the boundedness of the stress in r. Thus the shear rate is

$$\frac{du}{dr} = -\frac{\Delta P}{2L}r\eta^{-1}. \qquad\qquad (S.369)$$

The flow rate is

$$Q = 2\pi\int_0^R ru\,(r)\,dr$$

$$= -\pi\int_0^R r^2\frac{du}{dr}dr$$

$$= \frac{\pi\Delta P}{2L}\int_0^R r^3\eta^{-1}dr$$

Since $r = -2L\tau/\Delta P$, where $\tau = S_{rz}$, a change in variable yields

$$Q = 8\pi \left(\frac{L}{\Delta P}\right)^3 \int_0^{\tau_w} \tau^3 \eta^{-1} d\tau \qquad (S.370)$$

where $\tau_w = -\Delta P R/(2L)$ is the shear stress at the wall. In terms of the reduced discharge rate,

$$q = \frac{Q}{\pi R^3} = \tau_w^{-3} \int_0^{\tau_w} \tau^3 \eta^{-1} d\tau, \qquad (S.371)$$

and therefore

$$\frac{dq}{d\tau_w} = \frac{1}{\eta(\tau_w)} - \frac{3q}{\tau_w}, \qquad (S.372)$$

or

$$\eta^{-1}(\tau_w) = \frac{q}{\tau_w}\left(3 + \frac{d\ln q}{d\ln \tau_w}\right). \qquad (S.373)$$

Since $\eta(\tau_w)\dot\gamma_w = \tau_w$, or $\eta^{-1}(\tau_w) = \dot\gamma_w/\tau_w$

$$\dot\gamma_w = q(\tau_w)\left[3 + \frac{d\ln q}{d\ln \tau_w}\right]. \qquad (S.374)$$

The relation (S.374) is due to Rabinowitch and is the basis for capillary viscometry.

Problems in Chap. 7

Problem 7.1

The solution to (7.22) is given by (7.24), reproduced here

$$\dot{\mathbf{x}}(t) = \int_0^t \exp\left\{\mathbf{m}^{-1}\boldsymbol{\zeta}(t' - t)\right\} \mathbf{m}^{-1}\mathbf{F}^{(b)}(t') dt'. \qquad (S.375)$$

Now, the diffusivity is defined

$$\mathbf{D} = \lim_{t\to\infty} \frac{1}{2}\int_0^t \langle \dot{\mathbf{x}}(t)\dot{\mathbf{x}}(t-\tau) + \dot{\mathbf{x}}(t-\tau)\dot{\mathbf{x}}(t)\rangle d\tau. \qquad (S.376)$$

First, we form the correlation function

$$
\begin{aligned}
\mathbf{R}(\tau) &= \langle \dot{\mathbf{x}}(t+\tau)\,\dot{\mathbf{x}}(t)\rangle \\
&= \int_0^t \int_0^t \exp\left\{\mathbf{m}^{-1}\zeta\left(t'-t-\tau\right)\right\} \mathbf{m}^{-1}\left\langle \mathbf{F}^{(b)}\left(t'\right)\,\mathbf{F}^{(b)}\left(t''\right)\right\rangle \\
&\qquad\qquad \mathbf{m}^{-1}\exp\left\{\mathbf{m}^{-1}\zeta\left(t''-t\right)\right\} dt'\,dt'' \\
&= \int_0^t \exp\left\{\mathbf{m}^{-1}\zeta\left(t'-t-\tau\right)\right\} \mathbf{m}^{-1}2\mathbf{f}\mathbf{m}^{-1}\exp\left\{\mathbf{m}^{-1}\zeta\left(t'-t\right)\right\} dt' \\
&= e^{-\mathbf{m}^{-1}\zeta\tau}\,\langle \dot{\mathbf{x}}(t)\,\dot{\mathbf{x}}(t)\rangle .
\end{aligned}
\tag{S.377}
$$

Thus

$$
\begin{aligned}
\mathbf{D} &= \lim_{t\to\infty}\int_0^t e^{-\mathbf{m}\zeta\tau}\langle \dot{\mathbf{x}}\left(t'\right)\dot{\mathbf{x}}\left(t'\right)\rangle d\tau \\
&= \zeta^{-1}\mathbf{m}kT\mathbf{m}^{-1} \\
&= kT\zeta^{-1}.
\end{aligned}
\tag{S.378}
$$

This is the *Stokes–Einstein relation*, relating the diffusivity to the mobility of a Brownian particle.

Problem 7.2

In the limit $\mathbf{m}\to\mathbf{0}$, the Langevin equation (7.20) becomes (note that all the material matrices are symmetric)

$$
\dot{\mathbf{x}} = -\zeta^{-1}\cdot\mathbf{Kx} + \zeta^{-1}\mathbf{F}^{(b)}(t),
\tag{S.379}
$$

which has the solution

$$
\Delta\mathbf{x}(t) = -\zeta^{-1}\cdot\mathbf{Kx}\Delta t + \int_t^{t+\Delta t}\zeta^{-1}\mathbf{F}^{(b)}\left(t'\right)dt'.
\tag{S.380}
$$

From this,

$$
\begin{aligned}
\langle\Delta\mathbf{x}(t)\rangle &= -\zeta^{-1}\cdot\mathbf{Kx}\Delta t + \int_t^{t+\Delta t}\zeta^{-1}\langle\mathbf{F}^{(b)}\left(t'\right)\rangle dt' \\
&= -\zeta^{-1}\cdot\mathbf{Kx}.
\end{aligned}
\tag{S.381}
$$

In addition,

$$\langle \Delta \mathbf{x}\,(t)\,\Delta \mathbf{x}\,(t)\rangle = O\left(Deltat^2\right)$$
$$+ \int_{t}^{t+\Delta t} \int_{t}^{t+\Delta t} dt'' \zeta^{-1} \langle \mathbf{F}^{(b)}\,(t')\,\mathbf{F}^{(b)}\,(t'')\rangle \zeta^{-1} dt'$$
$$= 2 \int_{t}^{t+\Delta t} dt' \zeta^{-1} \mathbf{f}\,\zeta^{-1}$$
$$= 2kT \zeta^{-1} \Delta t. \tag{S.382}$$

Thus the Fokker–Planck equation is

$$\frac{\partial \phi}{\partial t} = \frac{\partial}{\partial \mathbf{x}} \cdot \left[\frac{\langle \Delta \mathbf{x} \Delta \mathbf{x}\rangle}{2\Delta t} \frac{\partial \phi}{\partial \mathbf{x}} + \frac{\langle \Delta \mathbf{x}\rangle}{\Delta t} \phi \right]$$

is simply

$$\frac{\partial \phi}{\partial t} = \frac{\partial}{\partial \mathbf{x}} \cdot \left[kT \zeta^{-1} \frac{\partial \phi}{\partial \mathbf{x}} + \zeta^{-1} \cdot \mathbf{K}\mathbf{x}\phi \right]. \tag{S.383}$$

Problem 7.3

In a non-homogeneous flow and using the dumbell model, the particle centre of gravity will migrate from the streamline according to

$$\langle \dot{\mathbf{R}}^{(c)} - \mathbf{u}^{(c)}\rangle = \frac{1}{8} \langle \mathbf{R}\mathbf{R}\rangle : \nabla\nabla \mathbf{u}^{(c)}. \tag{S.384}$$

In the planar Poiseuille flow,

$$\mathbf{u}^{(c)} = u\,(y)\,\mathbf{i}, \quad u\,(y) = U\left(1 - y^2/h^2\right)$$

$$\langle \dot{\mathbf{R}}^{(c)} - \mathbf{u}^{(c)}\rangle = \frac{1}{8} \left[\langle R_1^2\rangle \frac{\partial^2}{\partial x^2} + 2\langle R_1 R_2\rangle \frac{\partial^2}{\partial x \partial y} + 2\langle R_1 R_3\rangle \frac{\partial^2}{\partial x \partial z} \right.$$
$$\left. + \langle R_2^2\rangle \frac{\partial^2}{\partial y^2} + 2\langle R_2 R_3\rangle \frac{\partial^2}{\partial y \partial z} + \langle R_3^2\rangle \frac{\partial^2}{\partial z^2} \right] \mathbf{u}^{(c)}$$
$$= \frac{1}{8} \langle R_2^2\rangle \frac{\partial^2}{\partial y^2} \mathbf{u}^{(c)}$$
$$= -\frac{U}{4h^2} \langle R_2^2\rangle \mathbf{i}. \tag{S.385}$$

This must be solved together with the constitutive equation for $\mathbf{R}$ to obtain the migration velocity. In fact, putting $\mathbf{A} = \langle \mathbf{RR} \rangle$, and using the elastic dumbbell model, we have

$$A_{11} + \lambda \left(\dot{A}_{11} - 2u' A_{12} \right) = \frac{1}{3},$$
$$A_{12} + \lambda \left(\dot{A}_{12} - u' A_{22} \right) = 0,$$
$$A_{22} + \lambda \dot{A}_{22} = \frac{1}{3}.$$

Thus the migration is along a streamline, and of the amount

$$\langle \dot{\mathbf{R}}^{(c)} - \mathbf{u}^{(c)} \rangle = -\frac{U}{12h^2} \mathbf{i}. \tag{S.386}$$

Of course, this is a simplistic model, to have cross-streamline migration, a better model is required, see for example, Goh et al., J Chem Phys 81 (1985) 6259–6265 and some of the references cited thereon.

Problem 7.4

The average end-to-end vector of a linear dumbbell evolves in time according to

$$\langle \dot{\mathbf{R}} \rangle = \mathbf{L} \langle \mathbf{R} \rangle - 2H\zeta^{-1} \langle \mathbf{R} \rangle, \tag{S.387}$$

This has the integrating factor $e^{t/2\lambda} e^{\mathbf{L}t}$:

$$\frac{d}{dt} \left(e^{t/2\lambda} e^{-\mathbf{L}t} \langle \mathbf{R} \rangle \right) = e^{t/2\lambda} e^{-\mathbf{L}t} \left\langle \dot{\mathbf{R}} + \frac{1}{2\lambda} \mathbf{R} - \mathbf{L} \mathbf{R} \right\rangle$$
$$= 0,$$

where $\lambda = \zeta / (4H)$ is the relaxation time. This has the solution

$$\langle \mathbf{R}(t) \rangle = e^{-t/2\lambda} e^{\mathbf{L}t} \mathbf{R}_0. \tag{S.388}$$

Whether or not $\langle \mathbf{R} \rangle$ decays to zero depends on the eigenvalues of $\mathbf{L} - \mathbf{I}/2\lambda$, if this is positive, $\langle \mathbf{R} \rangle$ is a run-away process, and if this is negative, then $\langle \mathbf{R} \rangle$ will decay to zero. Thus a strong flow will result if

$$\text{eigen} \ (\mathbf{L}) \geq 1/2\lambda, \tag{S.389}$$

where eigen $(\mathbf{L})$ is the maximum eigenvalue of $\mathbf{L}$. Otherwise we have a weak flow.

Problem 7.5

The upper-convected Maxwell model is written as

$$\mathbf{S}^{(p)} + \lambda \left\{ \frac{d}{dt}\mathbf{S}^{(p)} - \mathbf{L}\mathbf{S}^{(p)} - \mathbf{S}^{(p)}\mathbf{L}^T \right\} = G\mathbf{I}, \qquad (S.390)$$

Now, consider the following integral model:

$$\mathbf{S}^{(p)}(t) = \frac{G}{\lambda} \int_{-\infty}^{t} e^{(s-t)/\lambda} \mathbf{C}_t(s)^{-1}\, ds, \qquad (S.391)$$

one has

$$\dot{\mathbf{S}}^{(p)}(t) = \frac{G}{\lambda}\mathbf{I} + \frac{G}{\lambda} \int_{-\infty}^{t} -\frac{1}{\lambda} e^{(s-t)/\lambda} \mathbf{C}_t(s)^{-1}\, ds$$

$$+ \frac{G}{\lambda} \int_{-\infty}^{t} e^{(s-t)/\lambda} \left[\mathbf{L}(t)\mathbf{C}_t(s)^{-1} + \mathbf{C}_t(s)^{-1}\mathbf{L}^T(t) \right] ds,$$

or that

$$\mathbf{S}^{(p)} + \lambda \left\{ \frac{d}{dt}\mathbf{S}^{(p)} - \mathbf{L}\mathbf{S}^{(p)} - \mathbf{S}^{(p)}\mathbf{L}^T \right\} = G\mathbf{I}.$$

We conclude that (S.391) indeed solves (S.390).

Problems in Chap. 8

Problem 8.1

In the shear reversal experiments of Gadala-Maria and Acrivos (1980), it was found that if shearing is stopped after a steady state has been reached in a Couette device, the torque is reduced to zero instantaneously. This is due to the insignificant inertia of the suspended particles, and the micromechanics are governed by the Stokes equation (8.3), only the present boundary conditions matter. When the flow stops, all the forces, including the torque, go to zero instantaneously.

If shearing is resumed in the same direction after a period of rest, then the torque would attain its final value that corresponds to the resumed shear rate almost instantaneously. This is due to the equilibrated configuration of the suspended particles has been achieved and frozen in place after the flow stops. If the flow starts in the same direction, the particles are happy to stay in their previously preferred configuration, and the forces and torques instantaneously assume their previous values.

However, if shearing is resumed in the opposite direction, then the particles find a different preferred configuration corresponding to the reversed shear, and the forces and torques go through a period of adjustment to their steady state values. Zero, fading or infinite memory is a convenient description - in this suspension case, it is irrelevant to think of fading memory - the whole rheology is what matters.

Problem 8.2

Jeffery's solution for a unit vector $\mathbf{p}$ directed along the major axis of a spheroidal supsended particle obeys

$$\dot{\mathbf{p}} = \mathbf{W} \cdot \mathbf{p} + \frac{R^2 - 1}{R^2 + 1} \left(\mathbf{D} \cdot \mathbf{p} - \mathbf{D} : \mathbf{ppp} \right) \tag{S.392}$$

$$= \mathbf{L} \cdot \mathbf{p} - \frac{2}{R^2 + 1} \mathbf{D} \cdot \mathbf{p} - \frac{R^2 - 1}{R^2 + 1} \mathbf{D} : \mathbf{ppp},$$

where R is the aspect ratio of the particle (major to minor diameter ratio), $\mathbf{L}$ is the velocity gradient, $\mathbf{D} = \left(\mathbf{L} + \mathbf{L}^T \right) / 2$ is the strain rate, and $\mathbf{W} = \left(\mathbf{L} - \mathbf{L}^T \right) / 2$ is the vorticity tensor. Denote the effective velocity gradient as

$$\mathcal{L} = \mathbf{L} - \frac{2}{R^2 + 1} \mathbf{D} = \mathbf{L} - \zeta \mathbf{D}, \tag{S.393}$$

then (S.392) can be simplified to

$$\dot{\mathbf{p}} = \mathcal{L} \cdot \mathbf{p} - \mathcal{L} : \mathbf{ppp}. \tag{S.394}$$

Now, consider the linear system

$$\dot{\mathbf{Q}} = \mathcal{L} \cdot \mathbf{Q}. \tag{S.395}$$

If we denote $\mathbf{Q} = Q\mathbf{p}$, where Q is the magnitude of $\mathbf{Q}$, and $\mathbf{p}$ is a unit vector, then

$$\dot{\mathbf{Q}} = \dot{Q}\mathbf{p} + Q\dot{\mathbf{p}} = Q\mathcal{L} \cdot \mathbf{p}.$$

Since $\mathbf{p} \cdot \dot{\mathbf{p}} = 0$,

$$\dot{Q} = Q\mathcal{L} : \mathbf{pp},$$

and thus

$$\dot{\mathbf{p}} = \mathcal{L} \cdot \mathbf{p} - \mathcal{L} : \mathbf{ppp}.$$

We conclude that (S.395) solves (S.394) and therefore solves (S.392). The parameter $\zeta = 2/(R^2 + 1)$ is a 'non-affine' parameter, representing the straining inefficiency of the flow.

Problem 8.3

In the start-up of a simple shear flow, the velocity gradient is

$$[\mathbf{L}] = \begin{bmatrix} 0 & \dot{\gamma} & 0 \\ 0 & 0 & 0 \\ 0 & 0 & 0 \end{bmatrix}, \quad [\mathcal{L}] = \begin{bmatrix} 0 & \left(1 - \frac{\zeta}{2}\right)\dot{\gamma} & 0 \\ -\frac{\zeta}{2}\dot{\gamma} & 0 & 0 \\ 0 & 0 & 0 \end{bmatrix},$$

where the shear rate is $\dot{\gamma}$, the process $\mathbf{Q}$ obeys

$$\dot{Q}_1 = \left(1 - \frac{\zeta}{2}\right)\dot{\gamma}Q_2,$$

$$\dot{Q}_2 = -\frac{\zeta}{2}\dot{\gamma}Q_1, \tag{S.396}$$

$$\dot{Q}_3 = 0.$$

This implies $Q_3 = Q_{30}$, a constant, and

$$\ddot{Q}_1 + \frac{\zeta}{2}\left(1 - \frac{\zeta}{2}\right)\dot{\gamma}^2 Q_1 = 0,$$

$$\ddot{Q}_2 + \frac{\zeta}{2}\left(1 - \frac{\zeta}{2}\right)\dot{\gamma}^2 Q_2 = 0.$$

The solutions are

$$Q_1 = Q_{10}\cos\omega t + \sqrt{\frac{2-\zeta}{\zeta}}Q_{20}\sin\omega t,$$

$$Q_2 = Q_{20}\cos\omega t - \sqrt{\frac{\zeta}{2-\zeta}}Q_{10}\sin\omega t,$$

where $\{Q_{10}, Q_{20}, Q_{30}\}$ are the initial components of $\mathbf{Q}$, and the frequency of the oscillation is

$$\omega = \frac{1}{2}\dot{\gamma}\sqrt{\zeta(2 - \zeta)} = \frac{\dot{\gamma}R}{R^2 + 1}.$$

From these results, the result for $\mathbf{p}$ can be obtained.

The stress is

$$\mathbf{\alpha} = 2\eta_s \mathbf{D} + 2\eta_s \phi \{A\mathbf{D} : \mathbf{pppp} + B (\mathbf{D} \cdot \mathbf{pp} + \mathbf{pp} \cdot \mathbf{D}) + C\mathbf{D}\}, \qquad \text{(S.397)}$$

In full we have the reduced viscosity:

$$\frac{\langle \sigma_{12} \rangle - \eta_s \dot{\gamma}}{\eta_s \dot{\gamma} \phi} = 2A p_1^2 p_2^2 + B (p_1^2 + p_2^2) + C, \qquad \text{(S.398)}$$

the reduced first normal stress difference:

$$\frac{N_1}{\eta_s \dot{\gamma} \phi} = 2A p_1 p_2 (p_1^2 - p_2^2), \qquad \text{(S.399)}$$

and the reduced second normal stress difference:

$$\frac{N_2}{\eta_s \dot{\gamma} \phi} = 2 p_1 p_2 (A p_2^2 + B). \qquad \text{(S.400)}$$

Thus, the particles tumble along with the flow, with a period of $T = 2\pi(R^2+1)/\dot{\gamma}R$, spending most of their time aligned with the flow.

Problem 8.4

In the start-up of an elongational flow with a positive elongational rate $\dot{\gamma}$, the velocity gradient is

$$[\mathbf{L}] = \begin{bmatrix} \dot{\gamma} & 0 & 0 \\ 0 & -\dot{\gamma}/2 & 0 \\ 0 & 0 & -\dot{\gamma}/2 \end{bmatrix}, \quad [\mathcal{L}] = (1 - \zeta) \begin{bmatrix} \dot{\gamma} & 0 & 0 \\ 0 & -\dot{\gamma}/2 & 0 \\ 0 & 0 & -\dot{\gamma}/2 \end{bmatrix},$$

$$\dot{Q}_1 = (1 - \zeta) \dot{\gamma} Q_1,$$
$$\dot{Q}_2 = -\frac{1}{2} (1 - \zeta) \dot{\gamma} Q_2,$$
$$\dot{Q}_3 = -\frac{1}{2} (1 - \zeta) \dot{\gamma} Q_3,$$

which have the solutions

$$Q_1 = Q_{10} \exp \{(1 - \zeta)\dot{\gamma}t\},$$
$$Q_2 = Q_{20} \exp \left\{-\frac{1}{2}(1 - \zeta)\dot{\gamma}t\right\},$$

$$Q_3 = Q_{30} \exp\left\{-\frac{1}{2}(1 - \zeta)\dot{\gamma}t\right\},$$

so that the particle is quickly aligned with the flow in a time scale $O(\dot{\gamma}^{-1})$, $\mathbf{p} \to (1, 0, 0)$. The stress is

$$\text{œ} = 2\eta_s \mathbf{D} + 2\eta_s \phi \{A\mathbf{D} : \mathbf{pppp} + B (\mathbf{D} \cdot \mathbf{pp} + \mathbf{pp} \cdot \mathbf{D}) + C\mathbf{D}\}, \qquad (S.401)$$

In full we have:

$$\frac{\sigma_{11}}{\eta_s \dot{\gamma}} = 2 + 2\phi \left[A \left(p_1^2 - \frac{1}{2}p_2^2 - \frac{1}{2}p_3^2\right) p_1^2 + 2Bp_1^2 + C\right]$$

$$\frac{\sigma_{22}}{\eta_s \dot{\gamma}} = -1 + 2\phi \left[A \left(p_1^2 - \frac{1}{2}p_2^2 - \frac{1}{2}p_3^2\right) p_2^2 - 2Bp_2^2 - \frac{1}{2}C\right]$$

$$\frac{\sigma_{33}}{\eta_s \dot{\gamma}} = -1 + 2\phi \left[A \left(p_1^2 - \frac{1}{2}p_2^2 - \frac{1}{2}p_3^2\right) p_3^2 - 2Bp_3^2 - \frac{1}{2}C\right]$$

This yields the reduced elongational viscosity

$$\frac{N_1 - 3\eta_s \dot{\gamma}}{\eta_s \dot{\gamma}\phi} = 2\left[A \left(p_1^2 - \frac{1}{2}p_2^2 - \frac{1}{2}p_3^2\right) (p_1^2 - p_2^2) + 2B \left(p_1^2 + p_2^2\right) + \frac{3}{2}C\right]$$

At a steady state, show that the reduced elongational viscosity is given by

$$\frac{N_1 - 3\eta_s \dot{\gamma}}{\eta_s \dot{\gamma}\phi} = 2A + 4B + 3C \approx 2A = \frac{R^2}{\ln 2R - 1.5}. \qquad (S.402)$$

This elongational viscosity could be several order of magnitudes greater than the shear viscosity, due to the term $O\left(R^2\right)$.

Problems in Chap. 9

Problem 9.1

We start with the 1-D system

$$\frac{dr}{dt} = v, \quad m\frac{dv}{dt} = F_c - \gamma w_D v + \sigma w_R \theta(t), \quad r(0) = r_0, \quad v(0) = v_0, \qquad (S.403)$$

$$\langle \theta(t) \rangle = 0, \quad \langle \theta(t) \theta(t + \tau) \rangle = \delta(\tau).$$

In the inertial time scale, the displacement, the force F_c *and* the wighting functions may be regarded as constant. We may re-define the velocity as

$$v = u + F_c/\gamma w_D,$$

thus eliminating F_c in the governing equation altogether, thus only deal with a linear system

$$m\frac{dv}{dt} = -\gamma w_D v + \sigma w_R \theta(t). \tag{S.404}$$

This system has the integrating constant $e^{m^{-1}\gamma w_D t}$

$$\frac{d}{dt}\left(e^{m^{-1}\gamma w_D t} v\right) = e^{m^{-1}\gamma w_D t}\left(\dot{v} + m^{-1}\gamma w_D v\right)$$

$$= e^{m^{-1}\gamma w_D t} m^{-1}\sigma w_R \theta(t),$$

which can be integrated to yield

$$v(t) = \int_0^t e^{m^{-1}\gamma w_D (t'-t)} m^{-1}\sigma w_R \theta(t') \, dt'. \tag{S.405}$$

Its mean square velocity is

$$\langle v(t) v(t) \rangle = \frac{\sigma^2}{m^2}\int_0^t \int_0^t e^{m^{-1}\gamma w_D(t'-t)} w_R^2 \langle \theta(t') \theta(t'') \rangle e^{m^{-1}\gamma w_D(t''-t)} dt' dt''$$

$$= \frac{\sigma^2}{m^2}\int_0^t e^{m^{-1}\gamma w_D(t'-t)} w_R e^{m^{-1}\gamma w_D(t'-t)} w_R dt',$$

$$= \frac{1}{2}m^{-1}\sigma^2 w_R^2 \gamma^{-1} w_D^{-1}. \tag{S.406}$$

Assuming the equi-partition principle

$$\frac{1}{2}m \langle v^2(t) \rangle = \frac{1}{2}k_B T, \tag{S.407}$$

leads directly to

$$\sigma^2 w_R^2 \gamma^{-1} w_D^{-1} = 2k_B T. \tag{S.408}$$

Problem 9.2

In the case of small inertia, our main stochastic system becomes

$$\frac{dr}{dt} = \gamma^{-1} w_D^{-1} F_c + \gamma^{-1} w_D^{-1} \sigma w_R \theta(t). \tag{S.409}$$

This may be solved in one time step Δt

$$\Delta r = \gamma^{-1} w_D^{-1} F_c \Delta t + \int_t^{t+\Delta t} \gamma^{-1} w_D^{-1} \sigma w_R \theta \left(t'\right) dt'. \tag{S.410}$$

From this, and the properties of white noise, the drift velocity is given by

$$\frac{\langle \Delta r \rangle}{\Delta t} = \gamma^{-1} w_D^{-1} F_c. \tag{S.411}$$

In addition

$$
\begin{aligned}
\langle \Delta r \Delta r \rangle &= O\left(\Delta t^2\right) + \gamma^{-2} w_D^{-2} \sigma^2 w_R^2 \int_t^{t+\Delta t} dt' \int_t^{t+\Delta t} dt'' \left\langle \theta \left(t'\right) \theta \left(t''\right) \right\rangle \\
&= O\left(\Delta t^2\right) + \gamma^{-2} w_D^{-2} \sigma^2 w_R^2 \Delta t \\
&= O\left(\Delta t^2\right) + 2 k_B T \gamma^{-1} w_D^{-1} \Delta t.
\end{aligned}
$$

In the last step, the equipartition principle has been used. Thus

$$\frac{\langle \Delta r \Delta r \rangle}{2\Delta t} = O\left(\Delta t\right) + k_B T \gamma^{-1} w_D^{-1}, \tag{S.412}$$

and the Fokker–Planck equation is

$$\frac{\partial}{\partial t} W\left(r, t\right) = \lim_{\Delta t \to 0} \frac{\partial}{\partial r} \left[\left(\frac{\langle \Delta r \Delta r \rangle}{2\Delta t} \frac{\partial}{\partial r} - \frac{\langle \Delta r \rangle}{\Delta t} \right) W\left(r, t\right) \right], \tag{S.413}$$

or

$$\frac{\partial}{\partial t} W\left(r, t\right) = \frac{\partial}{\partial r} \left[\left(\frac{k_B T}{\gamma w_D} \frac{\partial}{\partial r} - \frac{F_c}{\gamma w_D} \right) W\left(r, t\right) \right]. \tag{S.414}$$

Problem 9.3

In order to focus on events on the time scale τ_I we can regard the restoring force F_c as constant, so that it can be absorbed in a re-definition of the system state system:

$$\frac{dv}{dt} = -m^{-1} \gamma w_D v + m^{-1} \sigma w_R \theta \left(t\right). \tag{S.415}$$

Noting that

$$2vr = \frac{d}{dt} \left(r^2\right), \quad 2 \left(\dot{v} r + v^2\right) = \frac{d^2}{dt^2} \left(r^2\right),$$

we find, by multiplying (S.415) with r,

$$m \left(\dot{v} r + v^2 \right) - m v^2 - \gamma w_D v r = \sigma w_R \theta (t) r,$$

or,

$$\frac{1}{2} m \frac{d}{dt} \left(\frac{d}{dt} \langle r^2 \rangle \right) - m \langle v^2 \rangle + \frac{1}{2} \gamma w_D \frac{d}{dt} \langle r^2 \rangle = \sigma w_R \langle \theta (t) r \rangle . \qquad (S.416)$$

We now define

$$e = d \langle r^2 \rangle / dt.$$

From the temperature definition, $m \langle v^2 \rangle = k_B T$; furthermore, $\langle \theta (t) r \rangle = 0$ due to different time scales of $\theta (t)$ and r. Thus

$$\dot{e} + m^{-1} \gamma w_D e = 2 k_B T m^{-1}, \quad e (0) = 0.$$

This has the integration factor

$$\frac{d}{dt} \left(e \exp \left(m^{-1} \gamma w_D t \right) \right) = 2 k_B T m^{-1} \exp \left(m^{-1} \gamma w_D t \right) ,$$

which has the solution, for the assumed initial condition,

$$e = \frac{d}{dt} \langle r^2 \rangle = 2 k_B T \gamma^{-1} w_D^{-1} \left[1 - e^{-m^{-1} \gamma w_D t} \right] . \qquad (S.417)$$

Consequently, if $\Delta t \gg \tau_I = O \left(m^{-1} \gamma w_D \right)$, writing $\Delta r = r (\Delta t)$,

$$\frac{\langle \Delta r \Delta r \rangle}{2 \Delta t} = k_B T \gamma^{-1} w_D^{-1} . \qquad (S.418)$$

Problem 9.4

Define the velocity correlation as

$$R (\tau) = \lim_{t \to \infty} \langle v (t + \tau) v (t) \rangle , \qquad (S.419)$$

where the limit refers to large time compared to the inertial time scale, but yet small compared to the relaxation time scale. A formal solution of (S.415) is

$$v (t) = \int_0^t e^{m^{-1} \gamma w_D (t' - t)} m^{-1} \sigma w_R \theta (t') dt' . \qquad (S.420)$$

From this solution, and for $\tau > 0$

$$
R(\tau) = \lim_{t \to \infty} \int_0^{t+\tau} dt' \int_0^t e^{m^{-1}\gamma w_D(t'-t-\tau)} \left(m^{-1}\sigma w_R\right)^2
$$
$$
\langle \theta(t')\,\theta(t'') \rangle\, e^{m^{-1}\gamma w_D(t''-t)} dt'' \tag{S.421}
$$
$$
= e^{-m^{-1}\gamma w_D \tau} \lim_{t \to \infty} \langle v(t)\, v(t) \rangle = e^{-m^{-1}\gamma w_D \tau} R(0)
$$
$$
= k_B T m^{-1} e^{-m^{-1}\gamma w_D \tau}. \tag{S.422}
$$

That is, the velocity correlation decays after an inertial time scale, after which the velocity is independent to its previous state.

Next, the diffusivity can also be defined as

$$
D = \lim_{t \to \infty} \frac{1}{2} \langle v(t)\, r(t) + r(t)\, v(t) \rangle.
$$

This is equivalent to

$$
D = \lim_{t \to \infty} \int_0^t \langle v(t)\, v(t+\tau) \rangle\, d\tau = \int_0^\infty R(\tau)\, d\tau
$$
$$
= k_B T \gamma^{-1} w_D^{-1},
$$

consistent with previous results.

Problem 9.5

Now, consider the Langevin system

$$
\frac{d\mathbf{r}_i}{dt} = \mathbf{v}_i, \quad m\frac{d\mathbf{v}_i}{dt} = \sum_j \mathbf{F}_{ij}. \tag{S.423}
$$

Here, $\mathbf{F}_{ij}$ is the pairwise additive interparticle force by particle j on particle i; this force consists of three parts, a conservative force $\mathbf{F}_{ij}^C$, a dissipative force $\mathbf{F}_{ij}^D$, and a random force $\mathbf{F}_{ij}^R$:

$$
\mathbf{F}_{ij} = \mathbf{F}_{ij}^C + \mathbf{F}_{ij}^D + \mathbf{F}_{ij}^R. \tag{S.424}
$$

From (S.423), the drift velocity of the process is

$$
\left\langle \frac{\Delta\mathbf{v}_i}{\Delta t} \right\rangle = -m^{-1}\sum_j \left(\mathbf{F}_{ij}^C + \mathbf{F}_{ij}^D\right) = -m^{-1}\sum_j \left(\mathbf{F}_{ij}^C - \gamma w_{ij}^D \mathbf{e}_{ij}\mathbf{e}_{ij} \cdot \mathbf{v}_{ij}\right). \tag{S.425}
$$

Furthermore,

$$\langle \Delta v_\alpha \Delta v_\beta \rangle = O\left(\Delta t^2\right)$$
$$+ \sum_{k,m} \int^{\Delta t} dt' \int^{\Delta t} dt'' \frac{\sigma^2}{m^2} w_{\alpha k}^R w_{\beta m}^R \left(\delta_{\alpha\beta}\delta_{km} + \delta_{\alpha m}\delta_{\beta k}\right) \delta\left(t' - t''\right) \mathbf{e}_{\alpha k} \mathbf{e}_{\beta m},$$

or

$$\langle \Delta v_\alpha \Delta v_\beta \rangle = O\left(\Delta t^2\right)$$
$$+ \Delta t \frac{\sigma^2}{m^2} \left[\delta_{\alpha\beta} \sum_k \left(w_{\alpha k}^R\right)^2 \mathbf{e}_{\alpha k}\mathbf{e}_{\beta k} + \left(w_{\alpha\beta}^R\right)^2 \mathbf{e}_{\alpha\beta}\mathbf{e}_{\beta\alpha}\right],$$

Using the fluctuation-dissipation theorem,

$$\langle \Delta v_\alpha \Delta v_\beta \rangle = O\left(\Delta t^2\right) + 2\frac{k_B T \gamma}{m^2} \Delta t \left[\delta_{\alpha\beta} \sum_k w_{\alpha k}^D \mathbf{e}_{\alpha k}\mathbf{e}_{\beta k} - w_{\alpha\beta}^D \mathbf{e}_{\alpha\beta}\mathbf{e}_{\alpha\beta}\right],$$

$$\left\langle \frac{\Delta v_\alpha \Delta v_\beta}{2\Delta t} \right\rangle = \frac{\gamma k_B T}{m^2} \left[\delta_{\alpha\beta} \sum_k w_{\alpha k}^D \mathbf{e}_{\alpha k}\mathbf{e}_{\beta k} - w_{\alpha\beta}^D \mathbf{e}_{\alpha\beta}\mathbf{e}_{\alpha\beta}\right]. \qquad \text{(S.426)}$$

The Fokker–Planck equation for the process is

$$\frac{\partial f}{\partial t} + \sum_i \frac{\partial}{\partial \mathbf{r}_i} \cdot (\mathbf{v}_i f) + \sum_i \frac{\partial}{\partial \mathbf{v}_i} \cdot \left(\left\langle \frac{\Delta \mathbf{v}_i}{\Delta t}\right\rangle f\right) \qquad \text{(S.427)}$$
$$= \sum_{i,j} \frac{\partial}{\partial \mathbf{v}_i} \cdot \left(\left\langle \frac{\Delta \mathbf{v}_i \Delta \mathbf{v}_j}{2\Delta t}\right\rangle \cdot \frac{\partial f}{\partial \mathbf{v}_j}\right),$$

where the limit $\Delta t \to 0$ is implied. Using the results obtained above for the drift and the diffusivity, we finally obtain the Fokker–Planck equation for the process

$$\frac{\partial f}{\partial t} + \sum_i \frac{\partial}{\partial \mathbf{r}_i} \cdot (\mathbf{v}_i f) + \sum_{i,j} \frac{\partial}{\partial \mathbf{p}_i} \cdot \left(\mathbf{F}_{ij}^C f\right) = \gamma \sum_{i,j} w_{ij}^D \mathbf{e}_{ij} \frac{\partial}{\partial \mathbf{p}_i} \cdot \left(\mathbf{e}_{ij} \cdot \mathbf{v}_{ij} f\right)$$
$$+ \gamma k_B T \sum_{i,j} w_{ij}^D \mathbf{e}_{ij} \cdot \frac{\partial}{\partial \mathbf{p}_i} \cdot \left(\mathbf{e}_{ij} \cdot \left(\frac{\partial f}{\partial \mathbf{p}_i} - \frac{\partial f}{\partial \mathbf{p}_j}\right)\right), \qquad \text{(S.428)}$$

where $\mathbf{p}_i = m\mathbf{v}_i$ is the linear momentum of particle i.

Problem 9.6

We note that the Hamiltonian of the associate system is

$$\mathcal{H} = \sum_i \frac{\mathbf{p}_i \cdot \mathbf{p}_i}{2m} + \frac{1}{2}\sum_{i,j} \varphi\left(r_{ij}\right), \tag{S.429}$$

noting

$$\frac{\partial\mathcal{H}}{\partial\mathbf{p}_\alpha} = \frac{\mathbf{p}_\alpha}{m} = \mathbf{v}_\alpha, \quad \frac{\partial\mathcal{H}}{\partial\mathbf{r}_{\alpha\beta}} = -\mathbf{F}_{\alpha\beta}^C.$$

Thus, the equilibrium distribution of the associate system to is

$$
\begin{aligned}
f_{eq}\left(\chi, t\right) &= \frac{1}{Z}\exp\left[-\frac{1}{k_B T}\left(\sum_i \frac{\mathbf{p}_i \cdot \mathbf{p}_i}{2m} + \frac{1}{2}\sum_{i,j}\varphi\left(r_{ij}\right)\right)\right] \tag{S.430}\\
&= \frac{1}{Z}\exp\left[-\frac{\mathcal{H}}{k_B T}\right],
\end{aligned}
$$

where Z is a normalizing constant.

We further note that,

$$\sum_i \mathbf{v}_i \cdot \frac{\partial f_{eq}}{\partial\mathbf{r}_i} = \frac{1}{k_B T}\sum_{i,j}\mathbf{v}_i \cdot \mathbf{F}_{ij}^C f_{eq}, \quad \sum_{i,j}\mathbf{F}_{ij}^C \cdot \frac{\partial f_{eq}}{\partial\mathbf{p}_i} = -\frac{1}{k_B T}\sum_{i,j}\mathbf{F}_{ij}^C \cdot \mathbf{v}_i f_{eq},$$

$$\gamma\sum_{i,j} w_{ij}^D \mathbf{e}_{ij}\frac{\partial}{\partial\mathbf{p}_i} \cdot \left(\mathbf{e}_{ij} \cdot \mathbf{v}_{ij} f_{eq}\right) = \gamma\sum_{i,j} w_{ij}^D\left(-\frac{1}{k_B T}\mathbf{e}_{ij} \cdot \mathbf{v}_i \mathbf{e}_{ij} \cdot \mathbf{v}_{ij} + \frac{1}{m}\right)f_{eq},$$

$$
\begin{aligned}
\gamma k_B T\sum_{i,j} w_{ij}^D \mathbf{e}_{ij} \cdot \frac{\partial}{\partial\mathbf{p}_i} \cdot \left(\mathbf{e}_{ij} \cdot \left(\frac{\partial}{\partial\mathbf{p}_i} - \frac{\partial}{\partial\mathbf{p}_j}\right)f_{eq}\right) \\
= \gamma\sum_{i,j} w_{ij}^D\left(-\frac{1}{m} + \frac{1}{k_B T}\mathbf{e}_{ij} \cdot \mathbf{v}_i \mathbf{e}_{ij}\mathbf{v}_{ij}\right),
\end{aligned}
$$

and conclude that f_{eq} is also a stationary solution (i.e., solution that is independent of time) of the Fokker–Planck equation (S.428).

Problem 9.7

The stress contributed from the conservative forces,

$$\mathbf{S}_C(\mathbf{r}, t) = -\frac{1}{2} \int d\mathbf{R} \mathbf{F}^C \mathbf{R} \left\{ 1 - \frac{1}{2} \mathbf{R} \cdot \nabla + \cdots \right\} \bar{f}_2(\mathbf{r} + \mathbf{R}, \mathbf{r}, t). \qquad \text{(S.431)}$$

Marsh used a different technique in deriving the stresses, which involves expressing the delta function as an integral. From this, the stresses from the conservative forces is expressed as,

$$\mathbf{S}_C(\mathbf{r}, t) = -\frac{1}{2} \int d\mathbf{v}' \int d\mathbf{v}'' \int \mathbf{F}^C \mathbf{R} \bar{W}_2(\chi', \chi'', t; \mathbf{r}) \, d\mathbf{R}, \qquad \text{(S.432)}$$

where

$$\bar{W}_2(\mathbf{r}', \mathbf{v}', \mathbf{r}'', \mathbf{v}'', t; \mathbf{r}) = \int_0^1 f_2(\mathbf{r} + \lambda\mathbf{R}, \mathbf{r} - (1 - \lambda)\mathbf{R}, \mathbf{v}', \mathbf{v}'', t) d\lambda. \qquad \text{(S.433)}$$

Expressing $f_2(\mathbf{r} + \lambda\mathbf{R}, \mathbf{r} - (1 - \lambda)\mathbf{R}, \mathbf{v}', \mathbf{v}'', t)$ as a function of $(\mathbf{r} - \varepsilon\mathbf{R})$, where $\varepsilon = 1 - \lambda$,

$$f_2 = f_2(\mathbf{r} + \mathbf{R} - \varepsilon\mathbf{R}, \mathbf{r} - \varepsilon\mathbf{R}, \mathbf{v}', \mathbf{v}'', t),$$

and taking a Taylor's series in ε,

$$f_2 = f_2(\mathbf{r} + \mathbf{R}, \mathbf{r}, \mathbf{v}', \mathbf{v}'', t) - \varepsilon\mathbf{R} \cdot \nabla f_2(\mathbf{r} + \mathbf{R}, \mathbf{r}, \mathbf{v}', \mathbf{v}'', t) + \cdots$$

Integrating

$$\bar{W}_2(\mathbf{r}', \mathbf{v}', \mathbf{r}'', \mathbf{v}'', t; \mathbf{r}) = \int_0^1 f_2(\mathbf{r} + \lambda\mathbf{R}, \mathbf{r} - (1 - \lambda)\mathbf{R}, \mathbf{v}', \mathbf{v}'', t) d\varepsilon$$

$$= \left\{ 1 - \frac{1}{2}\mathbf{R} \cdot \nabla + \cdots \right\} f_2(\mathbf{r} + \mathbf{R}, \mathbf{r}, \mathbf{v}', \mathbf{v}'', t).$$

This, substituted in (S.432) and the velocity spaces are integrated out, leads to (S.431). The remaining part of the question is demonstrated similarly.

Problem 9.8

The "heat flux" has been shown to be

$$\mathbf{q}_C = \left\langle \sum_{i,j} \frac{1}{4} \mathbf{F}_{ij}^C \cdot (\mathbf{v}_i + \mathbf{v}_j) \mathbf{r}_{ij} \left(1 - \frac{1}{2} \mathbf{r}_{ij} \cdot \nabla + \cdots \right) \delta(\mathbf{r} - \mathbf{r}_j) \right\rangle. \qquad \text{(S.434)}$$

This can be demonstrated, in an almost verbatim manner to the treatment of the stress, to be

$$\mathbf{q}_C\left(\mathbf{r},t\right) = \frac{1}{4}\int d\mathbf{R}\int d\mathbf{v}\int d\mathbf{v}'\mathbf{F}^C\left(\mathbf{R}\right)\cdot\left(\mathbf{v}+\mathbf{v}'\right)\mathbf{R}\left\{1-\frac{1}{2}\mathbf{R}\cdot\nabla+\cdots\right\}$$
$$\cdot f_2\left(\mathbf{r}+\mathbf{R},\mathbf{r},\mathbf{v},\mathbf{v}',t\right). \tag{S.435}$$

Problem 9.9

The stress contributed from the damping forces is,

$$\mathbf{S}_D\left(\mathbf{r},t\right) = -\frac{1}{2}\int d\mathbf{R}\int d\mathbf{v}\int d\mathbf{v}'\gamma w^D\left(R\right)\hat{\mathbf{R}}\hat{\mathbf{R}}\cdot\left(\mathbf{v}-\mathbf{v}'\right)\mathbf{R}\left\{1+O\left(R\right)\right\}$$
$$\cdot f_2\left(\mathbf{r}+\mathbf{R},\mathbf{r},\mathbf{v},\mathbf{v}',t\right).$$

For a homogeneously shear flow, the velocity gradient is constant. By taking the particles' velocities as the fluid's velocities at the particles' locations: $\mathbf{v}-\mathbf{v}' = \mathbf{LR}$, where $\mathbf{L}$ is the velocity gradient, together with Groot and Warren's approximation, $f_2 = n^2\left(1+O\left(R\right)\right)$, the stress contributed by the damping forces is

$$S_{D,\alpha\beta}\left(\mathbf{r},t\right) = -\frac{\gamma n^2}{2}\left\langle\hat{R}_\alpha\hat{R}_\beta\hat{R}_i\hat{R}_jL_{ij}\int R^2w^D\left(R\right)4\pi R^2dR\right\rangle$$
$$= -\frac{2\pi\gamma n^2}{15}\left(\delta_{\alpha\beta}\delta_{ij}+\delta_{\alpha i}\delta_{\beta j}+\delta_{\alpha j}\delta_{\beta i}\right)L_{ij}\int R^4w^D\left(R\right)dR.$$

Here we have assume a completely isotropic distribution for the structure tensor:

$$\left\langle\hat{R}_\alpha\hat{R}_\beta\hat{R}_i\hat{R}_j\right\rangle = \frac{1}{15}\left(\delta_{\alpha\beta}\delta_{ij}+\delta_{\alpha i}\delta_{\beta j}+\delta_{\alpha j}\delta_{\beta i}\right).$$

For the standard weighting function (9.88) adopted in DPD, one has

$$S_{D,\alpha\beta}\left(\mathbf{r},t\right) = -\frac{2\pi\gamma n^2}{15}\left(L_{\alpha\beta}+L_{\beta\alpha}+L_{ii}\delta_{\alpha\beta}\right)\int_0^{r_c}R^4\left(1-R/r_c\right)^2dR$$
$$= \frac{2\pi\gamma n^2r_c^5}{1575}\left(L_{\alpha\beta}+L_{\beta\alpha}+L_{ii}\delta_{\alpha\beta}\right), \tag{S.436}$$

and consequently the damping-contributed viscosities are given by

$$\eta_D = \frac{2\pi\gamma n^2r_c^5}{1575}, \quad \zeta_D = \frac{5}{3}\eta_D = \frac{2\pi\gamma n^2r_c^5}{945}. \tag{S.437}$$

For the modified weighting function (9.125) with $s = 1/2$,

$$S_{D,\alpha\beta}(\mathbf{r}, t) = \frac{512\pi\gamma n^2 r_c^5}{51975} \left(L_{\alpha\beta} + L_{\beta\alpha} + L_{ii}\delta_{\alpha\beta}\right) \tag{S.438}$$

and consequently

$$\eta_D = \frac{512\pi\gamma n^2 r_c^5}{51975}, \quad \zeta_D = \frac{5}{3}\eta_D = \frac{512\pi\gamma n^2 r_c^5}{31185}. \tag{S.439}$$

Problem 9.10

By using the kinetic theory or by considering the case of a homogeneously shear flow (as shown above), one can obtain an estimate for the viscosity and diffusivity

$$\eta = \eta_K + \eta_D = \frac{\rho D}{2} + \frac{\gamma n^2 \left[R^2 w^D\right]_R}{30},$$

$$D = \frac{3k_B T}{n\gamma \left[w^D\right]_R},$$

where

$$\left[w^D\right]_R = \int_0^{r_c} 4\pi R^2 w^D (R) \, dR, \quad \left[R^2 w^D\right]_R = \int_0^{r_c} 4\pi R^4 w^D (R) \, dR .$$

With the standard weighting function,

$$\left[w^D\right]_R = \frac{2\pi}{15}r_c^3, \quad \left[R^2 w^D\right]_R = \frac{4\pi}{105}r_c^5,$$

which leads to

$$\eta = \frac{45mkT}{4\pi\gamma r_c^3} + \frac{2\pi\gamma n^2 r_c^5}{1575}, \quad D = \frac{45k_B T}{2\pi\gamma nr_c^3}.$$

Problem 9.11

With the modified weighting function ($s = 1/2$),

$$\left[w^D\right]_R = \frac{64\pi}{105}r_c^3, \quad \left[R^2 w^D\right]_R = \frac{1024\pi}{3465}r_c^5,$$

which leads to

$$\eta = \frac{315 m k_B T}{128 \pi \gamma r_c^3} + \frac{512 \pi \gamma n^2 r_c^5}{51975}, \quad D = \frac{315 k_B T}{64 \pi \gamma n r_c^3}.$$

Problem 9.12

```
%%%%%%%%%%%%%%%%%%% INITILISATION %%%%%%%%%%%%%%%%%%
% Array to store the number of times that the particles cross boundaries
ncc=zeros(2,nFreeAtom);
% Array to store the mean square displacements against time
MSDs=zeros(stepSample,2);


%%%%%%%%%%% COMPUTATION %%%%%%%%%%%%%%%%%%%%%%%%%%%%%
    for i=1:nFreeAtom
        if r(1,i) > 0.5*Lx
            r(1,i)=r(1,i)-Lx; ncc(1,i) = ncc(1,i)+1;
        elseif r(1,i) < -0.5*Lx
            r(1,i)=r(1,i)+Lx; ncc(1,i) = ncc(1,i)-1;
        end
        if r(2,i) > 0.5*Ly
            r(2,i)=r(2,i)-Ly; ncc(2,i) = ncc(2,i)+1;
        elseif r(2,i) < -0.5*Ly
            r(2,i)=r(2,i)+Ly; ncc(2,i) = ncc(2,i)-1;
        end
    end

    if stepCount == stepEquil
        % Actual positions of particles at the thermal equilibrium
        r0=r+[Lx*ncc(1,:);Ly*ncc(2,:)];
    end

    if stepCount > stepEquil
        timeNow=timeNow+deltaT;
        % Actual positions of particles at the current time t
        rt = r+[Lx*ncc(1,:);Ly*ncc(2,:)];
        % Relative positions
        DeltaR = rt-r0;
        % Mean Square Displacement
        MSD = sum(sum(DeltaR.^2,1))/nFreeAtom;
        MSDs(round(timeNow/deltaT),:) = [timeNow,MSD];
    end

%%%%%%%%%%%%%%%%%%% OUTPUT %%%%%%%%%%%%%%%%%%%%%%%%
% Plot MSD and Diffusivity of DPD particles
figure; plot(MSDs(:,1),MSDs(:,2),'b-')
xlabel('Time'); ylabel('MSD')
title('Mean Square Displacements of DPD particles');
figure; plot(MSDs(:,1),(1/4)*(MSDs(:,2)./MSDs(:,1)),'b-');
xlabel('Time'); ylabel('D');
title('Self-diffusion coefficient of DPD particles');
```

Problem 9.13

```
%%%%%%%%%%%%%%%%%%%% INPUT %%%%%%%%%%%%%%%%%%%%%%%%%%
% Number of annular elements and the domain size
nRDF=100; rRDF=1;

%%%%%%%%%%%%%%%%%%% INITILISATION %%%%%%%%%%%%%%%%%%
% Array to store RDF values against time
RDFs=zeros(stepSample,nRDF);

%%%%%%%%%%%% COMPUTATION %%%%%%%%%%%%%%%%%%%%%%%%%%%%
    if stepCount > stepEquil
        timeNow=timeNow+deltaT;

        % Pick up particles in the core region for computing their RDF
        id = find(abs(r(1,:))<Lx/2-rRDF & abs(r(2,:))<Ly/2-rRDF);
        rs = r(:,id); ns=size(rs,2); RDF=zeros(ns,nRDF);
        for i=1:ns
            DeltaR = [r(1,:)-rs(1,i);r(2,:)-rs(2,i)];
            DistR = sqrt(sum(DeltaR.^2,1));
            for j=1:nRDF
                Rmin=(j-1)*(rRDF/nRDF); Rmax=j*(rRDF/nRDF);
                id=find(DistR>Rmin & DistR<=Rmax); RDF(i,j)=length(id);
            end
        end
        RDFs(round(timeNow/deltaT),:) = mean(RDF,1);
    end

%%%%%%%%%%%%%%%%%% OUTPUT %%%%%%%%%%%%%%%%%%%%%%
% Plot RDF of the DPD particles
R = mean(RDFs,1); dq=(rRDF/nRDF); q=dq^(1:nRDF);
R(1)=R(1)/(numDen*pi*dq^2);
R(2:end)=R(2:end)./(numDen*2*pi*(q(2:end)-dq)*dq);
figure; plot(q,R,'b-o')
xlabel('Distance'),ylabel('RDF')
title('Radial distribution function for DPD particles')
```

References

1. M. Arienti, W. Pan, X. Li, G. Karniadakis, J. Chem. Phys. **134**, 204114 (2011)
2. H. Ashi, *Numerical Methods for Stiff Systems*. Ph.D thesis, University of Nottingham (Nottingham 1998)
3. E.B. Bagley, *Food Extrusion Science and Technology* (Marcel Dekker, New York, 1992)
4. G.K. Batchelor, J. Fluid Mech. **44**, 545–570 (1970)
5. X. Bian, M. Ellero, Comput. Phys. Commun. **85**(1), 53–62 (2014)
6. R.B. Bird, R.C. Armstrong, O. Hassager, *Dynamics of Polymeric Liquids. Fluid Mechanics*, vol. 1, 2nd edn. (Wiley, New York, 1987)
7. E. Bodewig, *Matrix Calculus* (North-Holland Publishing, Amsterdam, 1956)
8. E. Boek, P. Coveney, H. Lekkerkerker, J. Phys. Cond. Matt. **8**(47), 9509–9512 (1996)
9. D.V. Boger, R. Binnington, Trans. Soc. Rheol. **21**, 515–534 (1977)
10. D.V. Boger, K. Walters, *Rheological Phenomena in Focus* (Elsevier, Amsterdam, 1993)
11. J.F. Brady, Int. J. Multiphase Flow **10**(1), 113–114 (1984)
12. F. Brauer, J.A. Nohel, *Qualitative Theory of Ordinary Differential Equations* (Dover, New York, 1989)
13. C. Bustamante, J.F. Marko, E.D. Siggia, S. Smith, Science **265**, 1500–1600 (1994)
14. S. Chandrasekhar, Rev. Mod. Phys. **15**, 1–89 (1943)
15. S. Chen, N. Phan-Thien, B.C. Khoo, X.-J. Fan, Phys. Fluids **18**, 103605 (2006)
16. B.D. Coleman, Arch. Rat. Mech. Anal. **9**, 273–300 (1962)
17. W.O. Criminale Jr., J.L. Ericksen, G.L. Filbey Jr., Arch. Rat. Mech. Anal. **1**, 410–417 (1957)
18. M. Doi, S.F. Edwards, *The Theory of Polymer Dynamics* (Oxford University Press, Oxford, 1988)
19. D. Duong-Hong, N. Phan-Thien, X.-J. Fan, Comput. Mech. **35**, 24–29 (2004)
20. A. Einstein, in *Investigations on the Theory of the Brownian Movement*, ed. by R. Fürth, R., Transl. A.D. Cowper (Dover, New York, 1956)
21. J.L. Ericksen, in *Viscoelasticity - Phenomenological Aspects*, ed. by J.T. Bergen (Academic Press, New York 1960)
22. P. Español, P. Warren, Europhys. Lett. **30**(4), 191–196 (1995)
23. E. Fahy, G.F. Smith, J. Non-Newt, Fluid Mech. **7**, 33–43 (1980)
24. X.-J. Fan, N. Phan-Thien, S. Chen, X. Wu, T.Y. Ng, Phys. Fluids **18**(6), 063102 (2006)
25. X.-J. Fan, N. Phan-Thien, R. Zheng, J. Non-Newtonian Fluid Mech. **84**(2–3), 257–274 (1999)
26. P.J. Flory, *Statistical Mechanics of Chain Molecules* (Wiley Interscience, New York, 1969)
27. F.P. Folgar, C.L. Tucker Jr., J. Reinf, Plast. Compos. **3**, 98–119 (1984)
28. U. Frisch, B. Hasslachcher, Y. Pomeau, Phys. Rev. Lett. **56**, 1505–1508 (1986)

© Springer International Publishing AG 2017

N. Phan-Thien and N. Mai-Duy, *Understanding Viscoelasticity*,

Graduate Texts in Physics, DOI 10.1007/978-3-319-62000-8

29. R.M. Füchslin, H. Fellermann, A. Eriksson, H.-J. Ziock, J. Chem. Phys. **130**, 214102 (2009)
30. F. Gadala-Maria, A. Acrivos, J. Rheol. **24**, 799–814 (1980)
31. H. Giesekus, Rheol. Acta **3**, 59–81 (1963)
32. A.E. Green, R.S. Rivlin, Arch. Rat. Mech. Anal. **1**, 1–21 (1957)
33. R.D. Groot, P.B. Warren, J. Chem. Phys. **107**, 4423–4435 (1997)
34. M.E. Gurtin, *An Introduction to Continuum Mechanics* (Academic Press, New York, 1981)
35. J. Happel, H. Brenner, *Low Reynolds Number Hydrodynamics* (Noordhoff International Publishing, Leyden, 1973)
36. E.J. Hinch, J. Fluid Mech. **72**, 499–511 (1975)
37. E.J. Hinch, L.G. Leal, J. Fluid Mech. **52**, 683–712 (1972)
38. P. Hoogerbrugge, J. Koelman, Europhys. Lett. **19**(3), 155–160 (1992)
39. R.R. Huilgol, N. Phan-Thien, *Fluid Mechanics of Viscoelasticity: General Principles, Constitutive Modelling, Analytical and Numerical Techniques* (Elsevier, Amsterdam, 1997)
40. J.H. Irving, J.G. Kirkwood, J. Chem. Phys. **18**, 817–829 (1950)
41. G.B. Jeffery, Proc. R. Soc. Lond. A **102**, 161–179 (1922)
42. E.E. Keaveny, I.V. Pivkin, M. Maxey, G.E. Karniadakis, J. Chem. Phys. **123**, 104107 (2005)
43. O.D. Kellogg, *Foundations of Potential Theory* (Dover, New York, 1954)
44. Y. Kong, C. Manke, W. Madden, A. Schlijper, J. Chem. Phys. **107**(2), 592–602 (1997)
45. I.M. Krieger, T.J. Dougherty, Trans. Soc. Rheol. **3**, 137–152 (1959)
46. W. Kuhn, Kolloid-Zeit. **68**, 2–15 (1934)
47. L.D. Landau, E.M. Lifshitz, *Fluid Mechanics*, translated by J.B. Sykes, W.H. Reid (Pergamon Press, New York 1959)
48. R.G. Larson, *Constitutive Equations for Polymer Melts and Solutions* (Butterworth Publishers, Boston, 1988)
49. R.G. Larson, H. Hu, D.E. Smith, S. Chu, J. Rheol. **43**(2), 267–304 (1999)
50. R.G. Larson, T.T. Perkins, D.E. Smith, S. Chu, Phys. Rev. E **55**(2), 1794–1797 (1997)
51. L.G. Leal, E.J. Hinch, Rheol. Acta **12**, 127–132 (1973)
52. Y.K. Lin, *Probabilistic Theory of Structural Dynamics* (McGraw-Hill, New York, 1967)
53. A.S. Lodge, *Elastic Liquids* (Academic Press, New York, 1964)
54. A.E.H. Love, *A Treatise in the Mathematical Theory of Elasticity, Note B* (Dover, New York, 1944)
55. N. Mai-Duy, N. Phan-Thien, B.C. Khoo, Comput. Phys. Commun. **189**, 37–46 (2015)
56. C. Marsh, *Theoretical Aspects of Dissipative Particle Dynamics*. Ph.D thesis, University of Oxford (Oxford 1998)
57. A.B. Metzner, J. Rheol. **29**, 739–775 (1985)
58. M. Mooney, J. Appl. Phys. **11**, 582–592 (1940)
59. F.A. Morrison, *Understanding Rheology: Topic in Chemical Engineering* (Oxford Unversity Press, New York, 2001)
60. P. Ninkunen, M. Karttunen, I. Vattulainen, Comput. Phys. Commun. **153**, 407–423 (2003)
61. W. Noll, J. Ratl. Mech. Anal. **4**, 2–81 (1955)
62. W. Noll, Ratl. Mech. Anal. **2**, 197–226 (1958)
63. L. Novik, P. Coveney, Int. J. Mod. Phys. C **8**(4), 909–918 (1997)
64. J.G. Oldroyd, Proc. Roy. Soc. Lond. A **200**, 523–541 (1950)
65. D. Pan, N. Phan-Thien, N. Mai-Duy, B.C. Khoo, J. Comput. Phys. **242**, 196–210 (2013)
66. W. Pan, B. Caswell, G.E. Karniadakis, Langmuir **26**(1), 133–142 (2010)
67. W. Pan, I.V. Pivkin, G.E. Karniadakis, EPL **84**, 10012 (2008)
68. N. Phan-Thien, S. Kim, *Microstructures in Elastic Media: Principles and Computational Methods* (Oxford University Press, New York, 1994)
69. N. Phan-Thien, N. Mai-Duy, B.C. Khoo, J. Rheol. **58**, 839–867 (2014)
70. A.C. Pipkin, *Lectures on Viscoelasticity Theory*, 2nd edn. (Springer, Berlin, 1986)
71. A.C. Pipkin, R.S. Rivlin, Arch. Rational Mech. Anal. **4**, 129–144 (1959)
72. A.C. Pipkin, R.I. Tanner, in *Mechanics Today*, vol. 1, ed. by S. Nemat-Nasser (Pergamon, New York, 1972), pp. 262–321

73. I.V. Pivkin, B. Caswell, G.E. Karniadakis, Reviews in Computational Chemistry **27**, 85–110 (2011)
74. B. Rabinowitch, Z. Physik, Chem. A **145**, 1–26 (1929)
75. D. Rapaport, *The Art of Molecular Dynamics Simulation* (Cambridge University Press, Cambridge, 1995)
76. L.E. Reichl, *A Modern Course in Statistical Physics* (University of Texas Press, Austin, Texas, 1980)
77. M. Revenga, I. Zuñiga, P. Español, Int. J. Mod. Phys. C **9**(8), 1319–1328 (1998)
78. R.S. Rivlin, J.L. Ericksen, J. Ration. Mech. Anal. **4**, 323–425 (1955)
79. G. Ryskin, J.M. Rallison, J. Fluid Mech. **99**, 513–529 (1980)
80. H. See, J. Phys. D Appl. Phys. **33**, 1625–1633 (2000)
81. L.F. Shampine, C.W. Gear, SIAM Rev. **21**(1), 1–17 (1979)
82. A.J.M. Spencer, in *Continuum Physics*, vol. 1, ed. by A.C. Eringen (Academic Press, New York, 1971)
83. R.I. Tanner, Phys. Fluids **9**, 1246–1247 (1966)
84. R.I. Tanner, J. Polym. Sci. A **2**(8), 2067–2078 (1970)
85. R.I. Tanner, *Engineering Rheology*, Revised edn. (Oxford University Press, Oxford, 1992)
86. R.I. Tanner, K. Walters, *Rheology: An Historical Perspective* (Elsevier, Amsterdam, 1998)
87. L.R.G. Treloar, *The Physics of Rubber Elasticity*, 3rd edn. (Clarendon Press, Oxford, 1975)
88. L.E. Wedgewood, D.N. Ostrov, R.B. Bird, J. Non-Newtonian Fluid Mech. **40**, 119–139 (1991)
89. L. Weese, Comput. Phys. Commun. **77**, 429–440 (1993)
90. H. Weyl, *Classical Groups* (Princeton University Press, New Jersey, 1946)
91. S. Willemsen, H. Hoefsloot, P. Iedema, Int. J. Mod. Phys. C **11**(5), 881–890 (2000)
92. W.-L. Yin, A.C. Pipkin, Arch. Rat. Mech. Anal. **37**, 111–135 (1970)

Index

© Springer International Publishing AG 2017
N. Phan-Thien and N. Mai-Duy, *Understanding Viscoelasticity*,
Graduate Texts in Physics, DOI 10.1007/978-3-319-62000-8

Printed in the United States
By Bookmasters